Contents

Contents   v

vi   *Contents*

INTRODUCTION TO

PHYSICAL CHEMISTRY

M.F.C.LADD & W.H.LEE

Department of Chemistry, University of Surrey

The right of the
University of Cambridge
to print and sell
all manner of books
was granted by
Henry VIII in 1534.
The University has printed
and published continuously
since 1584.

WITHDRAWN

CAMBRIDGE UNIVERSITY PRESS

Cambridge

New York   New Rochelle

Melbourne   Sydney

Published by the Press Syndicate of the University of Cambridge
The Pitt Building, Trumpington Street, Cambridge CB2 1RP
32 East 57th Street, New York, NY 10022, USA
10 Stamford Road, Oakleigh, Melbourne 3166, Australia

First published 1986
Reprinted (with corrections) 1988

Printed in Great Britain at the University Press, Cambridge

British Library cataloguing in publication data

Ladd, M. F. C.
Introduction to physical chemistry.

1. Chemistry, Physical and theoretical
I. Title   II. Lee, W. H.
541.3   QD453.2

Library of Congress cataloguing in publication data

Ladd, M. F. C. (Marcus Frederick Charles)
Introduction to physical chemistry.

Includes index.
1. Chemistry, Physical and theoretical. I. Lee,
W. H. (Walter Henry)   II. Title.
QD453.2.L33   1986   541.3   85-30918

ISBN 0 521 26448 0 hard covers
ISBN 0 521 26995 4 paperback

Preface

This book has been developed from an earlier publication by the authors, now brought up to date and intended to meet the requirements of students in their first year of a degree course in chemistry, or in those sciences of which chemistry forms a major part. The mathematical treatments used are not difficult, and should be found to lie within the scope of any chemistry degree student. Some mathematical arguments are developed in appendices, because their inclusion in the text might distract readers from the development of the subject. Competence in mathematics, at least to a good A-level standard, is essential for any student of modern chemistry.

Each chapter has been provided with a set of problems of varying degrees of difficulty, which the reader is encouraged to solve because they will assist in gaining familiarity with the themes of the book, and in testing one's ability to apply these themes to new situations. A suggested scheme for solving problems, and a listing of values of fundamental constants used in the text, appear before the opening chapter. Solutions to the problems are provided at the end of the book.

The SI system of units has been used throughout and, in several instances, conversions between the SI and cgs systems have been indicated. It is not yet, and will not be for some time, possible to neglect the cgs system. Competency in more than one system should not be despised: indeed, such ability will enhance a chemistry student's appreciation of both his subject and its literature. Some useful conversion factors are tabulated in the introductory matter that appears immediately before Chapter 1.

In studying stereochemistry and solids, one needs an understanding of the three-dimensional character of molecules and crystals. This aspect of structure is not so much intrinsically difficult as it is unfamiliar, despite the nature of the world in which we live. In order to assist in an appreciation of three-dimensional structure, many of the illustrations are provided as stereoscopic views, and directions for observing them are given in an appendix.

It is our pleasure to record gratitude to Dr S. Motherwell of Cambridge University for the use of the program PLUTO, with which the stereoscopic illustrations have been prepared, to publishers and authors for permission to reproduce those figures that carry appropriate acknowledgements, to numerous students who have worked through problems to our advantage and, we hope, to theirs and to the publishers for enabling this work to be brought to a state of completion.

Department of Chemistry   M. F. C. Ladd
University of Surrey   W. H. Lee
                                                              July 1985

Reprinting with corrections, 1988
It is with sadness that I record that, while the corrections for this reprinting were being prepared, my colleague and co-author Dr W. H. Lee died at his home after an illness.

                                                              M. F. C. Ladd

Physical constants

The following data form a self-consistent, least-squares adjusted set, and details of their compilation may be found in *Atomic Masses and Fundamental Constants 5*, ed. J. H. Sanders & A. H. Wapstra, Plenum Press, NY, 1976 and in references therein, particularly E. R. Cohen & B. N. Taylor, *J. Phys. Chem. Ref. Data* (1973), **2**, 663–734. Each datum is quoted to its known precision although rarely shall we need to use the full precision of any constant: the figure in parentheses is the uncertainty in the final digit of the constant.

Speed of light in a vacuum	c	$2.997924590(8) \times 10^{8}$	m s^{-1}
Permittivity of a vacuum	ε_0	$8.85418782(7) \times 10^{-12}$	$\text{J}^{-1}\,\text{C}^2\,\text{m}^{-1}$
Atomic mass unit	u	$1.660566(9) \times 10^{-27}$	kg
Mass of hydrogen atom	m_{H}	$1.673560(8) \times 10^{-27}$	kg
Rest mass of proton	m_{p}	$1.672649(9) \times 10^{-27}$	kg
Rest mass of neutron	m_{n}	$1.674954(9) \times 10^{-27}$	kg
Rest mass of electron	m_{e}, m	$9.10953(5) \times 10^{-31}$	kg
Elementary charge	e	$1.602189(5) \times 10^{-19}$	C
Classical electron radius	r_{e}	$2.817938(7) \times 10^{-15}$	m
Boltzmann constant	k	$1.38066(4) \times 10^{-23}$	J K^{-1}
Planck constant	h	$6.62618(4) \times 10^{-34}$	J s
Bohr radius	a_0	$5.294664(4) \times 10^{-11}$	m
Rydberg constant	R_{∞}	$1.09737314(1) \times 10^{7}$	m^{-1}
Rydberg constant for hydrogen	R_{H}	$1.09677582(1) \times 10^{7}$	m^{-1}
Bohr magneton	μ_{B}	$9.27408(4) \times 10^{-24}$	A m^2
Avogadro constant	L, N_{A}	$6.022094(6) \times 10^{23}$	mol^{-1}
Gas constant	R	$8.3160(2)$	$\text{J K}^{-1}\,\text{mol}^{-1}$
Ice-point temperature	T_{ice}	$273.1500(1)$	K
Faraday	F	$9.64865(3) \times 10^{4}$	C mol^{-1}
Reduced mass of (proton +electron)	μ	$9.10457(5) \times 10^{-31}$	kg

Conversion of units

There will be, for a very long time, a need to convert from cgs to SI units and vice versa. While the conversion factors can always be determined from first principles, and it is good practice to do so occasionally, a table of conversion factors is useful for quick reference. We set out here a selection of such data.

Physical quantity	cgs unit ×	conversion factor →	SI unit
Length, l	cm	10^{-2}	m
	Å	10^{-10}	m
Volume[a], V	cm^3 ml	10^{-6}	m^3
Molar volume, V_m	$cm^3\,mol^{-1}$	10^{-6}	$m^3\,mol^{-1}$
Velocity, v	$cm\,sec^{-1}$	10^{-2}	$m\,s^{-1}$
Wavelength, λ	cm	10^{-2}	m
Wavenumber v	cm^{-1}	10^2	m^{-1}
Mass, m	g	10^{-3}	kg
Density ρ	$g\,cm^{-3}$	10^3	$kg\,m^{-3}$
Force, F	dyne	10^{-5}	N
Pressure[b], P	$dyne\,cm^{-2}$	10^{-1}	$N\,m^{-2}$
Energy, U	cal	4.184	J
Work, W	erg	10^{-7}	J
Quantity of electricity, q	esu	3.3356×10^{-10}	C
Current, I	$esu\,sec^{-1}$	3.3356×10^{-10}	A
Charge density, σ	$esu\,cm^{-2}$	3.3356×10^{-6}	$C\,m^{-2}$
Electric field intensity, E	$dyne\,esu^{-1}$	2.9979×10^4	$V\,m^{-1}$
Electric potential, V	$erg\,esu^{-1}$	299.79	V
Capacitance, C	$esu^2\,erg^{-1}$	1.1127×10^{-12}	$J^{-1}\,C^2$
Permittivity ε_0	1	8.8542×10^{-12}	$J^{-1}\,C^2\,m^{-1}$
Dipole moment[c], p	esu cm	3.3356×10^{-12}	$C\,m$

Physical quantity	cgs unit $\times$	conversion factor $\rightarrow$	SI unit
Resistance, R	ohm	1	Ω
Resistivity, ρ	ohm cm	10^{-2}	$\Omega\,m$
Conductivity, σ	$ohm^{-1}\,cm^{-1}$	10^{2}	$\Omega^{-1}\,m^{-1}$
Molar conductivity, Λ	$ohm^{-1}\,cm^{2}\,mol^{-1}$	10^{-4}	$\Omega^{-1}\,m^{2}\,mol^{-1}$
Ionic mobility, u	$cm^{2}\,V^{-1}\,sec^{-1}$	10^{-4}	$m^{2}\,V^{-1}\,s^{-1}$
Specific heat capacity, s	$cal\,g^{-1}\,deg^{-1}$	4184	$J\,kg^{-1}\,K^{-1}$
Molar refraction, R_m	$cm^{3}\,mol^{-1}$	10^{-6}	$m^{3}\,mol^{-1}$
Molar polarization, P_m	$cm^{3}\,mol^{-1}$	10^{-6}	$m^{3}\,mol^{-1}$

[a] $1\,l = 1\,dm^{3} = 10^{-3}\,m^{3}$.

[b] $1\,atm = 101325\,N\,m^{-2} = 760\,Torr$; $1\,mmHg = 133.322\,N\,m^{-2}$.

[c] $1\,D$ (Debye unit) $= 10^{-18}$ esu cm.

Prefixes for SI units

The following prefixes are the approved indicators of decimal fractions or multiples of the basic SI units (e.g. metre), or the derived SI units (e.g. joule).

		Prefix	Symbol
Fraction	10^{-1}	deci	d
	10^{-2}	centi	c
	10^{-3}	milli	m
	10^{-6}	micro	μ
	10^{-9}	nano	n
	10^{-12}	pico	p
	10^{-15}	femto	f
	10^{-18}	atto	a
Multiple	10	deka	D
	10^{2}	hecto	h
	10^{3}	kilo	k
	10^{6}	mega	M
	10^{9}	giga	G
	10^{12}	tera	T

Solution of numerical problems

Introduction

Numerical problems are essential to a study of chemistry, because they relate experimental observations to theoretical models. The insertion of magnitudes into a given equation is a common scientific activity: it should be mastered and, however trivial, never despised.

The solving of problems leads to an appreciation of several important features:

(*a*) the orders of magnitude in physical and chemical quantities;

(*b*) the need for an understanding of units;

(*c*) the value of checking dimensional homogeneity;

(*d*) the sources of physical and chemical data;

(*e*) the precision of the data and its transmission to the result.

Most problems involve algebraic manipulation. It is essential to obtain a clear picture of the chemistry and physics involved in the problem before embarking on a series of mathematical processes. It is often useful to obtain an explicit algebraic expression before inserting numerical values. There are several advantages in so doing:

(*f*) the expression can be checked dimensionally;

(*g*) the possible cancellation of terms may improve the precision of the result;

(*h*) the chemical or physical significance of the result may be more apparent;

(*i*) similar problems with other magnitudes can be solved with little additional effort;

(*j*) if the result is erroneous, it is easy to check whether the error is in the deduction or in the arithmetic;

(*k*) in examinations, the derivation of a correct explicit expression will score marks, even though the arithmetic may be in error.

If the data is inserted into an expression in the form of numbers between one and ten, multiplied by the appropriate powers of ten, it is easy to estimate an approximate answer: calculators and other aids can go wrong; they can even be manipulated incorrectly. Suppose that we have for a relative permittivity, ε_r,

$$(\varepsilon_r - 1) = Np^2/9\varepsilon_0 kT \tag{1}$$

N (number of molecules per unit volume) $= 2.461 \times 10^{25}\,\mathrm{m}^{-3}$
p (dipole moment) $= 5.11 \times 10^{-30}\,\mathrm{C\,m}$
ε_0 (permittivity of a vacuum) $= 8.8542 \times 10^{-12}\,\mathrm{J}^{-1}\,\mathrm{C}^2\,\mathrm{m}^{-1}$
k (Boltzmann constant) $= 1.3807 \times 10^{-23}\,\mathrm{J\,K}^{-1}$
T (absolute temperature) $= 298.15\,\mathrm{K}$

We can see that the expression is dimensionally correct; the right-hand side of (1) has the units

$$\frac{\mathrm{m}^{-3} \times \mathrm{C}^2\,\mathrm{m}^2}{\mathrm{J}^{-1}\,\mathrm{C}^2\,\mathrm{m}^{-1} \times \mathrm{J\,K}^{-1} \times \mathrm{K}}$$

which is dimensionless. Inserting the magnitudes, we have

$$(\varepsilon_r - 1) = \frac{2.461 \times 10^{25} \times (5.11)^2 \times 10^{-60}}{9 \times 8.8542 \times 10^{-12} \times 1.3807 \times 10^{-23} \times 298.15} \tag{2}$$

We can see that $(\varepsilon_r - 1) \approx 60/(100 \times 300)$ or 2×10^{-3}. Thus, when the expression is evaluated, we can write with confidence, $(\varepsilon_r - 1) = 1.96 \times 10^{-3}$.

Approach to problems

There are different ways of tackling problems, so that these notes are offered only as a guide. Sometimes a recommended stage may be changed or bypassed. Elegant derivations are often concise: the converse is not necessarily true, and failure to justify a stage in a derivation may indicate a lack of judgement or a lack of confidence. On the other hand, overelaboration of trivial detail or of arithmetic manipulation may be equally unacceptable in a polished answer to a problem. Some degree of subjective judgement is involved in the solution of problems, and in the marking of such solutions in examinations. Few examiners would give high marks for a completely correct numerical answer in the absence of satisfactory evidence of the method used.

Procedure

(a) Read the problem carefully. If you think that it contains an ambiguity (which can happen sometimes), assume the simplest interpretation of the ambiguity, and comment on it.

(b) Summarize the given information by appropriate means, such as:
 (i) labelled drawings;
 (ii) energy-level diagrams;
 (iii) sketch-graphs, correctly labelled;

(iv) defining symbols used in diagrams and formulae;

(v) listing numerical values with units.

(c) State the answers required, defining quantities involved together with their units and symbols.

(d) Indicate relevant laws and equations which are to be used in developing the problem, at least initially.

(e) State the method to be used, for example 'take ln of both sides of Equation (1)'.

(f) Where appropriate, attempt to formulate an explicit equation before inserting numerical data. Look for cancellations of terms, and indicate any physical or functional approximations.

(g) Do *not* make needless numerical approximations, but state any approximations which are made, and include an estimate of the probable error as far as you are able.

(h) For convenience, substitute a new symbol for a complex group of symbols in deriving an expression.

(i) Check the dimensions of both quantities and expressions for consistency. Remember that exponents (and log terms) are dimensionless.

(j) Insert numerical values into expressions carefully. Determine an approximate result by 'gross cancellation', as in the example above.

(k) Think about the answer in terms of your knowledge of chemistry, and comment on it in the light of the question. If you feel doubtful about the validity of the result, check your arithmetic and deductions. If you still have some reservations about your answer indicate their nature.

(l) Keep a neat format in your answer. In an examination, allocate a numerical question only its fair share of time.

Example problem

The diffusion coefficient (D) of carbon in α-iron, as a function of temperature, follows the equation

$$D = D_0 \exp(-U/RT) \tag{3}$$

Values of D are as follows:

T/K	300	500	700	900	1100
$D/m^2\,s^{-1}$	4.73×10^{-21}	3.35×10^{-15}	1.08×10^{-12}	2.66×10^{-11}	2.05×10^{-10}

Verify the above equation, and find values for both the activation energy (U) for the diffusion process, and D_0. Comment briefly on the results.

$$(R = 8.316 \, J \, K^{-1} \, mol^{-1})$$

The following solution attempts to illustrate some of the points discussed above.

Solution

The presence of RT in the exponent and its units indicate that U is expected in J mol^{-1}. D_0 is the (hypothetical) limiting value of D as $T \to \infty$. Taking ln of both sides of (3) gives

$$\ln D = \ln D_0 - U/RT \tag{4}$$

If the graph of $\ln D$ against $1/T$ is a straight line, (3) would be verified. The slope of the line, $\Delta(\ln D)/\Delta(1/T)$, is $-U/R$, and the intercept, $\ln D$ at $1/T=0$, is $\ln D_0$.

$D/m^2\,s^{-1}$	$\ln(D/m^2\,s^{-1})$	T/K	$(10^3/T)/K^{-1}$
4.73×10^{-21}	-46.800	300	3.3333
3.35×10^{-15}	-33.330	500	2.0000
1.08×10^{-12}	-27.554	700	1.4286
2.66×10^{-11}	-24.350	900	1.1111
2.05×10^{-10}	-22.308	1100	0.90909

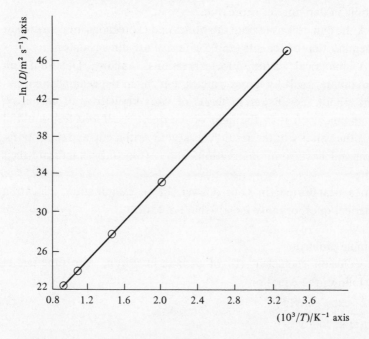

The straight-line graph verifies (3) for the diffusion of carbon in α-iron. From a least-squares fit to (4), the slope is -10.103 K, giving $U = 84.017$ kJ mol^{-1}; $D_0 = 1.999 \times 10^{-6}$ $m^2\,s^{-1}$. The sign and magnitude of U seem reasonable, since work must be done on the system to cause diffusion, and the energies of such processes are usually of the order of 1 eV (1 eV $= 1.6022 \times 10^{-19}$ J) per atom. D increases with T, and if (3) continues to hold, would tend to a limiting value of 1.000×10^{-6} $m^2\,s^{-1}$. However, at such high temperatures the material would melt and vaporize. The usefulness of D_0 here is mainly in connection with the evaluation of D from (3), over the temperature range studied.

Precision

In fitting a straight line to the experimental data by the method of least squares (Appendix A2), we obtain estimated standard deviations in both the slope and the intercept of the line. They are propagated as equivalent measures in the derived results for both U and D_0. Following the methods of Appendix A4, and assuming that errors other than in the experimental values of D are negligible, we obtain for the estimated standard deviations, σ (slope) and σ (intercept), the values of 0.00085 K and 0.00166 respectively. Hence, from (A4.5), $\sigma(U) = 0.007$ kJ mol^{-1} and $\sigma(D_0) = 0.003 \times 10^{-6}$ m^2 s^{-1}, and we may write the final results in the form $U = 84.017(7)$ kJ mol^{-1} and $D_0 = 1.999(3) \times 10^{-6}$ m^2 s^{-1}.

1

Introduction

Physical chemistry is concerned with the structures of chemical compounds, the mechanisms by which these compounds react, and the energy changes accompanying their reactions. The investigations of these fundamental facets of the subject are based principally upon experiment and measurement but, increasingly, theoretical and computer-simulation studies are developing to an extent where they can provide powerful alternative methods of investigation.

1.1 Structure

The term structure embraces a wide range of properties among which we may include the lengths of bonds between atoms, the angles between pairs of bonds, the distribution of electrons in atoms and molecules and the contact, or non-bonded, distances between them in the condensed state.

In the water molecule H_2O, for example, each hydrogen atom is linked by a bond of length 0.096 nm to the oxygen atom, and the H–O–H bond angle is 104.3°. The distance between the two hydrogen atoms, the interproton distance, is 0.152 nm. We may note here that a bond length depends to a small extent on the method used to measure it. If we take it to be the distance between the maxima of the electron density of two bonded atoms, as measured by X-ray diffraction, and compare it with the distance between the corresponding nuclei, as measured by neutron diffraction, there will be a small but significant difference in the results.

The hydrogen and oxygen atoms have, formally, one and eight electrons respectively, but there is a shift of electronic charge when the atoms are combined, as in the water molecule. Theoretical calculations have shown that the effective charges on the hydrogen and oxygen atoms are $0.16e$ and $-0.32e$ respectively, where e is the charge on an electron; the water molecule is said to be polar.

In the liquid state, the polarity of the water molecules leads to an association between them. Relatively strong hydrogen bonds are set up: a hydrogen atom acts as an electrostatic link between two oxygen atoms, one in the same molecule as the hydrogen atom and the other in an adjacent molecule. The hydrogen bonds in water are continually breaking and re-forming, and there exist constantly changing regions of water molecules with localized tetrahedral environments.

In ice at 90 K, there is nearly total hydrogen-bond formation between the water molecules, leading to a four-coordinate tetrahedral structure for ice (Figure 1.1); a discussion on stereoviewing is given in Appendix A1. At 313 K, the average number of hydrogen bonds is one-half of the maximum number. The structure of ice is relatively open, or loosely packed, and a volume contraction occurs on melting. The distance between oxygen atoms in adjacent, hydrogen-bonded molecules is 0.276 nm.

The average energy of hydrogen bonds in water is 18.8 kJ mol^{-1}. These bonds are much stronger among elements involving the first short row of the periodic table, particularly nitrogen, oxygen and fluorine. The hydrogen bonds in water are responsible for many of the properties of this substance that are anomalous when compared with those of the related hydrides of group VIB of the periodic table, as shown in Table 1.1. In this table, the quantity $\Delta H_f^{\ominus}(g)$ refers to the (standard) enthalpy change of the process of forming water (gas) from its

Table 1.1. *Some properties of the hydrides of group VIB elements*

	H_2O	H_2S	H_2Se	H_2Te
M p/K	273	190	210	225
B p/K	373	211	230	269
$\Delta H_f^{\ominus}(g)$/kJ mol^{-1}	-242	-22	$+66$	$+42$

Figure 1.1. Stereoview of the unit cell of the ice structure at 90 K; circles, in decreasing order of size, are oxygen and half-hydrogen (statistical representation).

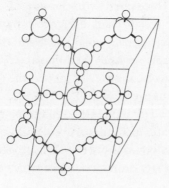 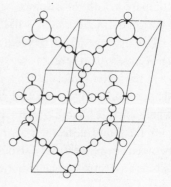

elements at 298 K and 1 atm. The important biological and chemical functions of water are related to its ability to form strong hydrogen bonds.

Many compounds do not exist in the form of discrete molecules. Sodium chloride, for example, exists as ions in the solid state, with no two ions particularly closely related to each other. The formula NaCl expresses the molar proportions of sodium and chlorine in the solid. If sodium chloride is heated, it melts at 1074 K to form a clear liquid containing sodium and chloride ions. On stronger heating, it boils at 1686 K, and the vapour consists of NaCl and Na_2Cl_2 molecules.

In the structure of sodium chloride (Figure 1.2), sodium and chloride ions occupy special positions in a face-centred cubic unit cell. There are four formula-entities associated with the unit cell of side 0.564 nm; the closest distances of ions are 0.282 nm for $Na^+–Cl^-$ and 0.399 nm for $Na^+–Na^+$ or $Cl^-–Cl^-$. The macroscopic crystal of sodium chloride can be considered to be formed by packing unit cells together such that each face is common to two adjacent unit cells.

Calcium sulphate dihydrate (gypsum) is a quite different structure (Figure 1.3). The structural units are calcium ions, sulphate ions and water molecules. The sulphate ion is an essentially covalently bonded entity in itself. It has a tetrahedral geometry that is preserved, with only minor variation, in the structures of all sulphates: S–O bond distance = 0.16 nm, O–S–O bond angle = 109.5°. The water molecules are linked in the structure by hydrogen bonds, and the cohesion in one direction arises solely from these bonds.

1.2 Kinetics

The mechanisms of chemical reactions, which include rate, molecularity and reaction pathway, are often derived from a study of their kinetics and of factors that control the kinetics. For example, the formation of methoxyethane

Figure 1.2. Stereoview of the unit cell of the sodium chloride NaCl structure; circles, in decreasing order of size, are chloride ions and sodium ions.

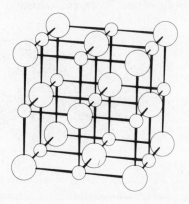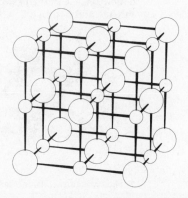

from iodomethane and sodium ethanoate occurs as a result of collisions between molecules of these reactants.

$$CH_3I + CH_3CH_2ONa \rightarrow CH_3OCH_2CH_3 + NaI \tag{1.1}$$

The rate of this simple reaction depends on how often the reactants meet each other, and these events are proportional to their concentrations. Not all collisions bring about reaction, but only those of sufficient energy to drive the process forward. Reaction (1.1) is bimolecular: there are two molecular species involved in the reaction process. It is also of second order kinetics: the rate is proportional to the concentrations of both reactants.

The apparently straightforward reaction between hydrogen and chlorine

$$\tfrac{1}{2}H_2(g) + \tfrac{1}{2}Cl_2(g) \rightarrow HCl(g) \tag{1.2}$$

is very slow unless activated by radiation of sufficient energy:

$$\tfrac{1}{2}Cl_2(g) + h\nu \rightarrow Cl\cdot(g) \tag{1.3}$$

$h\nu$ represents a photon, a packet of radiant energy, where h is the Planck constant and ν is the frequency of the radiation, in this case in the ultraviolet (uv) region of the spectrum (Figure 1.4); $Cl\cdot$ is a free radical. A chain mechanism is set up which, if uncontrolled, will proceed at an explosive rate.

$$Cl\cdot(g) + H_2(g) \rightarrow HCl(g) + H\cdot(g) \tag{1.4}$$

$$H\cdot(g) + Cl_2(g) \rightarrow HCl(g) + Cl\cdot(g) \tag{1.5}$$

Reactions (1.1) and (1.2) differ in the way in which they acquire the energy necessary to initiate reaction, (1.1) by warming in alcoholic solution and (1.2) by uv irradiation.

The bond dissociation energies of hydrogen and chlorine molecules, the energies needed to break the molecules down to atoms, are $436 \, kJ \, mol^{-1}$ and $242 \, kJ \, mol^{-1}$ respectively. It follows that (1.3) will begin if supplied with energy somewhat greater than $242 \, kJ \, mol^{-1}$, or $242 \times 10^3/(6.0221 \times 10^{23})$† per molecule;

Figure 1.3. Stereoview of the unit cell of the structure of calcium sulphate dihydrate $CaSO_4 \cdot 2H_2O$ (gypsum) structure; hydrogen bonds are shown by double lines. Circles, in decreasing order of size, are oxygen, calcium, sulphur and hydrogen.

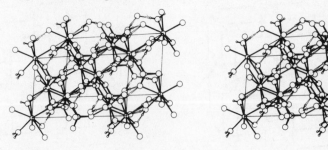

† The Avogadro constant/mol^{-1}.

this amount of energy corresponds to a radiation of wavelength 402 nm. Hence, uv radiation brings about a ready reaction.

The decay of the nuclei of elements of atomic number greater than 83 (bismuth) is spontaneous, and energy is emitted as α or β radiation (Figure 1.5). The rate of radioactive decay depends only on the concentration of the radioactive species present: it is a first-order process.

1.3 Energetics

Chemical reactions take place with differing degrees of completeness, ranging from hardly to all, such as the dissolution of silver iodide in water at

Figure 1.4. Radiation spectrum frequencies and wavelengths.

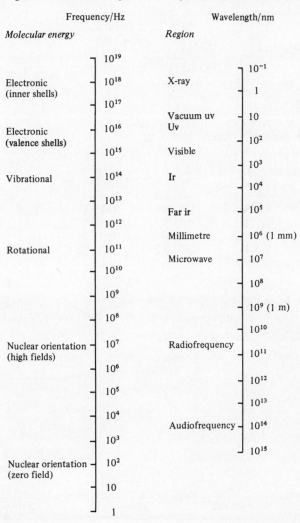

298 K, to completely for all practical purposes, such as the oxidation of metallic sodium in chlorine. Between these two extreme examples, all other stages can be encountered with reacting systems. As an example, consider the synthesis of ammonia by the Haber process:

$$\tfrac{1}{2}N_2(g) + \tfrac{3}{2}H_2(g) \rightleftharpoons NH_3(g) \tag{1.6}$$

where the $\rightleftharpoons$ sign indicates that an equilibrium is attained at any given temperature and pressure. Tables 1.2 and 1.3 list some experimental results for

Table 1.2. *Nitrogen–hydrogen–ammonia equilibrium: percentage of ammonia at equilibrium*

T/K	P/atm		
	10	50	100
623	10.4	25.1	37.1
673	3.85	15.1	24.9
723	2.04	9.17	16.4
773	1.20	5.58	10.4

Table 1.3. *Nitrogen–hydrogen–ammonia equilibrium: equilibrium constant K_P/atm^{-1}*

T/K	P/atm		
	10	50	100
623	0.0266	0.0278	0.0288
673	0.0129	0.0130	0.0137
723	0.00659	0.00690	0.00725
773	0.00381	0.00387	0.00402

Figure 1.5. Radioactive decay in the ^{238}U series: $-\alpha$ and $-\beta$ indicate emission of the corresponding radiation and the half-lives of the parent species are also given.

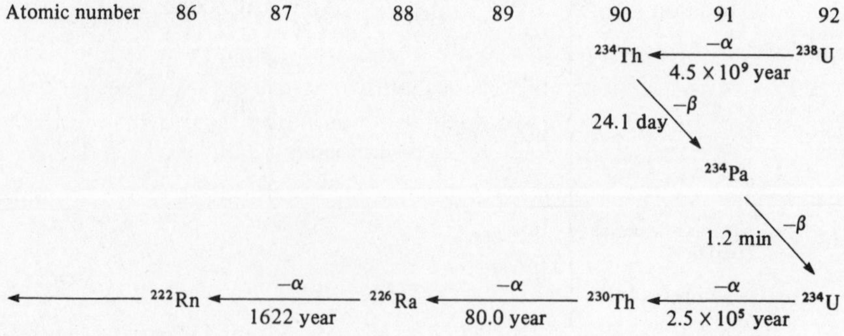

this equilibrium. We can see from the results that the equilibrium constant depends only on the temperature, and decreases with an increase in this variable, whereas the amount of ammonia formed increases with an increase in pressure but decreases with an increase in temperature. The relatively small change in K_P with P at constant T arises from the departure of the gases from ideality. It is important to distinguish between changes in the equilibrium constant and changes in the extent of reaction. The variations illustrated by Tables 1.2 and 1.3 are in accord with Le Chatelier's principle, which states that *a system at equilibrium, when subjected to a perturbation, responds in a way that tends to annul the effect of the perturbation.*

The production of ammonia, the forward reaction in (1.6), is accompanied by a change in the free energy of the reacting system.

$$\Delta G = \Delta H - T \, \Delta S \tag{1.7}$$

Thus, in (1.7) a negative value for the change in free energy ΔG is taken to define a thermodynamically spontaneous reaction. The other terms in (1.7) represent the change in enthalpy ΔH, the temperature T and the change in entropy ΔS. At the temperatures considered in Tables 1.2 and 1.3, ΔG is actually positive; the amount of ammonia formed at equilibrium is quite small.

We can conclude from Tables 1.2 and 1.3 that the amount of ammonia formed at 298 K would be much greater than that at, say, 723 K. This view is supported by the van't Hoff equation relating the equilibrium constant to the standard free energy change:

$$\Delta G^{\ominus} = - RT \ln K \tag{1.8}$$

Hence, from Table 1.3 $\Delta G^{\ominus}(723 \text{ K}, 10 \text{ atm})$ is 30.2 kJ mol^{-1}. Now $\Delta G^{\ominus}(298$ K, 10 atm) is -16.5 kJ mol^{-1} and the corresponding value of the equilibrium constant is 7.8×10^2. However, the rate of reaction at 298 K is immeasurably slow, and we see how considerations of kinetics are introduced. Notwithstanding the energetics are favourable at 298 K, the reactants must acquire a certain amount of energy, the activation energy, before reaction takes place at a measurable rate. The term spontaneous, applied to a reaction, means that it is accompanied by a decrease in the free energy of the reacting system (it is thermodynamically feasible), not necessarily that it takes place immediately the reactants are brought together.

Since kinetic studies involve the mechanisms of reactions, and mechanisms are often dependent on structure, we can see now how it is that the fundamental topics of structure, kinetics and energetics are very closely linked in the physical chemistry of reactions. In the ensuing chapters, we shall consider these subjects, and related matters, in some detail.

Problems 1

1.1 In the molecule of ClO$_2$, the Cl–O bond distance is 149 pm and the distance between the oxygen atoms is 254 pm. Calculate the O–Cl–O bond angle.

1.2 Assuming that no hydrogen bonding existed in water, what would be its approximate melting and boiling points (see Table 1.1)?

1.3 In the sulphate ion, the excess charge on each oxygen atom is $-0.95e$. What would be the corresponding charge on the sulphur atom?

1.4 What is the second largest sodium–chlorine distance in the sodium chloride structure?

1.5 The energy needed to initiate a reaction is approximately 400 kJ mol^{-1}. What is the wavelength of the equivalent radiation source, and in what region of the electromagnetic spectrum would such radiation occur?

1.6 The volume of helium, at 273 K and 1 atm, liberated by the radioactive disintegration of radium is $0.043 \text{ cm}^3 \text{ gm}^{-1} \text{ year}^{-1}$. In the same time, the number of α particles emitted in the disintegration of the same amount of radium is 116×10^{16}. Calculate the Avogadro constant N_A, assuming that each α particle yields one atom of helium.

2

Atoms, molecules and bonding

In this chapter, we shall consider some theoretical aspects of atoms and molecules, and the manner in which they interact to form solids. Chemical structure and reaction are governed by the behaviour of electrons in atoms and molecules, so that it is not surprising to find that the electron figures prominently in a discussion of chemical combination.

2.1 Bohr's atomic theory

When hydrogen gas is subjected to an electric discharge, the radiation given out, the emission spectrum, includes light of four discrete frequencies in the visible spectrum (Table 2.1); it is a line spectrum. Balmer showed, empirically, in 1895 that the spectrum could be represented by the equation

$$\tilde{v} = R_H[(1/n_1^2) - (1/n_2^2)] \tag{2.1}$$

where $\tilde{v}$ is the wavenumber, or reciprocal wavelength $1/\lambda$, of a spectral line; R_H is the Rydberg constant for hydrogen, n_1 is two for this series of lines – the Balmer series – and n_2 is another integer, greater than two.

It was a feature of the Bohr theory that it explained these results, as the spectrum of atomic hydrogen, in terms of transitions of electrons between two energy levels. An electron moving from an energy level E_2 to a lower, more negative, energy level E_1 emitted radiation of frequency v given by

$$v = (E_2 - E_1)/h \tag{2.2}$$

where h is the Planck constant.

From studies on black-body radiation, Planck had postulated in 1900 that an oscillating electron radiated energy only in certain amounts, or quanta, the single quantum of radiant energy being given by

$$E = hv \tag{2.3}$$

Bohr's theory applied classical mechanics to a moving electron, but with Planck's quantum condition imposed in order to explain experimental results. It departed from classical theory in requiring energy to be emitted or absorbed by a moving electron only when it changed from one energy state to another. In a given energy state, an electron was allowed to move without change of energy on a circular (Bohr) or elliptical (Sommerfeld) orbit; such energy states were called stationary states. Each orbit was specified by two integers, n a principal quantum number, and l an azimuthal, or angular momentum, quantum number. The allowed values of n were integral, to correspond with the quantum condition, and l was assigned the values $0, 1, 2, \ldots, (n-1)$ for a given value of n.

The simplest kinds of chemical species contain one electron, as in H, He$^+$, Li^{2+}, and are called 'hydrogen-like' species. Their energies depend only on n and the nuclear charge Z, and are independent of l. The lowest energy state is that for the orbit closest to the nucleus, and it corresponds to $n=1$, the ground state of the atom. In the next highest energy level n is 2, and so on. The radius of the first orbit, corresponding to $n=1$, is known as the Bohr radius for hydrogen a_0, and is given by

$$a_0 = h^2 \varepsilon_0 / \mu e^2 \pi \tag{2.4}$$

where ε_0 is the permittivity of a vacuum and μ is the reduced mass of the electron, $m_e m_p / (m_e + m_p)$ where m_e and m_p are respectively the masses of the electron and the proton; reduced mass is discussed in Appendix A5.

In the Bohr theory, the first four successive energy levels were denoted by the letters K, L, M and N, and could accommodate 2, 8, 18 and 32 electrons respectively. For hydrogen-like species, energy values can be calculated from the Bohr theory, using the equation

$$E_n = -\mu e^4 Z^2 / 8 n^2 h^2 \varepsilon_0^2 \tag{2.5}$$

as given in Table 2.2. The zero of energy corresponds here to the ionized atom (H$^+$ + e$^-$) at 0 K, so that the ionization energy of hydrogen is 1312.03 kJ mol^{-1}. If the electron changes from the energy of $n=3$ to that of $n=2$ in hydrogen, the energy is radiated at a wavelength of 656.5 nm (Table 2.1).

For atoms containing more than one electron, the energy of an electron depends on n, Z and l. For example, if n is 4, then l may take the values 0, 1, 2 and 3, and the electrons in these states are denoted by the letters s, p, d and f respectively. However, in applying the Bohr theory to species with two or more electrons,

Table 2.1. *Lines in the visible spectrum of atomic hydrogen*

	Red/orange	Blue	Blue/violet	Violet
$10^{-14}\, \nu/\text{Hz}$	4.566	6.165	6.905	7.307
λ/nm	656.5	486.3	434.2	410.3
$10^{-2}\, \bar{\nu}/\text{m}^{-1}$	15232	20563	23031	24372

certain objections arise. The theory requires us to specify, simultaneously, both the position and the momentum of an electron. This requirement cannot be achieved precisely; it was shown later that the uncertainties in position Δx and momentum Δp are governed by Heisenberg's uncertainty principle

$$\Delta x \, \Delta p \approx h \tag{2.6}$$

We cannot define the position of an electron precisely and follow its orbit. That it can be done with planets rather than with electrons depends upon the different sizes of these bodies in relation to the methods used for their detection and measurement; large bodies are not disturbed by the process of observation.

The Bohr theory could not explain the spectra of helium, the next species to hydrogen in the periodic table. Less still could it hope to account for, say, how methane has four equivalent bonds of C–H length 0.109 nm and H–C–H angle 109.7°, yet in ethene each carbon atom forms three bonds, two C—H and one C=C: these explanations are typical requirements of a good theory of atomic and molecular structure.

Experiments on the diffraction of electrons by Davisson and Germer in 1925 showed that, as well as the particle-like properties demonstrated by the photoelectric effect and Thompson's cathode-ray experiments, electrons had the properties of waves. If, then, we are not to obtain success based on treating the electron as a particle, it seems not unreasonable to consider its behaviour as a wave, and so we turn our attention to the wave mechanics of atomic structure.

2.2 Wave-mechanical theory and covalent bonding

The properties of the electron are now represented by a wave function ψ, which has special solutions known as standing, or time-independent, waves; they have positions of zero amplitude, or nodes. In the case of a stretched string, these standing waves represent the fundamental and overtone vibrations of the string, and may be labelled $0, 1, 2, \ldots$, depending upon the number of nodes between the fixed ends.

The wave equation for a total electronic energy E, given by Schrödinger in 1926, may be written as

$$\nabla^2 \psi + (8\pi^2 m_e/h^2)(E - V)\psi = 0 \tag{2.7}$$

Table 2.2. *Energy levels for hydrogen*

n	$-10^{-19}E_n/\text{J}$	$-E_n/\text{kJ mol}^{-1}$
∞	0	0
⋮	⋮	⋮
4	1.3617	82.00
3	2.4208	145.78
2	5.4468	328.01
1	21.787	1312.03

In this equation, ∇^2 is the Laplacian operator for three-dimensional space, $(\partial^2/\partial x^2)+(\partial^2/\partial y^2)+(\partial^2/\partial z^2)$, and V is the potential energy of the electron. If we consider the electron no longer at a point, but spread out to form a cloud, or electron density distribution, then $|\psi|^2$ measures the electron density at any point. Where the wave function is a complex quantity, $|\psi|^2$ is the product of ψ and its complex conjugate ψ^*. Hence, we may think of ψ as an amplitude of the electron wave, varying in space and, more generally, in time. More precisely, we may say that $|\psi|^2 \, d\tau$ represents the probability that an electron will be found in the elemental volume of space $d\tau$. It follows that

$$N^2 \int_{\substack{\text{all} \\ \text{space}}} |\psi|^2 \, d\tau = 1 \tag{2.8}$$

The wave equation is said now to be normalized, and N is the normalizing constant.

The wave equation can be solved without approximation for the hydrogen atom. It leads to values for the energy states of this species that are given by an equation equivalent to (2.5) and, hence, with the same values as given in Table 2.2. We still use the constant a_0 as the Bohr radius, a fundamental atomic unit of length, but its significance is changed. It is not now thought of as a precise value of the smallest radius of the hydrogen atom ($n = 1$), but rather as the most probable distance of the electron from the nucleus in the ground state of that atom. The discrete values for the energies are introduced explicitly in the solutions for the energies instead of in the empirical manner of the Bohr theory.

The solutions of (2.7) for ψ contain, in general, a radial part $R_{n,l}(r)$ and an angular part $Y_{m,m_l}(\theta, \phi)$. The radial part depends on the nuclear charge and the quantum numbers n and l, whereas the angular part depends on the quantum numbers l and m_l: like l, m_l arises through applying quantum restrictions to the angular momentum of the electron, and may have the $2l + 1$ integral values l, $(l-1), (l-2), \ldots, (-l)$. The symbols r, θ and ϕ are the polar spatial coordinates that conveniently replace x, y and z. A fuller discussion of the basis of m_l is to be found in *Physical Chemistry*, P. W. Atkins, Oxford University Press, Oxford, 1982.

2.3 Atomic orbitals

The solutions of the wave equation for the hydrogen atom may be called atomic orbitals. They are labelled according to the energy state of the corresponding electron (Table 2.3). From the table, we deduce that s orbitals have no angular dependence ($l=0$); they are spherically symmetrical. The three p orbitals have the same energy, and are said to be degenerate. In the case of the hydrogen atom, the ns and the three np orbitals have the same energy, as is evident from (2.5). This result is particular and, in general, the energy depends also upon l.

From the probability interpretation of $|\psi|^2$, it is clear that the orbitals do not terminate at any particular value of the distance r from the nucleus: there is a finite probability of finding some electron density at all distances from the nucleus. It is convenient to consider three-dimensional surfaces for the orbitals that are drawn so that a given fraction, say 99%, of the total electron density is enclosed by them. The shape of the orbital surface will depend on Y_{l,m_l} and its size on $R_{n,l}$. For the 99% surface, the radius of a carbon atom in its normal state is approximately that of its closest, non-bonded distance, the van der Waals' radius of 0.37 nm.

We think of the 1s, 2p and 3d atomic orbitals as shown in Figure 2.1. The position of the electron is considered in a probabilistic manner. Thus, the probability of finding the 1s electron lying within a spherical shell defined by radii r and $(r+dr)$ is $4\pi r^2|\psi|^2(1s)\,dr$. The 1s wave function for hydrogen is

$$\psi_{1,0,0}=N\exp(-r/a_0) \tag{2.9}$$

Hence, it follows that the maximum probability is given for $r=a_0$, that is, when the electron is distant from the nucleus by the Bohr radius.

So far we have seen that an electron may be represented by a wave function ψ from which the various electron energies may be calculated. They depend upon a set of quantum numbers: n determines its size, l determines its geometry and m_l determines which of the degenerate orbitals of the same value of l is being considered (see again Figure 2.1 and Table 2.3). In addition, each electron has a property referred to as spin, which arises from a full (relativistic) wave-mechanical treatment of the electron. There are two allowed values for the spin, given by the spin quantum numbers s of $\pm\frac{1}{2}$; two electrons with the same spin are said to be unpaired, or parallel; otherwise, they are paired, or antiparallel.

Table 2.3. *Electron notation*

n	l	m_l	State
1	0	0	1s
2	0	0	2s
2	1	0	
2	1	1	2p
2	1	-1	
3	0	0	3s
3	1	0	
3	1	1	3p
3	1	-1	
3	2	0	
3	2	1	
3	2	-1	3d
3	2	2	
3	2	-2	

The set of four quantum numbers is governed by the Pauli principle, which states that no two electrons in the same atom can have the same four quantum numbers. Hence, an orbital can accommodate a maximum of two electrons, having the same values of n, l and m_l, provided that their spins are paired.

2.4 Aufbau principle

The description of atomic orbitals is closely linked to the periodic table of the elements. Starting with hydrogen, we can feed electrons into the orbitals, one at a time and in accordance with the Pauli principle, so as to balance the nuclear charge. Figure 2.2 shows, diagrammatically, the results of carrying out this process for atoms up to argon. In the figure, each orbital is represented by a rectangular box, and paired spins are marked ↑↓. The elements hydrogen to beryllium present no new problem. In boron, the 2p electron can occupy any one

Figure 2.1. Atomic orbitals illustrated by surfaces of constant $|\psi|^2$: (*a*) 1s, (*b*) 2p, (*c*) 3d. The notation for the p orbitals is straightforward. With the d orbitals the lobes of d_{z^2} are directed along the z axis and those of $d_{x^2-y^2}$ are along x and y. The lobes in d_{xz}, d_{yz} and d_{xy} lie in the corresponding planes.

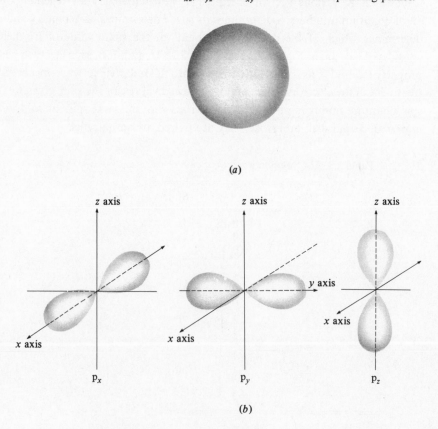

(*a*)

(*b*)

of three degenerate orbitals. It is convenient to designate these orbitals as $2p_x$, $2p_y$ and $2p_z$ according to their directional character (see Figure 2.1). With carbon, there are two arrangements to consider; that shown, and another in which any of the 2p orbitals contain two spin-paired electrons. The dilemma is resolved through Hund's rule, which states that degenerate orbitals tend to be occupied singly as far as possible; this arrangement leads to a minimum electronic energy for the species.

At this stage, we may recall that the stability of the inert-gas electronic configuration was the pillar of Lewis's electron-pair bond theory: it is given a quantitative realization here through the system of occupied atomic orbitals, up to and including those of the principal quantum number n, that is, a penultimate electronic configuration of $(ns)^2(np)^6$. The energy of an atomic orbital can be associated with the ionization energy of an atom; the value of the first ionization energy of each atom in Figure 2.2 is given in parentheses. (We make no apology for the occasional use of non-SI units, here the electron volt eV. Reference to the

Figure 2.1. (*c*) 3d.

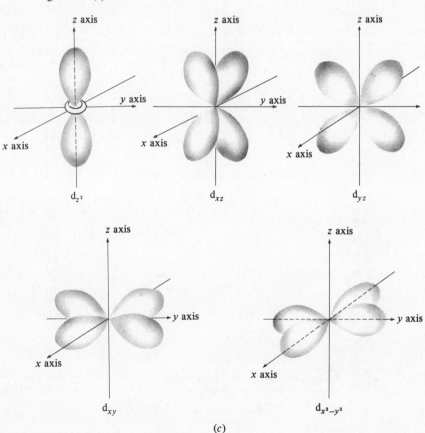

(*c*)

Figure 2.2. Illustrating the aufbau principle: ground state electron configurations of some elements, with first ionization energies in electron volts (1 eV = 96.487 kJ mol^{-1}).

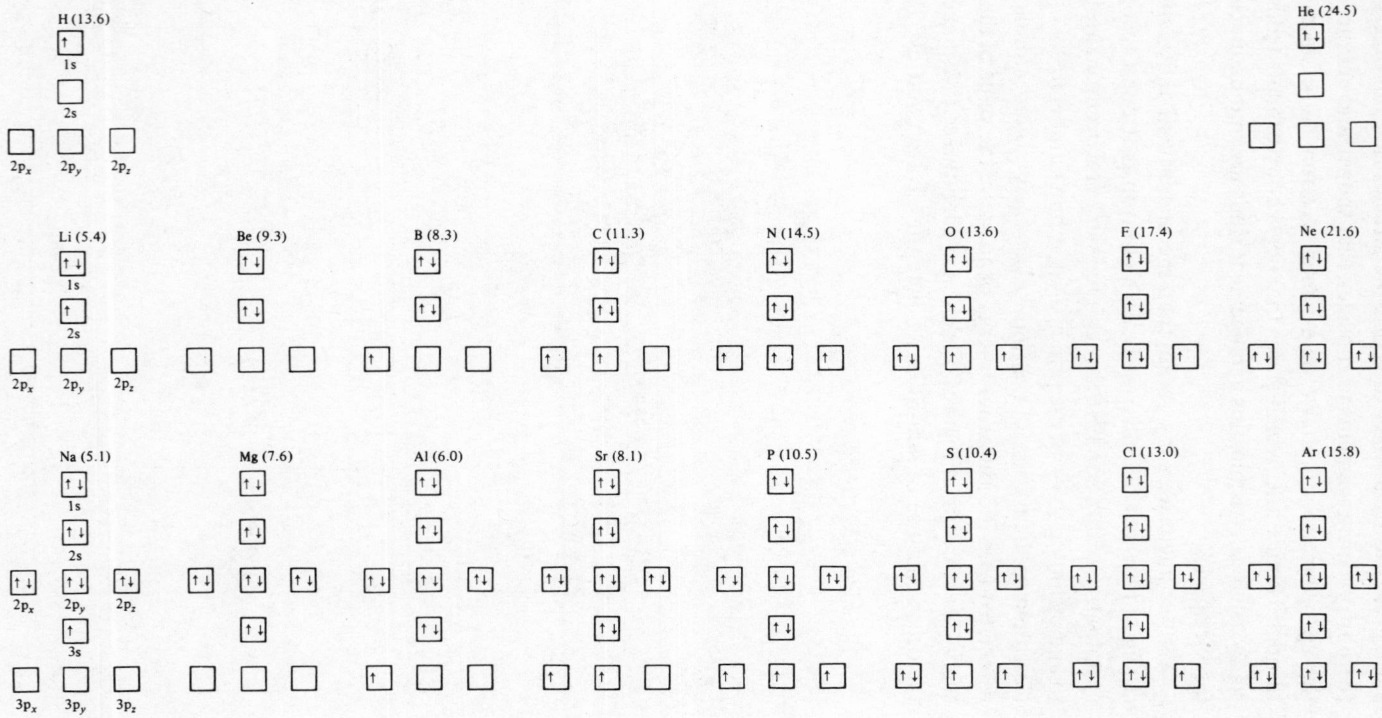

H (13.6)
1s
2s
2p$_x$
2p$_y$
2p$_z$

Li (5.4)
1s
2s
2p$_x$
2p$_y$
2p$_z$

Be (9.3)

B (8.3)

C (11.3)

N (14.5)

O (13.6)

F (17.4)

Ne (21.6)

Na (5.1)
1s
2s
2p$_x$
2p$_y$
2p$_z$
3s
3p$_x$
3p$_y$
3p$_z$

Mg (7.6)

Al (6.0)

Sr (8.1)

P (10.5)

S (10.4)

Cl (13.0)

Ar (15.8)

He (24.5)

literature will show that SI units have not attained universal acceptance, and the conscientious student will, for as long as we can forsee, need to be able to make the simple conversions between different units for the same quantity.) With an obvious notation, we can now write the electronic configurations of atoms; Table 2.4 lists them for some species from hydrogen to potassium .

The periodic table is not completed exactly in the manner indicated by Table 2.4. After calcium, the 3d energy level, which is not occupied in the ground state of atoms of atomic number less than 21, begins to fall below that of the 4s (Figure 2.3), and between calcium, $(Ar)(4s)^2$, and zinc, $(Ar)(3d)^{10}(4s)^2$, the 3d orbitals are progressively filled, giving rise to the first series of transition-metal elements.

Table 2.4. *Application of the aufbau principle*

H	$(1s)^1$
He	$(1s)^2$
Li	$(1s)^2(2s)^1$
$\vdots$	
C	$(1s)^2(2s)^2(2p)^2$ or $(1s)^2(2p_x)^1(2p_y)^1$
$\vdots$	
Ne	$(1s)^2(2s)^2(2p)^6$
Na	$(1s)^2(2s)^2(2p)^6(3s)^1$ or $(Ne)(3s)^1$
$\vdots$	
Si	$(Ne)(3s)^2(3p)^2$
$\vdots$	
Ar	$(Ne)(3s)^2(3p)^6$
K	$(Ar)(4s)^1$
K^+	(Ar)

Figure 2.3. Variation of atomic energy levels with atomic number Z.

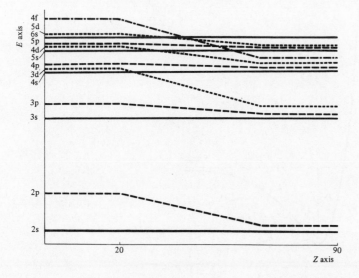

2.5 Molecules: valence-bond method

The wave equation for two or more electrons cannot be solved without approximations; this condition applies, therefore, to all molecules. A common approach considers a molecular wave function to be set up by a linear combination of atomic orbitals (LCAO):

$$\Phi = c_1\psi_1 + c_2\psi_2 + \cdots + c_n\psi_n \tag{2.10}$$

where $\psi_1 \ldots \psi_n$ form a basis set of atomic orbitals, and the constants $c_1 \ldots c_n$ are chosen to as to minimize the total electronic energy of the molecule.

Figure 2.4 shows the variation of potential energy V with interatomic distance for a pair of atoms capable of forming a molecule. The curve is quite well represented by an equation due to Morse:

$$V(r) = D_e\{1 - \exp[-a(r - r_e)]\}^2 \tag{2.11}$$

where D_e is the theoretical dissociation energy of the molecule, and a is a constant. At infinite separation the potential energy is zero. We can think of two atoms, say hydrogen atoms, represented by functions $\psi_A(1)$, with electron 1 on atom A, and $\psi_B(2)$, with electron 2 on atom B. As the atoms are brought together, we can imagine that they retain much of their individual electronic character, and that a molecular bond is formed through an overlap of atomic orbitals once the two atoms are close enough to each other. The potential energy of the pair of atoms

Figure 2.4. Variation in potential energy V with interatomic distance r. At the equilibrium distance r_e, the coulombic forces of attraction balance the forces of repulsion. The energy D_e is the theoretical dissociation energy of the paired atoms.

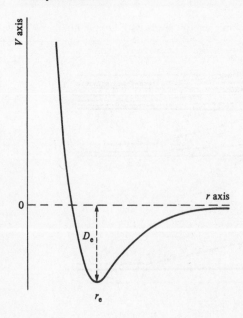

decreases as the molecule is formed, with a minimum in $V(r)$ at the equilibrium interatomic distance r_e. As r is decreased below r_e, the curve rises rapidly because of internuclear and electron–electron repulsion.

Once the molecule is formed, electron 1 may be found closer to nucleus B, and vice versa. Because of the indistinguishability of electrons, we formulate this result as

$$\Phi = c_1\psi_A(1)\psi_B(2) + c_2\psi_A(2)\psi_B(1) \tag{2.12}$$

where, again, c_1 and c_2 are chosen to produce a minimum energy conformation. From symmetry, in the case of the hydrogen molecule, and because probabilities are proportional to $|\psi|^2$, we have $c_1^2 = c_2^2$, whence

$$\Phi_\pm = \psi_A(1)\psi_B(2) \pm \psi_A(2)\psi_B(1) \tag{2.13}$$

Heitler and London used this type of wave function, the valence-bond wave function, and obtained energies E_+ and E_- (Figure 2.5): Φ_+, corresponding to paired electrons, leads to bond formation, whereas Φ_-, corresponding to parallel spins, does not. The value of E_+ was $-303\,\text{kJ mol}^{-1}$ (experimental, $-458\,\text{kJ mol}^{-1}$) and that of r_e was $0.087\,\text{nm}$ (experimental, $0.074\,\text{nm}$). Sophisticated modifications of (2.13), which will not concern us here, led to results in excellent agreement with the experimental data, so giving support to the theory and its mode of application. Pictorially, electron 1 is paired with electron 2 in the molecule of hydrogen. We see here the evidence for the electron-sharing concept of covalency put forward by Lewis, but now with a firm, quantitative foundation.

Before leaving this section, we shall consider one simple extension to (2.13). It is possible to consider the inclusion of ionic contributions to the bonding in molecular hydrogen, in which both electrons are around a single hydrogen atom,

Figure 2.5. Valence-bond energies for the hydrogen molecule; E_+, E_- and E_{true} are shown as a function of internuclear distance r.

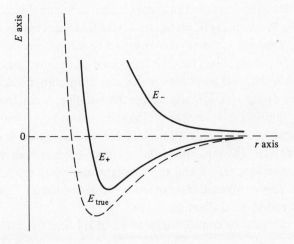

A or B. Thus, we might write the following possible structures: $H_A^1 H_B^2$, $H_A^2 H_B^1$, $H_A^{1,2} H_B$ and $H_A H_B^{1,2}$. Following the same notation, we may formulate the wave function for the bonding situation as

$$\Phi_+ = \psi_A(1)\psi_B(2) + \psi_A(2)\psi_B(1) + \lambda[\psi_A(1)\psi_A(2) + \psi_B(1)\psi_B(2)] \qquad (2.14)$$

The right-hand side of (2.14) may be considered as $\psi_{\text{covalent}} + \lambda\psi_{\text{ionic}}$, with λ selected so as to minimize the energy. The structures postulated above for the hydrogen molecule are of commensurate energy, and are known as canonical structures. The compounding of wave functions of similar energy is known as resonance. It does not express any actual state, but rather our attempt to obtain a formulation, in terms of possible and reasonable canonical structures, that is as closely representative of the true state as is possible within the terms of the given theory.

2.6 Molecules: molecular-orbital method

An alternative approach is termed the molecular-orbital method. It is of more general applicability to chemical problems, and forms a more logical extension of the atomic orbital treatment of atoms than does the valence-bond method. It takes the ion H_2^+ as the foundation structure, rather than the bond in molecular hydrogen used by the valence-bond method. The distribution of the electron is now given by

$$\Phi = \psi(1s_A) + \psi(1s_B) \qquad (2.15)$$

When the electron is close to nucleus A, the amplitude of $\psi(1s_B)$ is small and Φ approximates to $\psi(1s_A)$, and vice versa. The wave function (2.15) is like an atomic orbital, but it encompasses the whole molecule; it is a molecular orbital. The probability of finding the electron at any point is $|\Phi|^2 \, d\tau$, where $d\tau$ is the infinitesimal volume element around the given point. From (2.15), we see that this probability is given by $[|\psi|^2(1s_A) + |\psi|^2(1s_B) + 2\psi(1s_A)\psi(1s_B)] \, d\tau$. In the region of nucleus A, this function resembles the atomic orbital for hydrogen, and similarly near to B. In the intermediate region there is an enhancement, or accumulation, of electron density, given by the term $2\psi(1s_A)\psi(1s_B) \, d\tau$, above that expected for the hydrogen atoms alone. The reason lies in the constructive interference of the electron waves in the intermediate region between the nuclei: each contribution is of the same phase, so that the total amplitude is the sum of its components.

The molecular orbital $\psi(1s_A) + \psi(1s_B)$, responsible for bonding, is called a σ bonding orbital: it has cylindrical symmetry about the internuclear axis. Another molecular orbital, the next highest energy solution of the wave equation, is the combination $\psi(1s_A) - \psi(1s_B)$. In this case, the atomic orbitals interfere destructively in the intermediate region and the resultant molecular orbital is called antibonding (σ^*). These two molecular orbitals are illustrated by Figure 2.6, a molecular-orbital energy-level diagram.

Molecular hydrogen can now be considered in terms of H_2^+ and the aufbau principle. Two electrons must be fed into the molecular orbital of lowest energy. The configuration will now be $(1s\sigma)^2$, with the two electrons having paired spins. We can also explain the instability of the helium molecule, He_2. The first two electrons form $(1s\sigma)^2$: the next electron must enter the first antibonding orbital, so that the bonding is weakened. The fourth electron leads to $(1s\sigma^*)^2$, and the effects of $(1s\sigma)^2$ and $(1s\sigma^*)^2$ cancel; in fact, the antibonding effect is slightly the stronger. The configuration of dilithium, Li_2, is $(1s\sigma)^2(1s\sigma^*)^2(2s\sigma)^2$ and a bond is formed. Homonuclear diatomic molecules (X_2) are easily considered in this manner. Of course, as the number of atoms in the molecule increases, so does the complexity of the calculation, but the principles outlined here apply to all species.

When the second row elements are introduced into a molecule, p orbitals may be involved in bonding. Two 2p orbitals may overlap along the direction of the internuclear axis, as in the fluorine molecule. The bonding molecular orbital is called $2p\sigma$, and the corresponding antibonding molecular orbital is $2p\sigma^*$; both have cylindrical symmetry about the internuclear axis. Where p orbitals overlap in directions other than along the internuclear axis, the molecular orbitals so formed have reflexion symmetry across a nodal plane; they are called π molecular orbitals. Figure 2.7 illustrates some simple molecular orbitals.

In heteronuclear diatomic molecules (XY), similar considerations can be used. In hydrogen fluoride, for example, we have $\psi(1s)$ orbitals for hydrogen overlapping $\psi(2p)$ for fluorine, giving rise to pσ molecular orbitals of cylindrical symmetry.

2.7 Valence-bond and molecular-orbital methods compared

We will use the hydrogen fluoride molecule in order to compare the two methods that we have discussed. In the valence-bond approach, we can consider

Figure 2.6. Molecular-orbital energy-level diagram.

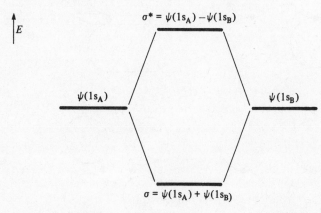

$\sigma^* = \psi(1s_A) - \psi(1s_B)$

E

$\psi(1s_A)$ $\psi(1s_B)$

$\sigma = \psi(1s_A) + \psi(1s_B)$

Atom A   Molecule   Atom B

Figure 2.7. Simple atomic and molecular orbitals (homonuclear species): (*a*)
s+s → sσ, (*b*) s−s → sσ*, (*c*) p+p → pσ, (*d*) p−p → pσ*, (*e*) p+p → pπ, (*f*)
p−p=pπ*. Here we have enhanced the notation so as to distinguish, for
example, between pσ and pπ molecular orbitals. In general, ψ, though not
$|\psi|^2$, being an amplitude has regions (lobes) of positive and negative values.

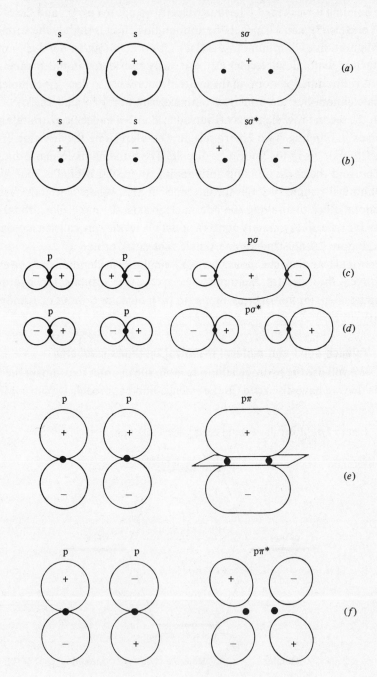

three canonical structures: H—F (covalent), H^+F^- (ionic) and H^-F^+ (ionic). We then set up the valence-bond function

$$\Phi_+ = \psi_{\text{covalent}} + \lambda_1 \psi_{\text{ionic}} + \lambda_2 \psi_{\text{ionic}} \tag{2.16}$$

Calculation shows that λ_2 is negligible, as might be expected. Hence,

$$\Phi_+ = \psi_{\text{covalent}} + \lambda \psi_{\text{ionic}} \tag{2.17}$$

and λ is chosen so as to obtain a minimum energy for the molecule.

In the molecular-orbital description, ten electrons are fed in, one at a time, into the molecular orbitals. The 1s and 2s electrons of fluorine take no part in the bonding, and may be said to form non-bonding molecular orbitals $1s\sigma$ and $2s\sigma$. A 2p electron of fluorine, say $2p_z$, forms a bonding $3p\sigma$ molecular orbital, and the remaining $(2p_x)^2$ and $(2p_y)^2$ electrons of fluorine form $2p\pi$ molecular orbitals. Thus, we may write hydrogen fluoride as $(1s\sigma)^2(2s\sigma)^2(2p\pi_x)^2(2p\pi_y)^2(3p\sigma)^2$. The bonding orbital is the $3p\sigma$; we must distinguish between molecular orbitals that take no part (non-bonding) and those that operate against bond formation (antibonding). The molecular orbital for hydrogen fluoride may be written formally as

$$\Phi_+ = c_1 \psi H_{1s} + c_2 \psi F_{2p_z} \tag{2.18}$$

2.8 Hybridization

So far we have regarded overlap as taking place between unmodified atomic orbitals. In the water molecule (see Chapter 1), we might expect the divalency of oxygen to lead to two bonds, with sp overlap, at 90° to each other. But we know from experiment that the bond angle is approximately 104.5°. Suppose that the oxygen atom makes use of its 2s and 2p atomic orbitals to produce a hybrid orbital:

$$\psi = \psi(p) + \lambda \psi(s) \tag{2.19}$$

In this modified atomic orbital, the density contributions of the s and p orbitals are in the ratio $\lambda^2:1$. It can be shown by calculation that for $\lambda = 0.5$ the angle between the axes of two hybrid orbitals is 104.5°. The linear combination of atomic orbitals (2.19) leads to important hybrid orbitals (Figure 2.8). As well as those involved in the bonding to the hydrogen atoms in the water molecule, there is a third hybrid orbital, containing a lone pair of electrons (non-bonding), directed along the negative bisector of the H–O–H angle, with a second lone pair in the $2p_z\pi$ orbital.

In methane, hybrids of the type $\psi(p) + \lambda \psi(s)$ can be formulated. For tetrahedrally directed bonds, $\lambda = 1/\sqrt{3}$, leading to sp^3 hybrid orbitals and four equivalent bonds (Figure 2.9). For the corner labelled (1,1,1) in the figure, the p orbital may be regarded as being resolved into three components p_x, p_y and p_z, so that we may write (2.19) as

$$\psi(sp^3) = \psi(p_x) + \psi(p_y) + \psi(p_z) + \lambda\psi(s) \tag{2.20}$$

(see also Problem 2.9).

In a similar manner we can derive sp^2 hybrid orbitals for the carbon atoms in ethene. They lead to three σ-bonds at 120° to one another, all in one plane, with one π-bond formed by overlap of the two $2p_z$ atomic orbitals (Figure 2.10). In

Figure 2.8. Water molecule: (*a*) bonding through $2p_x$ and $2p_y$ leads to an H–O–H bond angle of 90°, (*b*) $p + \lambda s$ hybrid orbital, (*c*) bonding with these hybrids leads to an H–O–H bond angle of 104.5° ($\lambda = 0.5$). The two hybrid orbitals are nearly independent of each other: if H_A is replaced by a methyl group, the O–H_B bond is almost unchanged, thus providing a basis for characteristic bond lengths, angles and energies.

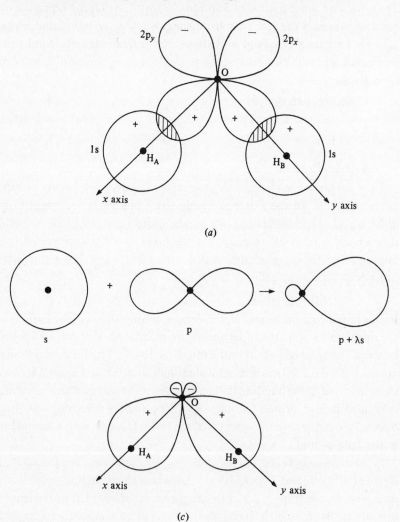

ethyne (Figure 2.11), there are sp hybrid orbitals on the carbon atoms: they lead to C–H σ-bonds which are linear, and two π-bonds obtained from the overlap of the two $2p_x$ and the two $2p_y$ atomic orbitals.

Benzene utilizes sp^2 hybrids, like ethene, to form six σ-bonds which take up the configuration of a planar regular hexagon (Figure 2.12). The six p electrons form a 'double-streamer' shaped molecular orbital, lying above and below the plane of the ring. These electrons are completely delocalized, that is, they are shared by, or are common to, all of the atoms in the molecule.

In all of these examples in which hybridization is invoked in order to obtain a good description of the geometry of the molecular bonds, energy is required to open the closed (paired) s orbitals prior to hybridization. However, this expenditure of energy is more than compensated by the energy subsequently released in forming the molecular bonds. It is important to remember that hybridization is a concept that we employ in order to provide a satisfactory model for chemical bonding; it does not necessarily imply the truth of that model.

2.9 Reactivity

The π-bond is weaker than the σ-bond, because the amount of overlap of atomic orbitals is smaller. Therefore, the electronic energy of a molecule may be lowered if a π-bond is disrupted and replaced by a further two σ-bonds, one on each atom:

$$H_2C{=}CH_2 + Br_2 \rightarrow H_2BrC{-}CBrH_2 \tag{2.21}$$

Ethyne is even more unsaturated than ethene, and even more reactive. Conjugated double bonds (alternate double and single bonds), particularly in cyclic compounds, have a special stability. The arrangement of atoms in benzene

Figure 2.9. Methane: sp^3 hybrid orbitals, showing their axes in relation to a cube. The carbon atom is imagined to be at the centre of the cube and the coordinates given refer to an origin at the carbon atom.

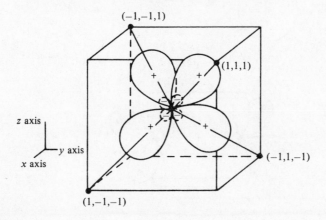

is very favourable for forming strong σ-bonds. Furthermore, application of the aufbau principle leads to only bonding orbitals being occupied by electrons.

The special stability of benzene is often called its resonance energy, resulting from a linear combination of its important canonical structures. The Kekulé formulations are shown in Figure 2.13(a). To show that benzene does not actually

Figure 2.10. Ethene: (a) formation of sp^2 hybrid orbitals from the 2s, 2p$_x$ and 2p$_y$ atomic orbitals of carbon, (b) overlap of orbitals; the 2p$_z$ orbital is directed normal to the xy plane, (c) σ and π molecular orbitals.

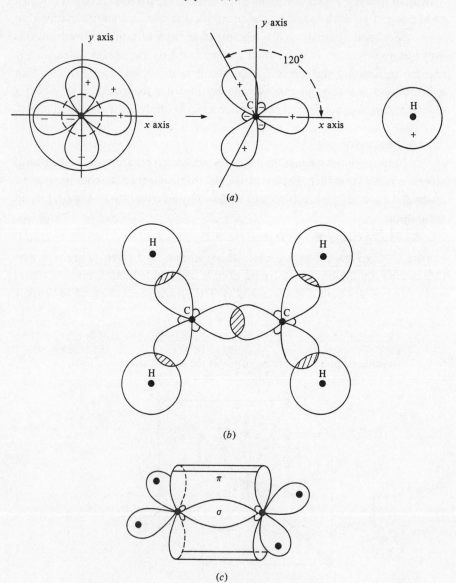

contain three double bonds, the resonating description in Figure 2.13(*b*) may be used. Again, we must be aware that the double-headed arrow does not imply any form of mixture of the two structures, except in a wave-mechanical sense, or indeed an oscillation between them. A better formulation is Figure 2.13(*c*), but again this is only an approximation to the true situation that can be described with precision only through the mathematics of the electron density distribution.

It is possible to obtain an experimental measure of the resonance energy of benzene by comparing its heat of hydrogenation with that of three times the amount for ethene or cyclohexane. The result is a value of approximately 150 kJ mol^{-1} for the resonance energy of benzene, its extra stability, which is a very significant quantity of energy.

2.10 Metallic bonding

Metals provide extreme examples of fully delocalized systems of electrons. The number of atoms present in a sample of metal is infinite, or at least very large, and the bonding electrons are delocalized over all of the atoms present.

Consider a simple metal, such as lithium. Each atom provides a relatively inert $(1s)^2$ core of electrons and a $(2s)^1$ electron for bonding. A second similar atom brought to the first leads to the formation of a bonding and an antibonding molecular orbital, the former being fully occupied by two electrons with antiparallel spins (Figure 2.14). A third atom brought up overlaps its nearest

Figure 2.11. Ethyne: (*a*) linear sp hybrid orbital on carbon, (*b*) σ and π molecular orbitals; the π molecular orbitals arise through the $2p_y$ and $2p_z$ atomic orbitals of carbon, which are mutually perpendicular to $2p_x$.

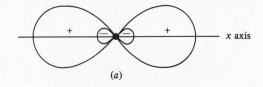

(*a*)

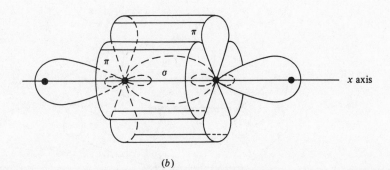

(*b*)

neighbour strongly and the other atom weakly. Each additional atom brought into combination spreads out the range of the energies of the orbitals, but also fills in the range obtained, that is, the energy gap between successive levels is reduced. In the combination of a very large number of atoms, the range of energies is filled and becomes a continuous band in which the lower orbitals are bonding and the upper orbitals are antibonding. The valence electrons from a total of N atoms in combination will occupy the lower $N/2$ orbitals of the metal. Hence, there are unfilled orbitals lying close to the uppermost filled orbitals (Figure 2.15), and only a small amount of energy is needed to excite the higher energy electrons so as to

Figure 2.12. Benzene: (*a*) σ(C–C) and σ(C–H) bonds forming a regular, planar hexagonal arrangement; the $2p_z$ atomic orbitals are directed normally to the ring (*xy*) plane, (*b*) a 'double streamer' π molecular orbital containing six delocalized electrons, common to the whole molecule.

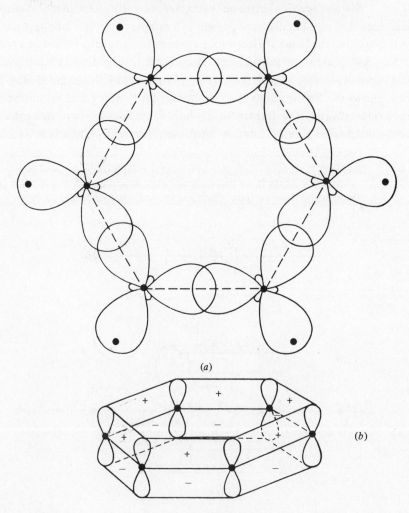

(*a*)

(*b*)

move to a vacant upper level: the electrons are mobile, and give rise to thermal and electrical conduction, which is characteristic of partially filled bands.

If each atom in the aggregate provides two electrons, the N orbitals are filled. If the next highest band is far removed in energy the element is an insulator, but if it is close it is a semi-conductor (Figure 2.15). In an element such as calcium, the two valence electrons fill the s band, but empty orbitals of the p band are available and under excitation, such as by the application of a potential difference, electrical conduction occurs.

2.11 d Electron compounds

The chemistry of the transition-metal elements, such as those lying between calcium and zinc in the periodic table, is distinguished by the d electron configuration. On ionization, a transition-metal element loses one or both of its outermost ns electrons, and thus the ion differs from the 'inert-gas' type of configuration shown by Na^+, Ca^{2+} or Cl^-. For example, Fe^{2+} may be written as $(Ar)(3d)^6$ and Cu^{2+} as $(Ar)(3d)^9$.

We shall confine our discussion to the octahedral complex ions, in which the central metal atom is attached to six other atoms or groups of atoms (ligands) so as to form an octahedral-shaped structure (Figure 2.16). A ligand that occupies only one of the six bonding positions is called a unidentate ligand, such as the ammonia or chloride ion species. If two of the ammonia ligands are replaced

Figure 2.13. Benzene: (*a*) classical formulae, (*b*) resonating (canonical) structures, (*c*) single formula, indicating a closer relationship to the molecular-orbital result.

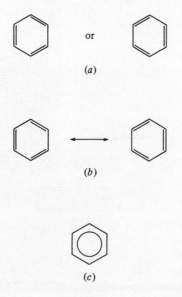

(*a*)

(*b*)

(*c*)

by chloride ions (Figure 2.17) the charge on the complex ion decreases from 3 + to 1 +. Furthermore, the replacement can be carried out in two stereochemically different ways: geometrical isomers, *cis* and *trans* exist, which differ only in the relative locations of the ligands.

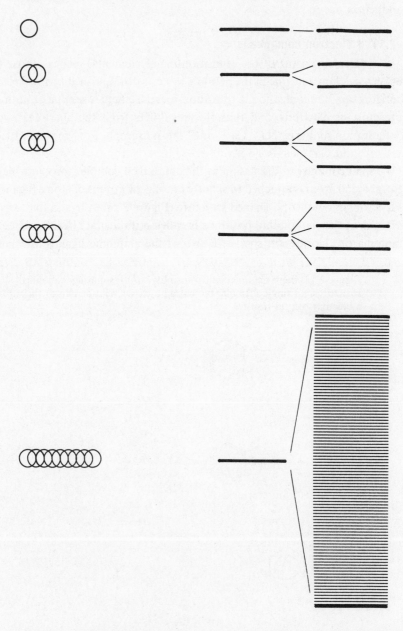

Figure 2.14. Band formation in a metal through the addition of atoms to a row.

The occupation of d orbitals by electrons leads to interesting magnetic properties. For convenience, we shall consider the five 3d orbitals of the first transition-metal elements as shown in Figure 2.18. The sequence Fe → Fe^{2+} needs no further explanation. Fe^{2+}* is the Fe^{2+} ion in a conceptual valence state, ready for bonding, in which the d electrons are paired and the remaining two

Figure 2.15. Band structures of energy levels: (*a*) metal, (*b*) insulator and (*c*) semi-conductor; e symbolizes electrons.

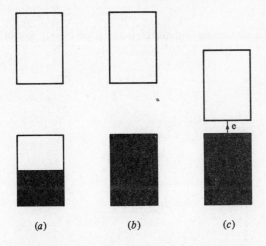

(*a*)   (*b*)   (*c*)

Figure 2.16. Hexamminocobalt(III) ion; the bond directions are shown in full lines and the edges of the octahedron in dashed lines; the overall charge is 3+.

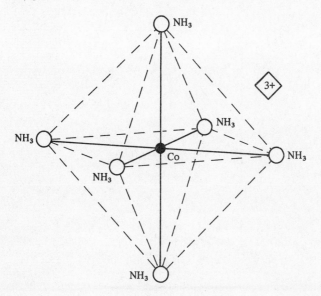

vacant 3d, the 4s and the three 4p orbitals form d^2sp^3 hybrid orbitals. These orbitals are equivalent, and their larger lobes are directed geometrically towards the six corners of an octahedron. Next, six cyanide ions donate their lone-pair electrons to the vacant hybrid orbitals, indicated in the figure by the symbol Υ. In the resulting hexacyanoferrate(II) ion, all electron spins are paired, and the ion is diamagnetic. A similar process carried out with the Fe^{3+} ion leads to the hexacyanoferrate(III) ion, which has one unpaired electron and is, thus, paramagnetic.

Figure 2.17. Dichlorotetramminecobalt(III) ion; charge $1+$ (*a*) *cis* and (*b*) *trans.*

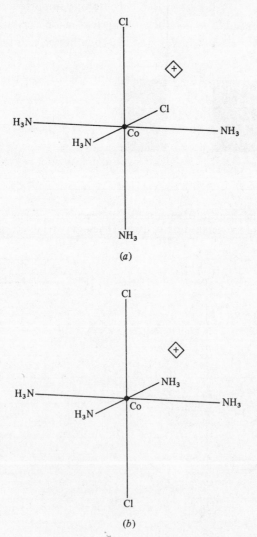

The bonds formed by the lone-pair electrons are sometimes called coordinate (dative) covalent, to indicate that both electrons in each orbital come from the same atom. However, electrons are indistinguishable, and once the bond is formed the term covalent is adequate – and probably more correct.

2.12 Ionic bonding

Ionic bonding involves an electron donor–acceptor mechanism among the participating atoms. The electrons in an ionic compound are localized in the atomic orbitals of ions, to a good first approximation, and the bonding between the species can then be considered as an electrostatic attraction between point charges. The attractive forces between the oppositely charged ions are balanced by strong electron–electron repulsion forces that become very important as the ions are considered to be brought close together; for this reason ionic crystals are not very compressible. The situation in an ionic solid is, qualitatively, not unlike that expressed by Figure 2.4.

The formation of ions from neutral atoms requires an expenditure of energy, because atoms are stable structures that do not ionize spontaneously. The

Figure 2.18. d Orbitals in hexacyanoferrate ions: (a) Fe metal, (b) Fe^{2+} ion, (c) Fe^{2+}_*, d^2sp^3 orbitals prepared for bonding, (d) hexacyanoferrate(II) ion – diamagnetic, (e) Fe^{3+} ion, (f) Fe^{3+}_*, d^2sp^3 orbitals prepared for bonding, (g) hexacyanoferrate(III) ion – paramagnetic.

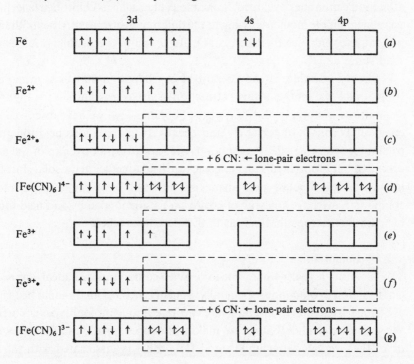

formation of certain negative ions, such as chloride ions, is an exothermic process, but the total energy for a pair of ions, such as Na^+ and Cl^- is strongly endothermic. This energy is more than recovered when an array of ions condenses to form an ionic crystal. We shall consider ionic crystals again in later chapters.

2.13 Van der Waals' bonding

The bonding between molecules such as benzene or sucrose takes place through forces that are relatively much weaker than those involved in covalent, ionic or metallic bonding; they are often referred to as van der Waals' forces. The simplest compounds that exhibit van der Waals' bonding are the inert gas elements in the solid state, such as solid argon. As atoms of argon are brought together it is not difficult to visualize electron–electron repulsion as the distance between the atoms is made very short, but what is the mechanism of attraction? The electron density of an atom is constantly changing with time. At any instant the centres of gravity of the positive and negative charge in the atom will not coincide, and a small, instantaneous dipole is created within the atom. This dipole will induce and interact with a similar dipole in a neighbouring atom, leading to a dipole–dipole, electrostatic attraction, albeit very small in magnitude. Van der Waals' forces of attraction exist between all atomic species, but they are of paramount importance in these compounds, where other, stronger forces of attraction cannot be developed. Some molecules, such as chlorobenzene, possess permanent dipole moments owing to the differing electronegativities of the atoms present; they enhance the electrostatic attraction between the molecules in the solid state.

Hydrogen bonding, already described in Chapter 1, may exist in molecular compounds; it is another form of attraction of an electrostatic nature. Figure 2.19 summarizes, pictorially, the four types of bonding that we have discussed. Ionic compounds consist of positively and negatively charged ions held together by coulombic forces; covalent compounds are formed by an overlap of the atomic orbitals of the atoms involved; molecules are held together in the solid state by the interaction of dipoles, permanent or induced, or both; and in metallic compounds, we have an array of positive ions surrounded by and held together by a sea of electrons that permeate the metal as a whole.

2.14 Compromise

A single (pure) bond type for a given compound is an ideal, approached closely in some compounds. Generally, a chemical compound should be regarded as bonded in a manner somewhat intermediate among the four types enumerated. We should remember that we are making an approximation to the true state in order that we can better understand the properties associated with the bond types. Diagrammatically, we might regard the bond type of a given compound as a point on a tetrahedron (Figure 2.20), thus expressing a wave-mechanical resonance between two or more structures. For example, in dilithium we might postulate the canonical forms Li–Li (covalent), Li^+Li^- (ionic), Li^-Li^+ (ionic), $Li^{\delta+}Li^{\delta-}$ (van der Waals') $Li^{\delta-}Li^{\delta+}$ (van der Waals'); the contribution of metallic bonding would be expected to be insignificant. The proportion of each canonical form would be determined, as always, by the requirement of a minimum energy conformation.

2.15 Electronegativity

The electronegativity x of an atom is an approximate measure of the tendency of that atom to attract electrons. Thus, the difference in electronegativity of two species should be a measure of the ionic character of a bond. In a bond A—B, the difference in electronegativities $\Delta x(AB)$ is given by the geometric mean of the dissociation energies of the A_2 and B_2 molecules. From such data, Pauling drew up a table of electronegativities (Table 2.5). The

Figure 2.19. Pictorial representation of the four principal bond types: (*a*) ionic, (*b*) covalent, (*c*) van der Waals' and (*d*) metallic.

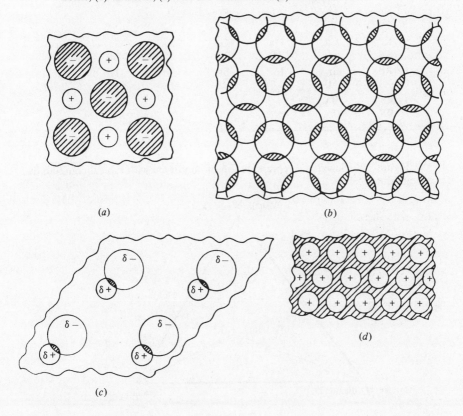

36   *Atoms, molecules and bonding*

relationship between Δx and fractional ionic character q is not easily quantified; the best equation is probably that of Hannay and Smith where

$$q = 0.16|\Delta x| + 0.035(\Delta x)^2 \tag{2.22}$$

with $q < 0.5$. Figure 2.21 shows the application of (2.22) to the halogen hydrides, using the data from Table 2.5.

2.16 Dipole moment

If a bond is formed between two dissimilar atoms, there will be a displacement of electronic charge towards the more electronegative species and a dipole will be formed. The dipole has a moment p given by

$$p = qed \tag{2.23}$$

where e is the charge on an electron and d is the distance between the effective

Table 2.5. *Pauling electronegativities of some elements*

H 2.2						
Li 1.0	Be 1.6	B 2.0	C 2.6	N 3.0	O 3.4	F 4.0
Na 0.9	Mg 1.3	Al 1.6	Si 1.9	P 2.2	S 2.6	Cl 3.2
K 0.8	Ca 1.0	Ga 2.0	Ge 2.0	As 2.2	Se 2.6	Br 3.0
Rb 0.8	Sr 1.0	In 1.8	Sn 2.0	Sb 2.1	Te 2.1	I 2.7
Cs 0.7	Ba 0.9					

Figure 2.20. Schematic representation of bond type with typical compounds indicated.

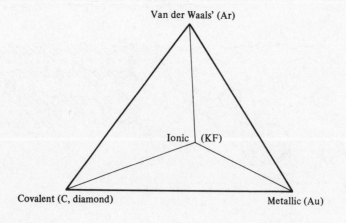

charges $\pm qe$ (Figure 2.22). The dipole moment can be obtained experimentally, through measurements of dielectric constant, and for hydrogen fluoride it is 6.07×10^{-30} C m. From (2.23), it follows that q is 0.41, and from (2.22) q is 0.40. We recall from (2.17) that the wave function for the hydrogen fluoride molecule may be written approximately as

$$\Phi_+ = \psi_{\text{covalent}} + \lambda\psi_{\text{ionic}} \tag{2.24}$$

Since the electron density is proportional to the square of a wave function, the covalent and ionic contributions are in the ratio of $1:\lambda^2$. Hence

$$q = \lambda^2/(1+\lambda^2) \tag{2.25}$$

from which it follows that $\lambda = 0.83$, if we take q as 0.41. Hence the wave function becomes

$$\Phi_+ = \psi_{\text{covalent}} + 0.83\psi_{\text{ionic}} \tag{2.26}$$

and it is clear that the H^+F^- canonical structure is very important. In the hydrogen iodide molecule, p is 1.50×10^{-30} C m, and a similar calculation shows that λ would be approximately 0.25 for this molecule.

Figure 2.21. Fractional ionic character q as a function of the electro-negativity difference $|\Delta x|$.

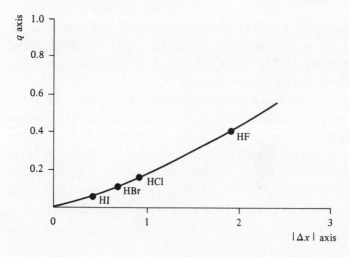

Figure 2.22. Effective dipole diagram for the hydrogen fluoride molecule.

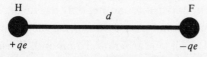

2.17 Physical and structural characteristics of the principal bond types

In Table 2.6 some of the physical and structural characteristics that may be associated with each of the principal types of chemical bonds are summarized.

2.17.1 *Mechanical properties*

The hardness of materials is often expressed by Moh's scale of scratch hardness, which ranges from one to ten, each material being capable of scratching those of lower hardness number:

1 talc	6 orthoclase
2 gypsum	7 quartz
3 calcite	8 topaz
4 fluorspar	9 corundum
5 apatite	10 diamond

Both electrovalent compounds, such as topaz $(Al_2SiO_4)(F, OH)_2$ and fluorspar CaF_2, and covalent compounds, such as quartz $(SiO_2)_n$ and corundum $(Al_2O_3)_m$, are represented within this scale, showing that hardness is not a distinguishing feature between these two types of bonding. Van der Waals' interactions are of relatively low energy and do not give rise to mechanically strong crystals, thus explaining the friability of organic crystals which are held together by forces of this type.

In metallic crystals, deformation of the crystal takes place by *gliding*, a process that occurs most easily along the planes of closely packed atoms in the metal crystal. The characteristic malleability of copper and the brittleness of tungsten are associated with the extent of close packing.

Another mechanical property of crystals that is intimately connected with the type of bond present is *cleavage*, the property of breaking cleanly in certain directions; it is not observed in metals, since they deform in the manner already described. The cleavage planes represent surfaces across which the binding forces are relatively weak, such as in calcite which cleaves parallel to the faces of a rhombohedron, and fluorspar which cleaves along planes parallel to the faces of an octahedron. Quartz lacks cleavage, because there is an even distribution of bonds in all directions.

2.17.2 *Thermal properties*

Electrovalent and covalent compounds generally have high melting points: sodium chloride (1074 K), magnesium oxide (3070 K), silicon carbide (2970 K), diamond (> 3770 K). Organic crystals are not covalent solids since the covalent molecules in the crystal are bound to one another only by van der Waals' forces. They have relatively low melting points, in contrast to the very high values for highly covalent compounds such as silicon carbide and diamond. The melting

Table 2.6. *Some properties of bond types*

Property	Bond type			
	Electrovalent	Covalent	Van der Waals'	Metallic
Mechanical	Strong and hard crystals	Strong and hard crystals	Weak and soft crystals	Variable strength; glide deformation
Thermal	High mp; low coefficient of thermal expansion; ions in the melt	High mp; low coefficient of thermal expansion; molecules in the melt	Low mp; high coefficient of thermal expansion	Variable mp; long liquid interval
Electrical	Insulators; conduction in the melt by ion transport	Insulators in the solid and in the melt	Insulators	Conduction by electron-transport
Optical	Absorption and related properties are those of component ions (similar in solutions)	Absorption different in solution (and in the gaseous state) from that of the solid	Absorption properties are those of the individual molecules and are similar in solution and in the gaseous state	Opaque; properties similar in the liquid state; transparent to high-energy radiation, such as X-rays
Structural	Non-directional; coordination number (CN) variable	Strongly directional; CN variable	Non-directional	Non-directional; CN high (8 or 12)

points of electrovalent compounds increase with increasing ionic charge, but for constant ionic charge the melting points decrease with increasing interionic distance r (Table 2.7).

The melting points of metals vary considerably: for example mercury 235 K, osmium 2990 K. All metals have a long liquid interval, the temperature range between melting point and boiling point: for example gallium 1953 K, copper 1512 K; electrovalent and covalent compounds have a liquid interval only about one-tenth that of metals.

2.17.3 *Electrical properties*

Electrovalent compounds conduct electricity in the molten state and in solution in a polar solvent by ion-transport, a characteristic that may be regarded as a criterion of the electrovalent bond. Covalent compounds and van der Waals' compounds are insulators in the solid state and in the melt; metals conduct readily owing to the presence of electrons in the conduction band which are free to move about the crystal as a whole. The small conductivity of semi-conductors is due to a smaller band gap than in insulators (see Figure 2.15).

2.17.4 *Optical properties*

The light absorption of ionic crystals is mainly in the uv region of the spectrum, so that they appear colourless. The optical properties of the crystal are the sum of the properties of the component ions since the electrons are localized in atomic orbitals. Ions of the transition-metal elements are usually coloured; optical absorption occurs in the visible region of the spectrum, because of electron transitions between the relatively close d energy levels.

In covalent compounds the electron orbitals overlap and hence the optical properties of these compounds are quite different from those of their constituent elements. It is noticeable that colour sometimes develops with an increasing percentage of covalent character, as in the silver halides. They also display an increased departure δ from additivity of the ionic radii, the difference between the interionic separation and the sum of the radii of the ions; this departure becomes larger with increasing polarizability of the anion (Table 2.8).

In van der Waals' compounds the optical properties are those of the component molecules, which means that they are similar in solution and in the

Table 2.7. *Melting points of electrovalent compounds*

	NaF	NaCl	NaBr	NaI	MgO
r/nm	0.231	0.282	0.298	0.324	0.211
mp/K	1261	1074	1028	924	3070

gaseous state. Visible light is scattered by electrons, so that metals appear opaque because their electrons are distributed over the entire crystal and therefore scatter all the light; to radiation of higher energy, such as X-rays, metals are transparent.

2.17.5 *Structural properties*

Only the covalent bond exhibits directional character. For example, a chloride ion experiences the same force of attraction in whichever direction it approaches a sodium ion during the formation of a sodium chloride crystal, whereas in methane a hydrogen atom must bond along the direction of one of the sp^3 orbitals of a carbon atom for bond formation to occur.

The number of nearest neighbours around any given atom or ion is called the *coordination number*. In covalent and electrovalent structures the coordination numbers are variable, but in metals they are usually eight to twelve. In molecular compounds the molecules pack together to obtain a minimum energy conformation, and the separation of nearest-neighbour atoms of different molecules is about 0.37 nm for many organic compounds. In certain crystals this may be reduced to as low a value as 0.25 nm where intermolecular hydrogen bonding is operative, the determination of such small distances being a frequent means of recognizing hydrogen bonds.

2.18 Elementary spectroscopy

At the beginning of this chapter we discussed the visible spectrum of atomic hydrogen, a set of spectral lines known as the Balmer series. Even for hydrogen, the simplest atom, there is more than this one series of lines that can be observed, all of which fit (2.1) but with different values of n_1 and n_2 (Table 2.9). The series are named after their discoverers. Each series terminates at $n_2 = \infty$, or $\tilde{v} = R_H/n_1^2$, in a continuum of energy levels, in contradistinction to the discrete lines that correspond to the spectral series.

The line spectra arise through electron transitions between two energy levels, according to (2.2). Now (2.5) gives the electron energy for atomic hydrogen, or for any 'hydrogen-like' species, in which the nuclear charge may vary while there is only one electron present. One is tempted to ascribe the spectra of atomic hydrogen to all possible transitions between energy levels of the hydrogen atom.

Table 2.8. *Departure from additivity of silver halides*

	AgF (colourless)	AgCl (colourless)	AgBr (pale yellow)	AgI (yellow)
r/nm	0.247	0.278	0.288	0.280
$[r(+)+r(-)]/nm$	0.259	0.307	0.322	0.346
δ/nm	0.012	0.029	0.034	0.066

However, the correct procedure invokes the conservation of angular momentum of the electron. The quantum number l, introduced in Section 2.2, specifies the angular momentum L of an electron:

$$L = [l(l+1)]^{\frac{1}{2}} \tag{2.27}$$

Because l has integral values, L is generally non-integral; but the component of L in an allowed direction, m_l, is again integral. A more detailed discussion of this topic lies outside the scope of this book, but it leads to important selection rules in spectroscopy which govern the allowed transitions between atomic energy levels:

$$\Delta n = \text{any integer} \tag{2.28}$$

$$\Delta l = \pm 1 \tag{2.29}$$

With this additional information, we can construct a diagram of allowed transitions; it is called a Grotrian diagram (Figure 2.23). We may remark here that once we leave the hydrogen-like species for more complex atoms, any two levels of the same n but of different l have differing energies. These other atoms produce similar series of spectra although there are many more lines in each spectrum, and (2.1) no longer predicts their wavenumbers.

Molecules absorb radiation in groups of very closely spaced lines that give a band-like appearance to their spectra; these bands are then absent from the transmitted radiation when it is analyzed. The radiation that is absorbed enhances the internal modes of motion of the molecule.

Consider a molecule containing n atoms. If these atoms were free, the system would possess $3n$ degrees of freedom of motion corresponding to components of translation along three mutually perpendicular x, y and z reference axes. For a complete molecule, wherein the atoms are bound by interatomic forces, there are only three degrees of translational freedom since the molecule must move as a single entity. The remaining modes of motion are divided between rotation of the molecule, and vibration (stretching or bending) of its interatomic bonds.

A non-linear molecule has three degrees of rotational freedom, corresponding to rotation about the x, y and z axes, so that the remaining $3n - 6$ modes must be assigned to vibrational degrees of freedom. In the case of a linear molecule, there are only two degrees of freedom for rotation since one of the reference axes can

Table 2.9. *Spectral series for atomic hydrogen*

Series	n_1	n_2	Spectral region
Lyman	1	2, 3, 4, ...	Uv
Balmer	2	3, 4, 5, ...	Visible
Paschen	3	4, 5, 6, ...	Ir
Brackett	4	5, 6, 7, ...	Ir
Pfund	5	6, 7, 8, ...	Ir

always be chosen to coincide with the molecular axis, and rotation about the molecular axis does not involve a displacement of the 'heavy' atom masses. Hence, the number of vibrational modes is then $3n-5$. The regions of the electromagnetic spectrum that are associated with the various molecular motions are shown in Figure 1.4.

2.18.1 *Nuclear magnetic resonance*

Nuclear magnetic resonance (nmr) arises through the small but finite magnetic moment of certain atomic nuclei, notably ^{1}H, ^{13}C and ^{14}N. The molecule absorbs energy from incident radiation of the radiofrequency region in aligning its nuclear dipole with the magnetic field of its environment. The specimen is rotated in a homogeneous magnetic field, of adjustable strength, and the spectrum is recorded at a constant value of the radiofrequency, while the magnetic field is swept over a range of values. Typical results are shown in Figure 2.24. The protons in differing environments in ethanol respond at different values of the field strength.

Figure 2.23. Portion of a Grotrian diagram for the hydrogen atom.

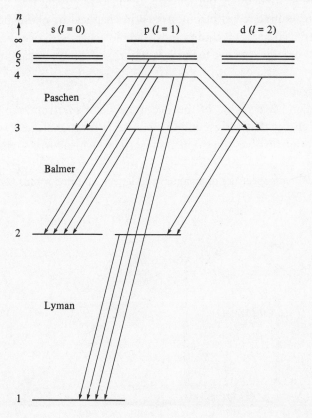

The magnetic field in which the nuclei move and align themselves is imposed by the external magnet, but modified by the magnetic environment created by the molecules of the specimen. Hence, the nmr spectrum provides information about this environment, and the technique has become a fundamental tool in the investigation of molecular structure.

From Figure 2.24, it may be inferred that the molecule contains a group of three hydrogen atoms similarly bonded, a second group of two hydrogen atoms, similarly bonded but in a different environment from the first group, and a single hydrogen atom uniquely bonded. Taken with the empirical formula and the relative molar mass, the molecular structure CH_3CH_2OH becomes apparent.

2.18.2 *Electron spin resonance*

Where an atom, molecule or ion contains an odd number of electrons, their spins cannot all be paired. The magnetic moment associated with resultant unpaired spin can give rise to electron spin resonance (esr) spectra.

An esr spectrum provides information rather similar to that given by nmr, but in the case of species with unpaired electrons absorption takes place in the microwave region of the electromagnetic spectrum, at convenient magnetic field strengths. It is found that esr is particularly useful for studying free radicals (which have a single unpaired electron) formed in chemical or photochemical reactions, many transition-metal compounds, and molecules with two unpaired electrons. The method is insensitive to the vast majority of spin-paired molecules.

2.18.3 *Microwave*

In the microwave region of the electromagnetic spectrum, molecules absorb quanta of energy that correspond to increases in rotational energy alone. The energy of rotation involves the moment of inertia which, in turn, is related to

Figure 2.24. Nmr spectrum of ethanol; bold letters indicate protons giving rise to resonance.

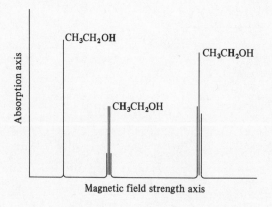

atomic masses and distances between them. Consequently, microwave spectra can be used to obtain values of bond lengths and angles. With small molecules, bond lengths have been obtained to a precision of ± 0.00002 nm. In O=C=S, for example, C=O$=0.11612$ nm and C=S$=0.15588$ nm.

2.18.4 *Visible*

Absorption in the visible region of the spectrum results in electronic excitation of molecules. Such molecules lead to coloured compounds; dyes and pigments absorb in this spectral region. The absorption of a photon by a molecule is centred on one or more *chromophoric groups*, such as C=C or C=O. The transition responsible for absorption by C=O arises through the non-bonding lone-pairs of electrons on the oxygen atom. Transition-metal compounds absorb light in d–d transitions. In other cases, absorption arises by transfer of an electron from a ligand atom to the central metal atom. The intense violet colour of the MnO_4^- ion is due to such a charge-transfer process.

2.18.5 *Ultraviolet*

Where absorption occurs in the uv region of the spectrum, the quanta of radiation are sufficiently large (equivalent to 300–500 kJ mol^{-1}) to cause dissociation of the molecule. The photochemical decomposition of hydrogen iodide, for example, occurs through the reactions:

$$HI + h\nu \xrightarrow{253\,nm} H\cdot + I\cdot \tag{2.30}$$

$$HI + H\cdot \rightarrow H_2 + I\cdot \tag{2.31}$$

$$I\cdot + I\cdot \rightarrow I_2 \tag{2.32}$$

Since process (2.31) is efficient, the absorption of one photon leads to the decomposition of two molecules of hydrogen iodide; the quantum efficiency of the process is two. In a chain reaction, which is a sort of chemical amplifier, quantum efficiencies may be of the order of 10^3–10^5. Hence, the uv region is appropriate for carrying out and investigating photolytic reactions in chemistry.

2.18.6 *Infrared*

In the infrared (ir) region, both rotational and vibrational energies are enhanced; the rotational fine structure spectrum is superimposed on the vibrational spectrum with the result that absorption bands appear rather than discrete lines. The absorption frequencies are often used to identify particular atomic groupings – the so-called fingerprint method of identification. Figure 2.25 is a chart of ir absorption 'frequencies'. In spectroscopy, one often speaks of frequency when meaning wavenumber: this piece of jargon is revealed, of course, by the units of the quantity.

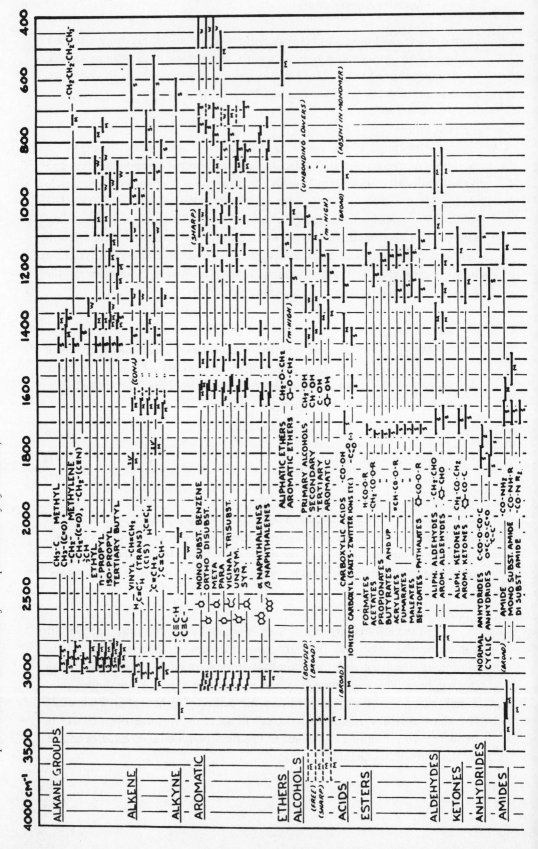

Figure 2.25. Infrared group 'frequencies'/cm^{-1}: S, strong intensity; M, medium intensity; W, weak intensity; 2ν, overtone vibration; Str, stretching frequency; Bend, bending frequency; Rock, rocking frequency (after N. B. Colthup, and reproduced by permission of the *Journal of the Optical Society of America*).

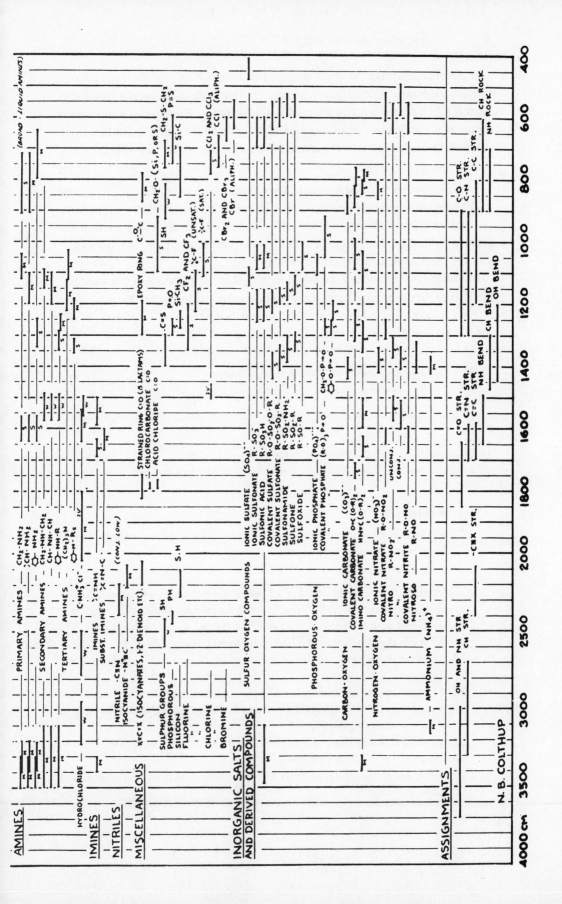

Ir spectra can also be used to provide information on the strengths of interatomic bonds. A bond force constant f is a measure of the energy required to stretch a bond between two atoms by unit distance. By treating two bonded atoms as a simple harmonic oscillator, (2.33) may be derived; v is the frequency (true) of the oscillator and μ is the reduced mass of the system (see Appendix A5).

$$v = (2\pi)^{-1}(f/\mu)^{\frac{1}{2}} \tag{2.33}$$

In the hydrogen chloride molecule, the reduced mass is 1.6275×10^{-27} kg and its absorption 'frequency' in the ir is 2989 cm^{-1}. From (2.33), the bond force constant is calculated as 516 N m^{-1}. The values for the hydrogen halides are listed in Table 2.10, together with data for the equilibrium interatomic distance r_e and the experimental dissociation enthalpy.

In concluding this section, we might remark that the literature of spectroscopy tends to be very specialized according to the spectral region under consideration. A good general first reference may be gained from *Physical Chemistry*, P. W. Atkins, Oxford University Press, Oxford, 1982.

Table 2.10. *Bond force constants in the hydrogen halides*

	HF	HCl	HBr	HI
$f/\text{N m}^{-1}$	989	516	405	308
r_e/nm	0.092	0.127	0.141	0.161
$D_0/\text{kJ mol}^{-1}$	564	428	363	295

Problems 2

2.1 The Balmer series in the hydrogen spectrum was analyzed originally in terms of wavelength λ through the equation

$\lambda = Kn^2/(n^2 - 4)$

where K is a constant and n is an integer greater than two. Show that this equation is equivalent to (2.1), and find K in terms of R_H. What is the energy, in J, associated with the spectral line nearest to the red end of the spectrum?

2.2 What are the four values of n_2 for the spectral lines in Table 2.1?

2.3 Calculate the values for the reduced mass of an electron and the Bohr radius of hydrogen.

2.4 A proton and an electron, taken as point charges, are held at a distance not greater than 10^{-15} m by coulombic forces. Is this hypothesis feasible in the light of the uncertainty principle?

2.5 Equation (2.9) may be completed by finding the value of N. Use (2.8) to obtain N.

2.6 Continue Table 2.2 up to krypton.

2.7 Give a description of Be$_2$ and LiH, in both valence-bond and molecular-orbital terms.

2.8 Given that an sp^x hybrid atomic orbital has a normalized wave function given by

$$\psi(\text{sp}^x) = (1+x)^{-\frac{1}{2}}[\psi(\text{s}) + x^{\frac{1}{2}}\psi(\text{p})]$$

determine the form of $\psi(\text{sp}^x)$ for the hybrid orbitals in Figure 2.9. Write down the other three hybrid atomic orbitals, equivalent to (2.20).

2.9 Arrange the following species in order of increasing numbers of unpaired electrons, and state these numbers: $[\text{Mn(CN)}_6]^{4-}$, Cr^{2+}, $\text{Co(NH}_3)^3\text{F}^3$, V^{3+}.

2.10 What geometrical isomers exist for the compound $\text{Co(NH}_3)_3\text{F}_3$?

2.11 The dipole moment for the hydrogen bromide molecule was measured as 2.77×10^{-30} C m. Calculate the fractional ionic character of the H–Br bond. How does the result compare with that obtained from the relative electronegativities? Complete the hydrogen bromide wave equation $\psi_{\text{covalent}} + \lambda\psi_{\text{ionic}}$ by giving the value of λ.

2.12 What is the electrostatic energy of a lithium ion Li^+ and a chloride ion Cl^- separated by 0.257 nm in vacuo?

2.13 The ir absorption of molecular oxygen is at 158036 m^{-1}. Calculate the force constant of the O–O bond.

2.14 The ir spectrum of an organic molecule, $\text{C}_4\text{H}_8\text{O}$, shows absorptions at 755, 945, 1080, 1180, 1370, 1425, 1725 and 2950 cm^{-1}. What is the probable structural formula for the molecule?

3

Thermochemistry and thermodynamics

Thermochemistry is concerned with the heat changes that accompany chemical reactions. It is based on the energy conservation law, which tells us that energy can be neither created nor destroyed. The expenditure of energy of one kind, such as work, involves the production of an equivalent amount of energy of other kinds, such as heat or light. The celebrated cannon-boring experiment of Rumford in 1798 introduced the idea of an equivalence between work and heat, the quantitative aspect of this relationship being demonstrated by Joule in 1840.

From the Einstein equation ($U = mc^2$), a change in energy ΔU is related to the corresponding mass change Δm for a quantity of matter by

$$\Delta U = \Delta mc^2 \tag{3.1}$$

where c is the speed of light in a vacuum. It implies not that matter contains heat or work energy, but rather that when matter is consumed in a reaction there arises an associated emission of an equivalent amount of energy: it is this energy that is manifested as heat, light or work, or some combination of them.

Thermodynamics is a much wider subject than is thermochemistry: it is concerned with all forms of energy changes accompanying chemical and physical processes. These energy changes may be used to determine the equilibrium position of a chemical system, and to show if a reaction is energetically feasible under a given set of conditions. Thermodynamic arguments are independent of atomic and molecular models and, thus, of changes in these models as new scientific ideas are introduced. Modern chemistry is, however, much concerned with models of systems, and an ability to relate thermodynamic quantities to molecular processes is useful in the study of chemistry, although not essential to the study of thermodynamics itself.

The equations of thermodynamics give no information about the rates of chemical processes. For example, a gaseous mixture of $\frac{1}{2}$ mol of hydrogen and $\frac{1}{2}$ mol of chlorine is thermodynamically unstable with respect to 1 mol of hydrogen chloride: the reaction

$$\tfrac{1}{2}H_2(g) + \tfrac{1}{2}Cl_2(g) \rightarrow HCl(g) \tag{3.2}$$

is accompanied by the release of energy as the more stable system, hydrogen chloride, is formed from the gaseous reactants. However, thermodynamics tells us neither that the mixture of hydrogen and chlorine can be kept indefinitely in a dark place, nor that the reaction takes place at an explosive rate in the presence of uv radiation. The conditions under which reactions occur at measurable rates, and the magnitudes of these rates, are studied in Chapter 8.

3.1 Conservation of energy and the first law of thermodynamics

The law of conservation of energy is based on practical experience: contrary processes, such as perpetual motion, have never been demonstrated, otherwise energy could be created continuously by such methods. The conservation of energy leads to the first law of thermodynamics: *the total energy of a system and its surroundings is constant*. The term *system* refers to a given quantity of matter. A system and its surroundings that cannot exchange energy with any other system and its surroundings is called an *isolated* system, so that another formulation of the first law is that *the energy of an isolated system is constant*.

To exemplify the first law of thermodynamics consider a gas in state 1, confined to a volume V_1 under an applied pressure P_1, being changed to state 2, of volume V_2 under a pressure P_2, by path I (Figure 3.1), and returning to its initial state by another route, II. The first law of thermodynamics requires that the energy of the gas in state 1 shall be the same after the complete cyclic process as it was initially: otherwise, the cycle would either create or destroy energy, which is contrary to

Figure 3.1. Forward and return paths for a gas between two states.

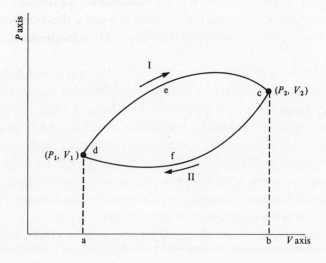

the law. The change in internal energy ΔU of the system in moving from state 1 to state 2 is given by

$$\Delta U = U_2 - U_1 \qquad (3.3)$$

3.2 State function

The notation of (3.3) is common in thermodynamics. A finite change in any property X of the system is represented by ΔX, where ΔX is the value of the property X after the change (X_2) *minus* the value of the property before the change (X_1). It follows from (3.3) that the change in internal energy for the complete cyclic path in Figure 3.1 is zero: ΔU is independent of the path followed between states 1 and 2. A property that behaves like the internal energy is called a *state function*. Physical properties, whether or no they are state functions, may be classified further. If a property depends upon the quantity of matter in the system, as does internal energy, it is called an *extensive* property. If it is independent of the quantity of matter in the system, as is density, the property is called *intensive*.

3.3 Heat and work

The system illustrated by Figure 3.1 may either lose or gain energy in going from state 1 to state 2, depending on the actual values of the variables V and P. The change in energy for the process may appear as heat q, or as work w, or as varying amounts of both heat and work. By convention, q is considered to be a positive quantity when the system *absorbs* heat from its surroundings; w is considered positive when it represents work done *on* the system by the surroundings. Thus, positive values of both q and w leads to an increase in the internal energy of a system, and for a change between two states we write

$$\Delta U = q + w \qquad (3.4)$$

Equations (3.3) and (3.4) may be regarded as the mathematical formulation of the first law of thermodynamics. The individual values of q and w depend upon the reaction path followed but their sum, the state function ΔU, is independent of the path.

As a further illustration of (3.4), consider Figure 3.2. We take as our system a brass weight of mass 1 kg attached to a string of negligible mass passing over a frictionless pulley. In (a), the weight is allowed to fall freely through 1 m to the ground level. The change in potential energy of the weight is -9.8 J; since no work is done, this energy is dissipated as heat, the heat generated by the impact of the weight with the ground. If all this heat were to pass into the brass weight, its temperature would increase by 0.025 K. In practice, however, it is more likely that part of the heat would pass into the weight and the remainder would be lost to the surroundings, but we will assume only the latter process.

In the second experiment (b), the falling weight does work in raising another weight of mass 0.5 kg through 1 m. The change in potential energy of the system (brass weight) is still -9.8 J, but this energy is now divided into -4.9 J of work done on the surroundings and -4.9 J of heat loss to the surroundings, assuming that no heat is taken up by the 1 kg weight, as before. We may note that the discussion would not be invalidated if some heat were taken up by the weight. If, for example, 1 J of the heat of impact were taken up by the 1 kg weight, then (3.4) would still hold, because the net change in potential energy of the system would be -8.8 J.

3.4 Reversibility

In thermodynamics, the term reversible has a meaning beyond that associated with equilibria such as

$$CH_3CO_2C_2H_5 + H_2O \rightleftharpoons CH_3CO_2H + C_2H_5OH \tag{3.5}$$

Let us consider first a mechanical process.

Figure 3.3 illustrates a spiral spring S carrying a pan P to which is attached a needle pointer N, all assembled on a baseboard B. When the pan is loaded with 0.8 kg, N is opposite the zero of the metre scale M. We shall assume that the spring and pan have negligible mass, and that the extension of the spring follows Hooke's law. Eight trays, T_1-T_8, are arranged to correspond with the positions of the load after successive changes of 0.1 kg loading. At the zero mark the pan is loaded with 0.8 kg, and we shall assume that each decrease of 0.1 kg causes the spring to contract by 0.1 m, following Hooke's law.

We begin with the system fully loaded with 0.8 kg, and consider how the energy stored in the spring can be caused to do work in raising masses above the working

Figure 3.2. Illustrating the relationship $\Delta U = q + w$.

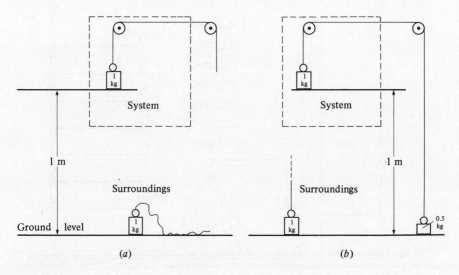

(a)   (b)

zero of energy, that corresponding to T_1, or when N is at zero. If the total of 0.8 kg is moved to T_1, it is clear that no work is done by the system on the weights: the spring contracts, and its stored energy is lost to the surroundings as heat. Next, let 0.4 kg be moved to T_1. The spring carries the remaining 0.4 kg upwards through a height of 0.4 m, and the 0.4 kg is now transferred to T_5. The work done w is that of raising 0.4 kg by 0.4 m against gravity, or -1.57 J (work $w =$ mass $m \times$ height $h \times g$); the negative sign shows that the work has been done *by* the system on the surroundings. The procedure can be carried out in several different stages, similar to those shown in Table 3.1.

It is evident that the work done by the system increases, numerically, as the mass transferred to the trays decreases, Figure 3.4 shows a plot of work done against mass transferred. It is linear and, if extrapolated to $\Delta m = 0$, intersects the ordinate at $w = -3.14$ J (this result can be confirmed by integration). Thus, we see that the maximum work obtainable from the system would be achieved for infinitesimally small values of Δm, that is, with successive infinitesimally small displacements of the spring from equilibrium. The experiment may be carried out in the reverse order, leading to similar but positive values for w.

The equilibrium condition of a system with respect to a process, which can be

Figure 3.3. Mechanical illustration of reversibility.

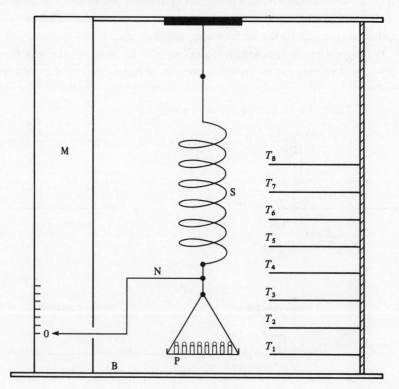

caused to move in either a forward or a backward direction by infinitesimal changes ($\pm \delta m$ in our experiment) constitutes the concept of *reversibility*. Of course, we cannot carry out our experiment in a truly reversible manner. We could approach reversibility closely by using, say 80 000 ten milligram weights

Table 3.1. *Mechanical analogue of reversibility*

Mass transferred, Δm/kg	Tray	Scale movement/m before transfer	Work done, w/J	
0.8	T_1	0	0	
0.4	T_1	0	0	
0.4	T_5	0.4	-1.57	-1.57
0.2	T_1	0	0	
0.2	T_3	0.2	-0.39	
0.2	T_5	0.4	-0.78	
0.2	T_7	0.6	-1.18	-2.35
0.1	T_1	0	0	
0.1	T_2	0.1	-0.10	
0.1	T_3	0.2	-0.20	
0.1	T_4	0.3	-0.29	
0.1	T_5	0.4	-0.39	
0.1	T_6	0.5	-0.49	
0.1	T_7	0.6	-0.59	
0.1	T_8	0.7	-0.69	-2.75

Figure 3.4. Variation of work done w with mass transferred Δm.

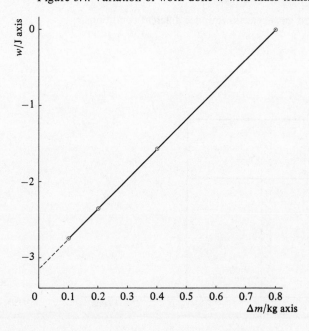

with, of course, 80 000 trays: it is left as an exercise to the reader to confirm this assertion. We shall see in Chapter 7 that the condition of reversibility is approached very closely in practice with the measurement of the electromotive force of an electrochemical cell. Next, we shall look at reversibility through the somewhat more chemical medium of a gas.

3.5 Expansion of a gas

There are several circumstances in which a gas may change its pressure and volume, and it is instructive to investigate some of them. We shall assume that our gas behaves *ideally*: by this term we imply that there are no intermolecular forces present, so that the internal energy of the gas is independent of its volume, depending only on the amount of substance and the temperature. If the amount of substance under examination is n mol, then the ideal gas equation is

$$PV = nRT \tag{3.6}$$

Consider the gas at a pressure P_1 and a volume V_1 (Figure 3.5, point d); the temperature is T_1. Let the gas be expanded to a volume V_2, but at the same pressure P_1 (point c). How may this process be carried out in practice? We can confine the gas to a cylinder by means of a perfect frictionless piston of cross-sectional area A (Figure 3.6), and supply heat to the gas to bring about its expansion. The mechanical work of expansion, done by the gas, utilizing the heat

Figure 3.5. Ideal gas expansions under different conditions.

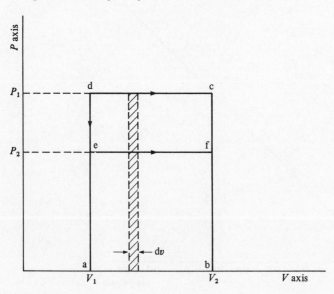

supplied, is given by

$$w = -P_1 \int A \, dl = -\int_{V_1}^{V_2} P_1 \, dV = -P_1(V_2 - V_1) \tag{3.7}$$

The negative sign shows that the work is done by the gas, and the magnitude of w is equivalent to the area abcd in Figure 3.5.

In another experiment, we may expand the gas isothermally from V_1 to V_2. In this case, because the temperature is constant, P_1 is changed to P_2 in accordance with (3.6). The work done is given by

$$w = -\int_{V_1}^{V_2} P \, dV = -P_2(V_2 - V_1) \tag{3.8}$$

since no work of expansion is done along the path de. The work given by (3.8) is the area abfe in Figure 3.5. Since the temperature is constant $\Delta U = 0$, and so $P_2(V_2 - V_1)$ is q, the heat absorbed by the gas in the isothermal expansion.

This expansion can be studied from another viewpoint, and it will transpire that abfe is not the maximum work derivable, at least theoretically, from this expansion. If the temperature is held constant, then from (3.6) the product PV is also constant (Boyle's law). The curve of P against V, at constant temperature, is a hyperbola (Figure 3.7). For an infinitesimal expansion dV at a constant pressure P, the work done is $-P \, dV$. From (3.6), we have

$$P \, dV = nRT \, dV/V \tag{3.9}$$

and the total work of expansion is given by

$$w = -nRT \int_{V_1}^{V_2} dV/V = -nRT \ln(V_2/V_1) = -nRT \ln(P_1/P_2) \tag{3.10}$$

Figure 3.6. Realization of non-isothermal expansion of an ideal gas at constant pressure.

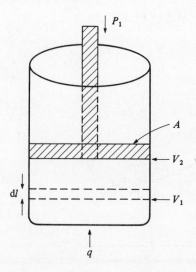

and this integral is the area abcd in Figure 3.7. How might this isothermal expansion be carried out in practice?

Again, the gas is confined to the cylinder (Figure 3.6). Starting at the point P_1, V_1 we decrease the pressure to some value P' at e (Figure 3.7) by cooling the gas. If there is no change in volume, no mechanical work is involved. However, since the gas must be cooled to move through the finite pressure step δP from d to e, it follows that q is negative, and from (3.4) U decreases. Thus, e does not lie in the constant temperature PV plane of the isotherm in Figure 3.7. A stepwise progression of non-isothermal, non-equilibrium steps, like def, would ultimately change the gas from P_1, V_1 to P_2, V_2. We see that a perfect isothermal process requires infinitesimal adjustments of the pressure and the volume, which cannot be achieved in practice.

The maximum work realizable from the gas by reversible expansion between states 1 and 2 is the area abcd (Figure 3.7). For any part of the complete change from state 1 to state 2, such as df, the maximum work is represented by the area agfd, and the practical amount by a lesser area, such as agfe. It should be remembered that any point not lying on the isothermal must be at either a lower or a higher temperature than that represented by the isotherm. If, however, the ratio $\delta P/P$ is small, an area such as agfd approximates to a rectangle. Although reversible conditions cannot be achieved in practice, the concept is important, and through it significant results can be deduced which are of practical value.

3.6 Exact differential

The values of w for the two paths in Figure 3.1 are different, being the areas abced (w_{I}) and abcfd (w_{II}) respectively. Since ΔU for the cycle $\mathrm{I}+\mathrm{II}$ is zero,

Figure 3.7. Isothermal expansion of an ideal gas.

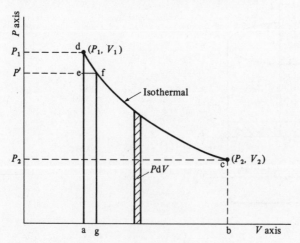

there are also two values for q. The quantities w and q are not state functions. They are not determined solely by the initial and final states of a system, and the infinitesimal amounts dw and dq cannot be integrated to give unique results. We shall indicate this condition by using the notation $\text{d}q$ and $\text{d}w$. Hence

$$\oint \text{d}w_I \neq \oint \text{d}w_{II} \text{ and } \oint \text{d}q_I \neq \oint \text{d}q_{II} \tag{3.11}$$

where the sign $\oint$ refers to the integral around a complete path, I and II in this case. But

$$\oint \text{d}w_I + \oint \text{d}q_I = -\oint \text{d}w_{II} - \oint \text{d}q_{II} \text{ and } \oint dU_I = -\oint dU_{II} \tag{3.12}$$

because U is a state function. For example, -2, 3, 4 and -5 could be self-consistent values for the first four integrals in (3.12); they show that ΔU_I, or $\oint dU_I$, is equal to $-\Delta U_{II}$, in accordance with (3.3). We call dU an exact differential; its integral is independent of the path of the corresponding reaction.

The energy of a real gas depends on any two of the variables P, V and T: we write for the exact differential dU

$$dU = (\partial U / \partial V)_T \, dV + (\partial U / \partial T)_V \, dT \tag{3.13}$$

Here, $(\partial U / \partial V)_T$ is the rate of change of U with respect to V while T is held constant, and similarly for the second term in this equation. We note that (3.13) does not refer to an ideal gas unless $(\partial U / \partial V)_T$ is zero (see Section 3.5).

We must be careful not to confuse the exact differential with the *total differential* of a function. Consider the function

$$V = f(T, P) \tag{3.14}$$

The total differential is defined as the sum of the partial differentials of all of the independent variables in the function

$$dV = (\partial f / \partial T)_P \, dT + (\partial f / \partial P)_T \, dP \tag{3.15}$$

For 1 mol of an ideal gas, $f = RT/P$, from (3.6). Hence

$$dV = (R/P) \, dT - (RT/P^2) \, dP \tag{3.16}$$

We need to be able to test whether or no dV is also exact. Let (3.15) be written as

$$dV = M(T, P) \, dT + N(T, P) \, dP \tag{3.17}$$

We know from calculus that for any exact function $f(x, y)$

$$\left[\frac{\partial}{\partial y} (\partial f / \partial x)_y \right]_x = \left[\frac{\partial}{\partial x} (\partial f / \partial y)_x \right]_y \tag{3.18}$$

Hence if dV is exact, applying (3.18) to (3.17), we have

$$\left[\frac{\partial}{\partial P} M(T, P) \right]_T = \left[\frac{\partial}{\partial T} N(T, P) \right]_P \tag{3.19}$$

and by substituting for the functions M and N from (3.16), we obtain $-R/P^2$ for both sides of (3.19), thus showing that the total differential dV in (3.16) is also exact.

3.7 Expansion under specified conditions

Our study of the expansion of a gas indicates certain lines for further consideration. A chemical process may be conducted at constant volume, at constant pressure, isothermally or adiabatically. We have considered the isothermal expansion, and we turn our attention now to other processes.

3.7.1 *Constant volume process*

At constant volume, the work of expansion of a gas is zero

$$w = -\int P\,dV = 0 \tag{3.20}$$

and from (3.4)

$$\Delta U = q_V \tag{3.21}$$

where q_V is the heat absorbed by the gas at constant volume. The heat capacity C of a body is defined as

$$C = đq/dT \tag{3.22}$$

Evidently C, like $đq$, depends upon the reaction path. The heat capacity at constant volume C_V is, however, unique, and is given by

$$C_V = (\partial U/\partial T)_V \tag{3.23}$$

3.7.2 *Constant pressure process*

Constant pressure processes are very common in chemistry. If we substitute $-P\,\Delta V$ for w in (3.4), we obtain

$$\Delta U = q_P - P\,\Delta V \tag{3.24}$$

where q_P is the heat absorbed by the gas at constant pressure. If ΔU and ΔV refer to a change from state 1 to state 2, then from (3.3)

$$q_P = (U_2 - U_1) + P(V_2 - V_1) \tag{3.25}$$

or

$$q_P = (U_2 + PV_2) - (U_1 + PV_1) \tag{3.26}$$

Since U, P and V are all state functions, $(U + PV)$ is also a state function; it is called the enthalpy H. Hence,

$$q_P = H_2 - H_1 = \Delta H \tag{3.27}$$

and

$$\Delta H = \Delta U + P\,\Delta V \tag{3.28}$$

where ΔH refers to a change at constant pressure P, and ΔU refers to the corresponding change under constant volume conditions. The difference between ΔH and ΔU is, therefore, the work of expansion at constant pressure.

The molar heat capacity of a gas at constant pressure C_P is given by

$$C_P = (\partial H/\partial T)_P \tag{3.29}$$

and for an ideal gas we may write

$$C_P = dH/dT \tag{3.30}$$

since, in this case, H is independent of pressure, as is U. From (3.23) and (3.30) we have, for an ideal gas,

$$C_P - C_V = (dH/dT) - (dU/dT) \tag{3.31}$$

or

$$C_P - C_V = \{(dU/dT) + [d(PV)/dT]\} - (dU/dT) \tag{3.32}$$

Using (3.6)

$$C_P - C_V = d(nRT)/dT = nR \tag{3.33}$$

In order to increase the temperature of a gas at constant volume, its internal energy must be increased, by supplying heat. To increase its temperature at constant pressure, heat must be supplied both to increase the internal energy and to provide for the work of expansion. From (3.33), we can see that the gas constant R represents the numerical value of the work done by one mole of an ideal gas in expanding against a constant pressure while energy is supplied to raise its temperature by 1 K. Equations (3.27) and (3.28) show that the heat gained or lost by a system under constant pressure conditions is the enthalpy change ΔH of the system, and generally ΔH is given by

$$\Delta H = \Delta U + nRT \tag{3.34}$$

in these circumstances. If ΔH is positive for a process, the system has absorbed heat from its surroundings (an *endothermic* reaction – for example, adding ammonium nitrate to water), whereas if ΔH is negative the system has given out heat to the surroundings (an *exothermic* reaction – for example, adding concentrated sulphuric acid to water).

3.7.3 *Adiabatic process*

In an adiabatic process, no heat enters or leaves the system; q is zero and ΔU is equal to w. Let us consider a sudden adiabatic expansion of an ideal gas against a constant pressure, this work being the only form involved. We have for dU

$$dU = (\partial U/\partial T)_V dT + (\partial U/\partial V)_T dV \tag{3.35}$$

Since the gas is ideal, $(\partial U/\partial V)_T$ is zero, and

$$dU = (\partial U/\partial T)_V dT = C_V dT = dw \tag{3.36}$$

Hence for a change between states 1 and 2

$$\Delta U = C_V(T_2 - T_1) \tag{3.37}$$

and because w is given also by $-P_2(V_2 - V_1)$, as for the isothermal expansion, we obtain

$$C_V(T_2 - T_1) = -P_2(V_2 - V_1) \tag{3.38}$$

Since $V_2 > V_1$, ΔT is negative, and the gas is cooled.

We can carry out the adiabatic expansion reversibly, at least in theory. From (3.36) we may write, recognizing the restriction of reversibility,

$$C_V \, dT = -P \, dV \tag{3.39}$$

Although we are working at constant pressure, we may use C_V by transforming the right-hand side of (3.39) to a variable in T. Dividing throughout by T and replacing P by RT/V, we obtain

$$C_V \, dT/T = -R \, dV/V \tag{3.40}$$

On integrating (3.40), we find

$$C_V \ln(T_2/T_1) = -R \ln(V_2/V_1) \tag{3.41}$$

Using (3.33) for $n = 1$, we have

$$T_2/T_1 = (V_1/V_2)^{(C_P - C_V)/C_V} = (V_1/V_2)^{\gamma - 1} \tag{3.42}$$

where γ is the ratio C_P/C_V. Since the ideal gas equation must be satisfied,

$$T_2/T_1 = P_2 V_2/P_1 V_1 \tag{3.43}$$

and

$$P_1 V_1^\gamma = P_2 V_2^\gamma \tag{3.44}$$

The bomb calorimeter operates under adiabatic, though not reversible, conditions.

Before leaving our consideration of the adiabatic process let us focus on q. Here it is zero: but q is not a state function; so is an adiabatic expansion unique, or is there another thermodynamic property involved here? Considering again the reversible adiabatic expansion of an ideal gas, we write

$$đq = dU - đw = C_V \, dT + P \, dV \tag{3.45}$$

or, for 1 mole,

$$đq = C_V \, dT + (RT/V) \, dV \tag{3.46}$$

Now $đq$ is inexact, as may be confirmed by differentiating the right-hand side of (3.46). However, on dividing by T we obtain

$$đq/T = (C_V \, dT/T) + (R \, dV/V) \tag{3.47}$$

The right-hand side of (3.47) is an exact differential, and so $\int đq/T$ is a state function: it is called the entropy change ΔS. Since $đq$ is zero for the reversible adiabatic expansion, it follows that

$$dS = đq/T = 0 \tag{3.48}$$

that is, the entropy change in an adiabatic process is zero. We shall consider the entropy function again later in this chapter.

3.8 Thermochemistry

In this section, we shall study enthalpy changes in chemical reactions. Let us consider first whether or no the fundamental conservation of energy is more significant than the conservation of mass for our purposes.

In a typical chemical reaction, a value for ΔU would usually be less than about 500 kJ mol^{-1}. From (3.1), taking the above value for energy, Δm is approximately 5.6×10^{-12} kg mol^{-1}, an amount not detectable by ordinary chemical means, so that for chemical processes the conservation of mass is a sufficiently precise concept. In contrast, consider a nuclear reaction, such as

$$^{14}_{7}N + {}^{2}_{1}H \rightarrow {}^{12}_{6}C + {}^{4}_{2}He \qquad (3.49)$$

where $^{14}_{7}N$ refers to the nitrogen isotope of mass number 14. The relative isotropic masses are, in order, 14.0037, 2.0140, 12.0000 and 4.00260. Hence Δm is $-0.01447u$ or -2.4026×10^{-29} kg. This mass is equivalent to -1.4470×10^{-5} kg mol^{-1}, and the corresponding energy change is -1.3×10^{9} kJ mol^{-1}; this nuclear process is highly exothermic.

Consider a chemical reaction generally:

$$A + B \rightarrow C + D \qquad (3.50)$$

It is likely that the sum of the enthalpies of the reactants is not equal to the sum of the enthalpies of the products. The reaction has an associated enthalpy change ΔH given by

$$\Delta H = \sum H \text{ (products)} - \sum H \text{ (reactants)} \qquad (3.51)$$

and it may be either positive or negative. Equation (3.51) forms the basis of thermochemistry.

We consider next the meaning of the terms on the right-hand side of (3.51). They are explained simply by

$$\sum H \text{ (products)} = H(C) + H(D) \qquad (3.52)$$
$$\sum H \text{ (reactants)} = H(A) + H(B) \qquad (3.53)$$

In a specific example†

$$Na(s) + \tfrac{1}{2}Cl_2(g) \rightarrow NaCl(s) \qquad (3.54)$$

we have, from (3.51)

$$\Delta H = H(NaCl,s) - [H(Na,s) + \tfrac{1}{2}H(Cl_2,g)] \qquad (3.55)$$

While we can measure ΔH for the reaction, we cannot measure the absolute values of H. A standard state is introduced so that a self-consistent basis is provided for enthalpic calculations. The standard state concept considers that the enthalpy (and free energy) of all elements in their normal state at 298.15 K and 1 atm is defined to be zero. Hence in (3.55)

$$\Delta H^{\ominus} = H^{\ominus}(NaCl,s) - (0 + 0) \qquad (3.56)$$

We refer to this particular $\Delta H^{\ominus}$ as the standard enthalpy of formation of sodium chloride, $\Delta H_f^{\ominus}(NaCl,s)$, that is, at 298.15 K and 1 atm. The superscript $^{\ominus}$ indicates a standard value of the property. The enthalpy of formation of a

---

† s, solid; l, liquid; g, gas.

compound from its elements is a particular example of the enthalpy change ΔH; other 'named' enthalpic reactions are given in Table 3.7.

Next, we study the oxidation of carbon (graphite), for which we can write three equations:

$$C\,(\text{graphite}) + O_2(g) \rightarrow CO_2(g) \quad \Delta H_f^{\ominus}(CO_2,g) = -394\,\text{kJ mol}^{-1} \tag{3.57}$$

$$C\,(\text{graphite}) + \tfrac{1}{2}O_2(g) \rightarrow CO(g) \quad \Delta H_f^{\ominus}(CO,g) \;\;= -111\,\text{kJ mol}^{-1} \tag{3.58}$$

$$CO(g) \qquad + \tfrac{1}{2}O_2(g) \rightarrow CO_2(g) \quad \Delta H^{\ominus} \qquad = -283\,\text{kJ mol}^{-1} \tag{3.59}$$

The total effect of (3.58) and (3.59) is equivalent to the single process (3.57). This situation arises from the fact that enthalpy is a state function, and is generalized as Hess's law: *if a reaction is carried out in stages, the algebraic sum of the enthalpies of the separate stages is equal to the enthalpy change for the complete reaction carried out in a single stage.* Mathematically

$$\sum \Delta H \,(\text{cycle}) = 0 \tag{3.60}$$

Hess's law is illustrated by Figure 3.8. From (3.60), taking the ΔH values in a cyclic manner, we obtain

$$\Delta H_2 + \Delta H_3 - \Delta H_1 = 0 \tag{3.61}$$

It is usually not possible to determine a heat of formation by direct combination of elements. Reaction (3.58) cannot be used to measure $\Delta H_f^{\ominus}(CO,g)$ experimentally because a mixture of carbon monoxide and carbon dioxide would be formed. We can, however, carry out both (3.57) and (3.59), and $\Delta H_f^{\ominus}(CO,g)$ can be obtained from the results (Figure 3.8).

As a second example, consider the evaluation of $\Delta H_f^{\ominus}(C_2H_5OH,l)$, which cannot be synthesized directly from its elements. The standard enthalpy of formation is still defined with respect to the elements in their standard states:

$$2\,C\,(\text{graphite}) + 3\,H_2(g) + \tfrac{1}{2}O_2(g) \rightarrow C_2H_5OH(l) \tag{3.62}$$

and the $\Delta H^{\ominus}$ for this reaction is the required $\Delta H_f^{\ominus}(C_2H_5OH,l)$. We need the following experimental data:

$$C\,(\text{graphite}) + O_2(g) \rightarrow CO_2(g) \quad \Delta H_f^{\ominus}(CO_2,g) = -394\,\text{kJ mol}^{-1} \tag{3.63}$$

Figure 3.8. Thermochemical cycle for carbon monoxide (gas).

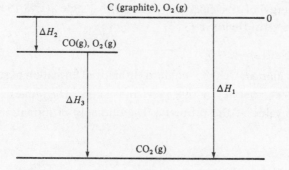

$$H_2(g) + \tfrac{1}{2}O_2(g) \rightarrow H_2O(l) \quad \Delta H_f^\ominus(H_2O,l) = -286 \text{ kJ mol}^{-1} \tag{3.64}$$

$$C_2H_5OH(l) + 3O_2(g) \rightarrow 2CO_2(g) + 3H_2O(l) \quad \Delta H^\ominus = -1373 \text{ kJ mol}^{-1} \tag{3.65}$$

From Figure 3.9, and using Hess's law, we have

$$2 \times (-394) + 3 \times (-286) - (1373) - H_f^\ominus(C_2H_5OH,l) = 0 \tag{3.66}$$

or

$$\Delta H_f^\ominus(C_2H_5OH^\ominus,l) = -273 \text{ kJ mol}^{-1} \tag{3.67}$$

The use of descriptive cycles, such as that in Figure 3.9, is preferable to treating processes such as (3.62)–(3.65) like algebraic equations, and the following hints may be useful in working with thermochemical cycles.

(a) Let the arrow indicate the direction of the reaction for which the corresponding ΔH, including its sign, is defined.

(b) Elements, in any required quantities, can be considered to be present on any enthalpy level, as they contribute zero enthalpy in standard state conditions. The number of moles of the products of the reactions of elements is, of course, significant.

(c) Sum the ΔH values around the cycle in a given sense, say anticlockwise, including the signs of the terms, but changing the sign of a term for which the arrow indicates a movement contrary to the direction of summing, as shown by (3.66).

3.8.1 *Dissolution*

Equation (3.64) represents the formation of liquid water. If the product had been water vapour, $\Delta H_f^\ominus$ would have been -242 kJ mol^{-1}. The difference, $-44.0 \text{ kJ mol}^{-1}$, represents the molar enthalpy of evaporation of water at 298.15 K and 1 atm, $\Delta H_e^\ominus(H_2O,l)$, and we see from this example the importance of quoting the physical state of each substance in a thermochemical reaction. Thus,

$$Na^+(aq) + OH^-(aq) + H^+(aq) + Cl^-(aq) \rightarrow Na^+(aq) + Cl^-(aq) + H_2O(l) \tag{3.68}$$

Figure 3.9. Thermochemical cycle for ethanol (liquid).

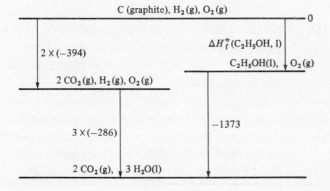

or

$$H^+(aq) + OH^-(aq) \rightarrow H_2O(l) \tag{3.69}$$

where aq represents the state of infinite dilution; this state needs some explanation.

When a substance, sodium hydroxide for example, dissolves in water an enthalpy change takes place that is dependent upon the amount x of water available for a given mass of sodium hydroxide. If x is sufficiently small, we may write

$$NaOH(s) + x\,H_2O(l) \rightarrow Na^+(y\,H_2O) + OH^-(z\,H_2O) + (x\text{-}y\text{-}z)H_2O(l) \tag{3.70}$$

and a certain enthalpy change, negative in this example, occurs. The x mol of water may be considered to be distributed as y mol associated with, or hydrating, the sodium ions and z *mol* hydrating the hydroxide ions. If more water is added to the solution, a further enthalpy change takes place, an enthalpy of dilution. The hydrated ions become further separated, and their energy of interaction is given up to the surroundings as heat. The standard enthalpy of dissolution $\Delta H_d^\ominus$ is referred to the state of infinite dilution, that is, it is the enthalpy change accompanying dissolution in such a quantity of water that further dilution produces no measurable enthalpy change. Generally, the value of $\Delta H_d^\ominus$ will be determined by observing the heats evolved at several given concentrations of the substance under consideration, and then extrapolating the results to zero concentration.

It is informative to try to understand some of these energy changes in terms of a model for the solution process. A solid dissolves because it interacts with the solvent. In the case of sodium hydroxide, an exothermic change is observed on adding a small quantity of the solid to water. The sodium and hydroxide ions that are present in the solid become hydrated by molecules of water, and the structure of the solid is disrupted. The ions become loosely bound to the water molecules by ion-dipole forces. In a small quantity of water, the ions may not be as fully hydrated as is energetically possible, because there is not enough water present to satisfy all of the ions present. It does not follow that some sodium hydroxide will not dissolve under these conditions. A sodium ion can be hydrated, on average, by six water molecules. We may think of a spherical region in the solution consisting of the central ion, sodium or hydroxide, surrounded by six water molecules forming a hydration sphere. The processes of initial dissolution and subsequent dilution involve electrical forces, and heat energy is exchanged between the solution and its surroundings.

Beyond a particular dilution, we may assume that complete hydration of the ions exists, with no additional interactions between the species present in solution because the ions are, on average, widely separated. Consequently, there is no exchange of energy on further dilution of the solution: the particular dilution may be taken to represent infinite dilution (aq) for all practical purposes.

It may be noted that sodium hydroxide dissolves endothermically in its near-saturated solution, and application of Le Chatelier's principle indicates that the solubility of sodium hydroxide increases with increasing temperature, as is known from experiment. The opposite, incorrect conclusion would be reached if the sign of $\Delta H_d^{\ominus}$ were considered instead.

The enthalpy data for the species involved in (3.68) are as shown in Table 3.2. Hence $\Delta H^{\ominus}$ for (3.68), and hence for (3.69), is given by

$$[\Delta H_f^{\ominus}(\text{NaCl,s}) + \Delta H_d^{\ominus}(\text{NaCl,aq}) + \Delta H_f^{\ominus}(\text{H}_2\text{O,l})]$$

$$- [\Delta H_f^{\ominus}(\text{NaOH,s}) + \Delta H_d^{\ominus}(\text{NaOH,aq}) + \Delta H_f^{\ominus}(\text{HCl,g}) + \Delta H_d^{\ominus}(\text{HCl,aq})] \quad (3.71)$$

which sums to $-56.2 \text{ kJ mol}^{-1}$. This value is very different from that for the reaction (3.64). The two results may be studied in the light of the cycle in Figure 3.10. Evidently,

$$\sum_{i=1}^{8} \Delta H_i^{\ominus} = \Delta H_e^{\ominus}(\text{H}_2\text{O,l}) - \Delta H_f^{\ominus}(\text{H}_2\text{O,g}) \quad (3.72)$$

where ΔH_8 is the enthalpy change for (3.69), and is a constant for all neutralizations of strong acids by strong bases when referred to infinite dilution.

In reacting a strong base with a weak acid at any non-zero concentration, a numerically smaller value for the enthalpy change corresponding to ΔH_8 is obtained, even at infinite dilution. In these circumstances, some of the enthalpy of (3.69) is absorbed by the system in bringing about dissociation of the weak acid. The value obtained for ΔH_8 in this situation is related to the dissociation constant for the weak acid, but this constant cannot be evaluated from the enthalpy change alone.

3.8.2 Enthalpy changes and temperature

Enthalpy changes vary with temperature. Differentiating (3.51) with respect to temperature, at constant pressure, leads to

$$(\partial \Delta H / \partial T)_P = [\partial H(\text{products}) / \partial T]_P - [\partial H(\text{reactants}) / \partial T]_P \quad (3.73)$$

By analogy with (3.29), we can write

$$(\partial \Delta H / \partial T)_P = \Delta C_P \quad (3.74)$$

Table 3.2. *Enthalpy data for species in Equation (3.68)*

	$\Delta H_f^{\ominus}/\text{kJ mol}^{-1}$		$\Delta H_d^{\ominus}/\text{kJ mol}^{-1}$
NaOH(s)	−427	NaOH(aq)	−42.3
HCl(g)	−92.5	HCl(aq)	−73.6
NaCl(s)	−411	NaCl(aq)	+5.4
H$_2$O(l)	−286		

where
$$\Delta C_P = C_P(\text{products}) - C_P(\text{reactants}) \tag{3.75}$$

Integrating (3.74) leads to the Kirchoff equation:

$$\Delta H_{T_2} - \Delta H_{T_1} = \int_{T_1}^{T_2} \Delta C_P \, dT \tag{3.76}$$

In general, C_P varies with temperature and may be represented as a function of temperature T:

$$C_P = a + bT + cT^2 \tag{3.77}$$

We shall illustrate these equations through the Haber process

$$\tfrac{1}{2}N_2(g) + \tfrac{3}{2}H_2(g) \rightleftharpoons NH_3(g) \tag{3.78}$$

The data listed in Table 3.3 apply between 273 K and 1500 K; $\Delta H_f^{\ominus}(NH_3, g)$ is $-46.1 \text{ kJ mol}^{-1}$, and the corresponding value at 598 K will be calculated.

Table 3.3. *Heat capacity constants for nitrogen, hydrogen and ammonia gases*

	$a/\text{J K}^{-1}\text{mol}^{-1}$	$10^3 b/\text{J K}^{-2}\text{mol}^{-1}$	$10^6 c/\text{J K}^{-3}\text{mol}^{-1}$
$\tfrac{1}{2}N_2$	13.49	2.96	-0.17
$\tfrac{3}{2}H_2$	43.60	-1.26	3.02
NH_3	25.89	32.58	-3.05

Figure 3.10. Formation of water by two routes.

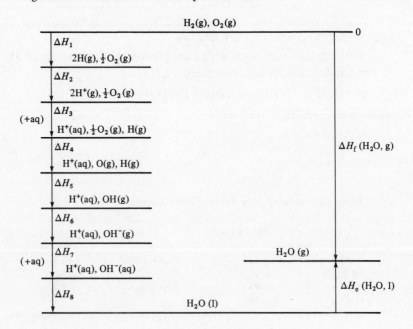

From (3.75) ΔC_P is given by

$$\Delta C_P = 25.89 - (13.49 + 43.60) + 10^{-3}T[32.58 - (2.96 - 1.26)]$$
$$+ 10^{-6}T^2[-3.05 - (-0.17 + 3.02]$$
$$= -31.20 + 30.88 \times 10^{-3}T - 5.90 \times 10^{-6}T \tag{3.79}$$

Using (3.76),

$$\Delta H_{598K} = -46.1 + \int_{298}^{598} (-31.20 + 30.88 \times 10^{-3}T - 5.90 \times 10^{-6}T^2)\,dT \tag{3.80}$$

The reader is invited to confirm the result that ΔH_{598K} is $-51.7\,\text{kJ mol}^{-1}$.

3.8.3 *Calorimetry*

Equation (3.65) represents the complete combustion of ethanol in oxygen. The measurement of the associated enthalpy change can be carried out with a bomb calorimeter (Figure 3.11). In this apparatus the calorimeter is insulated thermally from its surroundings. The combustion reaction is conducted both adiabatically and at constant volume, though not reversibly, and the energy change calculated represents the property ΔU for the reaction. From (3.28), the corresponding enthalpy change is given by

$$\Delta H = \Delta U + P\,\Delta V \tag{3.81}$$

since the change in pressure for the reaction is negligible under the conditions of very large oxygen pressure within the bomb. If we assume ideality for the gaseous components of combustion, then

$$\Delta H = \Delta U + \Delta nRT \tag{3.82}$$

where Δn is the number of moles of gaseous products minus the number of moles of gaseous reactants. The apparatus is calibrated with a standard substance, such as benzoic acid for which the enthalpy change on combustion is $-3226.4 \pm 0.3\,\text{kJ mol}^{-1}$ at 298 K. A precision of approximately 0.01% is attainable, and in a typical experiment on the combustion of ethanol, ΔU was found to be $-1370.7\,\text{kJ mol}^{-1}$; since Δn is -1, ΔH is $-1373.2\,\text{kJ mol}^{-1}$, from (3.82).

3.9 Entropy and free energy

In a mechanical system, such as the spring shown in Figure 3.3, the capacity for work lies in the energy stored in the extended spring. We showed how a transfer of effectively all of the energy might be achieved, through an infinite number of infinitesimally small stages of mass transfer (Figure 3.4). The internal energy of a chemical system cannot be transformed wholly into work, and equilibrium cannot be defined through ΔU, which is a distinction from our mechanical analogue.

From (3.4) we have for an ideal gas doing only work of expansion

$$dU = \text{\dj}q - P\,dV \tag{3.83}$$

Using (3.23)

$$C_V \, dT = đq - P \, dV \qquad (3.84)$$

and from (3.6), for 1 mole of the ideal gas,

$$C_V \, dT = đq - RT \, dV/V \qquad (3.85)$$

Dividing throughout by T and rearranging gives

$$đq/T = (C_V \, dT/T) + (R \, dV/V) \qquad (3.86)$$

By integrating the right-hand side of (3.86) between two states 1 and 2, we obtain

$$C_V \int_{T_1}^{T_2} dT/T + R \int_{V_1}^{V_2} dV/V = C_V \ln(T_2/T_1) + R \ln(V_2/V_1) \qquad (3.87)$$

assuming that C_V is constant between temperatures T_1 and T_2. The result can be seen to depend only on the initial and final states of the system. Hence $\int đq/T$ is a

Figure 3.11. Bomb calorimeter: a weighed sample A is placed in the crucible of a steel bomb B which is then filled with oxygen to a pressure of about 20 atm through the valve C, and then immersed in water at temperature equilibrium with the rest of the apparatus. An electric current is passed through the platinum wire spiral D which ignites a thread of cotton resting in the sample, and so starts the reaction. Heat is evolved, and the associated temperature rise in the water filling the air-jacketed calorimeter E is measured with the Beckmann thermometer F. The calorimeter is contained within a large vessel of water G, all agitated by the stirrers H. The temperature change in G is monitored by another Beckmann thermometer I. A cooling correction must be applied for the loss of heat from H to G. In the *adiabatic* bomb calorimeter, the thermometer I is replaced by an electric heater which is controlled by the difference in temperature between H and G, as measured by matched thermistors. Thus, the outer vessel G is kept at the same temperature as the calorimeter H, and no cooling correction is necessary.

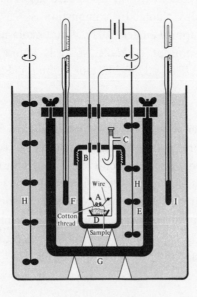

state function, and it is called the entropy change ΔS for the reaction: we anticipated a part of this result in Section 3.7.3.

Entropy provides for a measure of the unavailable work of a chemical reaction. At constant temperature, (3.4) shows that the infinitesimal amount of work done reversibly is given by

$$\text{đ}w_{rev} = dU - T\,dS \tag{3.88}$$

In taking the system from state 1 to state 2, ΔS is always the same by any path; it is equal to $\int \text{đ}q_{rev}/T$ only for a reversible process. Hence, $dS = \text{đ}q/T$ is a condition for equilibrium. For any observable change $dS > 0$, where dS refers to the combination of a system and its surroundings. For an isolated system at equilibrium, $dS = 0$, so that for any spontaneous process the total entropy of the system and its surroundings will increase. This argument leads to a statement of the second law of thermodynamics that *the entropy of the universe is tending to a maximum.*

3.9.1 *Expansion of a gas reconsidered*

Consider the isothermal expansion of an ideal gas from state 1 to state 2. From (3.10), we may write

$$w_{rev} = nRT \ln(V_2/V_1) \tag{3.89}$$

Hence $\quad \Delta S = q_{rev}/T = nR \ln(V_2/V_1) \tag{3.90}$

Equation (3.90) applies to both reversible and irreversible expansions. In a reversible expansion, $\Delta S_{total} = 0$ since the heat gained by the gas is equal to the heat lost by the surroundings. In an irreversible expansion, ΔS_{gas} is still $nR \ln(V_2/V_1)$, but $\Delta S_{total} > 0$. The entropy change for the gas is the same in each situation, and only an investigation of the surroundings will enable us to define the process from state 1 to state 2 fully. For this reason, entropy is not a useful function for defining equilibrium. Two factors arise in chemical systems, both of which tend to drive a reaction to equilibrium: there is the tendency to minimize the energy (ΔU or ΔH negative) and there is the tendency to maximize the entropy (ΔS positive). The interaction of these two factors is usually expressed by the free energy change ΔG for the reaction, discussed more fully presently.

3.9.2 *Statistical basis for entropy*

Entropy is related to the degree of order in a system, and we may illustrate this basis in the following manner. Let a system of five identical but distinguishable molecules be distributed among a set of energy levels ξ_0, $\xi_0 + \xi$, $\xi_0 + 2\xi$, $\xi_0 + 3\xi$ and $\xi_0 + 4\xi$, where ξ_0 is the lowest energy level (the ground state of the molecule). The total energy of the system E is clearly $5\xi_0 + 10\xi$: if, for convenience, we take ξ_0 as zero, then $E = 10\xi$.

Let the five molecules be distributed among the energy levels $p\xi$, ($p = 0, 1, \ldots, 4$) in state P. If there is no restriction on the number of molecules that may occupy a given energy level, we can show that the number of arrangements given in Table 3.4 each leads to a total energy of 10ξ. Of the 381 possible arrangements for the molecules, one distribution is four times more probable than the next most probable distribution. If the number of molecules increases, even to 10^6 (a very small number of molecules in real chemical systems), the most likely distribution becomes overwhelmingly probable, and the properties of this distribution can be taken to represent those of the actual system. Next, suppose that our system of five molecules undergoes a chemical reaction, without a change in total energy, leading to a state Q with a more restricted set of energy levels, 4ξ, 2ξ and 0.

Only 51 arrangements are now possible (Table 3.5), and the most probable of them is only one-quarter as probable as the most probable distribution in state P. Thus, on probability grounds, the equilibrium

$$P \rightleftharpoons Q \qquad\qquad (3.91)$$

will lie preponderantly on the side of P. In an actual equilibrium, some of the molecules will be in state Q but the large majority will be in state P. The equilibrium is, of course, dynamic, and the result that we have given represents

Table 3.4. *Distribution of five molecules among five energy states (state P)*

4ξ	3ξ	2ξ	ξ	0	Number of arrangements, W
1	1	1	1	1	$5!/1! = 120$
2		1		2	
2			2	1	
1	2			2	$5!/2!\,2! = 30$ each
1		2	2		
	2	2		1	
	2	1	2		
1	1		3		
1		3		1	
	3		1	1	$5!/3! = 20$ each
	1	3	1		
		5			$5!/5! = 1$

Table 3.5. *Distribution of five molecules among three energy levels (state Q)*

4ξ	2ξ	0	Number of arrangements, W
2	1	2	$5!/2!\,2! = 30$
1	3	1	$5!/3! = 20$
	5		$5!/5! = 1$

the average over a period of time that is long in relation to the time of interchange between state P and state Q for any molecule.

We identify the change in probability of a system with a change in entropy through the Boltzmann equation, given without proof here by

$$S = k \ln W \tag{3.92}$$

where k is the Boltzmann constant. Hence, for (3.91), using the most probable distributions of the energy states, ΔS is $N_A k \ln[W(Q)/W(P)]$, or $-11.5 \, \text{J mol}^{-1}$, which is in agreement with our deduction that state Q is the less probable.

3.9.3 *Absolute entropy*

Entropy is an absolute quantity. The zero of entropy is defined for a perfect, infinite crystal at 0 K, which is one statement of the third law of thermodynamics. The specification 'perfect' implies freedom from any effects that may inhibit perfect ordering, such as crystal defects or hydrogen bonding.

At constant pressure, from (3.90), we have

$$dS = dH/T \tag{3.93}$$

and using (3.30) we obtain

$$dS = (C_P/T) \, dT \tag{3.94}$$

Hence the entropy of a substance, throughout any range of temperature for which that substance remains homogeneous is given by

$$S = \int_{T_1}^{T_2} (C_P/T) \, dT \tag{3.95}$$

Generally, C_P will not be known analytically over a wide range of temperature. The variation of C_V with temperature for sodium chloride is known; the difference between C_P and C_V is very small for solids, because the heat capacity of a solid is determined by its vibrational energy, and this quality does not vary appreciably between constant pressure and constant volume conditions. Hence, the entropy of sodium chloride at 300 K can be calculated as the area under the curve (Figure 3.12) between $T = 0$ K and $T = 300$ K. The area may be evaluated with Simpson's rule, as follows.

The curve is divided into an even number a, of strips of equal width b, which lead to $a + 1$ ordinates to the curve of height $h_0, h_1, \ldots, h_a$. Then the area A is given by

$$A = b/3[h_0 + h_a + 2(h_2 + \cdots + h_{a-2}) + 4(h_1 + h_3 + \cdots + h_{a-1})] \tag{3.96}$$

From the curve given, the absolute entropy of sodium chloride, $S(\text{NaCl,s})$, is $75.1 \, \text{J K}^{-1} \, \text{mol}^{-1}$. An entropy change may be positive or negative, but an absolute entropy can only be positive.

3.10 Free energy

Chemical reactions are often studied at constant pressure. On raising the temperature of a given mass of gas, heat is absorbed and the enthalpy of the system increases. The entropy of the system also increases, because of the wider distribution, or greater randomness, of the speeds of the molecules (Figure 3.13).

From the point of view of the spontaneity of a reaction, the probable directions of the simultaneous changes in enthalpy and entropy are antithetical. A reaction is favoured by a negative ΔH, because the products are then lower in energy than are the reactants. A decrease in entropy for a reaction means that the products are less probable than the reactants – the number of available energy states has decreased. The total effect of the changes in enthalpy and entropy is expressed by another state function, the free energy G, such that

$$G = H - TS \tag{3.97}$$

For a reaction at constant temperature, we may write

$$\Delta G = \Delta H - T\,\Delta S \tag{3.98}$$

The quantity ΔG represents the energy obtainable from a chemical reaction at a constant total composition, other than work of expansion against a constant external pressure. The difference $\Delta H - \Delta G$ is the energy that cannot be converted into work; it depends on the temperature and the entropy change at that temperature. At equilibrium, $\Delta G = 0$: this should be understood as relating to a dynamic equilibrium, such as (3.91), where the amounts involved in the reaction are infinitesimally small, so that the reaction can be considered to be reversible.

For a finite change under isothermal conditions, we see from (3.93) that

$$\Delta S = \Delta H / T \tag{3.99}$$

Figure 3.12. Variation of C_V/T with T for sodium chloride.

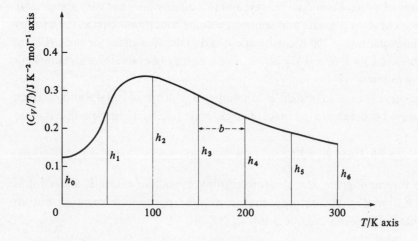

The reversible transition of ice to water at 273.15 K is represented by

$$H_2O(s) \rightleftharpoons H_2O(l) \tag{3.100}$$

and ΔH_t is found from experiment to be 6025.0 J mol^{-1}. Since ice and water are in equilibrium at 273.15 K, it follows that ΔS_t is 22.06 J K^{-1} mol^{-1}: ΔS_t is the entropy change for the system, and $-\Delta H_t/T$ that for the surroundings, because $\Delta S_{syst} + \Delta S_{surr} = 0$ for an equilibrium process. When liquid water is formed according to (3.64) $\Delta H_f^{\ominus}$ is -286 kJ mol^{-1}, and the corresponding entropy change is -163.2 J K^{-1} mol^{-1}; hence, $\Delta G^{\ominus}$ for (3.64) is -237 kJ mol^{-1}. When water vapour is formed by a similar reaction, $\Delta H_f^{\ominus}$ is -242 kJ mol^{-1} (it includes the enthalpy of evaporation) and $\Delta S_f^{\ominus}$ is -44.4 J K^{-1} mol^{-1}, so that $\Delta G_f^{\ominus}$ is now -229 kJ mol^{-1}. This example shows the effect of the higher entropy associated with the gaseous state compared to the condensed states of the same substance. In both cases, we see that some of the negative ΔH is

Figure 3.13. Probability distribution of molecular speeds. The shaded area under the curve represents the number of molecules that have speeds lying between v and $v + dv$.

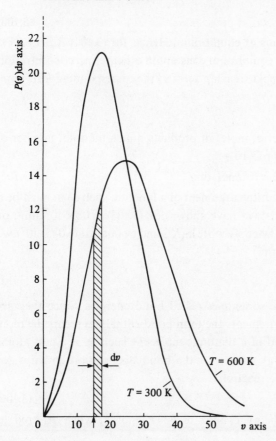

offset in ΔG by the negative ΔS. The decrease in ΔG for a reaction is the thermodynamic driving force causing it to take place. A negative value for ΔG indicates a spontaneous process, in the thermodynamic sense, under the conditions specified. The enthalpy change need not be negative in a spontaneous process. Consider the reaction

$$\tfrac{1}{2}I_2(s) + \tfrac{1}{2}Cl_2(g) \rightarrow ICl(g) \tag{3.101}$$

$\Delta H_f^{\ominus}(ICl,g)$ is 17.6 kJ mol^{-1}, and the entropies of $ICl(g)$, $Cl_2(g)$ and $I_2(s)$ are 247.4, 130.6 and 116.7 J K^{-1} mol^{-1} respectively. Thus, $\Delta S_f^{\ominus}(ICl,g)$ is 123.8 J K^{-1} mol^{-1}, and $\Delta G_f^{\ominus}(ICl,g)$ is -19.3 kJ mol^{-1}.

3.10.1 *Free energy change and equilibrium constant*

The equilibrium constant for reaction (3.101) is given, in terms of partial pressures, by

$$K_p = p(ICl,g)/p^{\frac{1}{2}}(Cl_2,g) \tag{3.102}$$

since the partial pressure of the solid iodine is a constant, incorporated into the value of K_p. The value of K_p is related to the standard free energy change by the equation

$$\Delta G^{\ominus} = -RT \ln(K_p) \tag{3.103}$$

a general equation for systems at equilibrium. Hence, for (3.101) K_p is 6.08×10^7 atm$^{\frac{1}{2}}$. We can express the equilibrium constant in other terms, concentration for example. In this case, we replace each p_i term by the corresponding RTc_i term, so that

$$K_p = K_c(RT)^{\Delta n} \tag{3.104}$$

where Δn is the total number of moles of products minus the total number of moles of reactants. Hence, for (3.101)

$$K_c = K_p/(RT)^{\frac{1}{2}} = 1.22 \times 10^6 \text{ mol}^{\frac{1}{2}} \text{ dm}^{-\frac{3}{2}} \tag{3.105}$$

We should note in passing that the argument of a function such as ln must be a dimensionless quantity. K_p is, we have shown for (3.101), 6.08×10^7 atm$^{\frac{1}{2}}$, or K_p/atm$^{\frac{1}{2}}$ is 6.08×10^7. Hence, when we write $\ln(K_p)$ and evaluate $\ln(6.08 \times 10^7)$, we should strictly write $\ln(K_p/\text{atm}^{\frac{1}{2}})$.

3.11 Bond enthalpies

Bond enthalpies are sometimes called bond energies. Since they are derived from enthalpy measurements the term bond enthalpies seems the more appropriate. The dissociation of a diatomic molecule, such as H_2, into atoms requires an amount of energy known as the bond dissociation enthalpy (or energy) D_0 (Figure 3.14). The reaction

$$H_2(g) \rightarrow 2H(g) \tag{3.106}$$

is accompanied by an enthalpy change of 436 kJ mol^{-1}, when referred to standard state conditions. If we take instead the reaction

$$CH_4(g) \rightarrow C(g) + 4H(g) \tag{3.107}$$

we can identify an average enthalpy of dissociation of the carbon–hydrogen single bond, $\Delta H_b(C–H)$; this quantity is sometimes called a bond enthalpy.

From Figure 3.15, and using principles that we have discussed already, we see that

$$\Delta H_b(C–H) = (872 + 717 + 74.8)/4 = 416 \text{ kJ mol}^{-1} \tag{3.108}$$

Figure 3.14. Variation of the bond dissociation energy of a diatomic molecule with interatomic distance. The energy is zero at infinite distance and $-D_0$ at the equilibrium interatomic distance r_e (see also Figure 2.4).

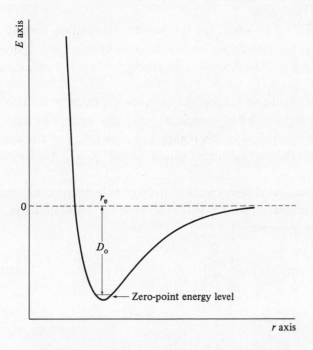

Figure 3.15. Estimation of ΔH_b for a C–H bond in methane (values are in kJ mol^{-1}).

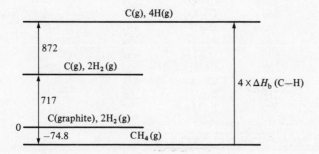

In a similar manner, the corresponding parameters for other bonds may be determined, and a selection of them is given in Table 3.6.

We distinguish between ΔH_b and D_o. A true bond dissociation enthalpy D_o has a definite value, associated with a given bond, but a bond enthalpy ΔH_b has a value for a given bond that is obtained as an average among a series of different dissociating species. Sometimes, as in the case of hydrogen molecules or hydrogen chloride, the two quantities become one and the same.

In the reaction

$$H_2O(g) \rightarrow H\cdot(g) + OH\cdot(g) \tag{3.109}$$

ΔH is $501.9\,kJ\,mol^{-1}$, whereas for

$$OH\cdot(g) \rightarrow H\cdot(g) + O\cdot(g) \tag{3.110}$$

ΔH is $423.4\,kJ\,mol^{-1}$. The value for the second dissociation enthalpy is influenced by the first dissociation: we are now dissociating an oxygen–hydrogen bond in $OH\cdot(g)$ and not in $OH_2(g)$, and it is the average of these two values that is listed in Table 3.6.

Bond enthalpies are evidently not precise quantities, and calculations based on them, while of the right order of magnitude, may not agree well with the corresponding experimental values, as Figure 3.16 indicates. The discrepancy between the experimental and calculated values for $\Delta H_f^\ominus(C_2H_6,g)$ is typical of such calculations.

In considering unsaturated compounds, enthalpies of hydrogenation tend to be more useful quantities for the following type of calculation. In the hydrogenation of cyclohexene

$$\bigcirc\!\!| \quad + \quad H_2(g) \quad \longrightarrow \quad \bigcirc \tag{3.111}$$

Table 3.6. *Average values of $\Delta H_b/kJ\,mol^{-1}$ for selected bonds*

H—H	436	C—C^a	717
C—C	348	C—O	358
C=C	610	C=O	745
C≡C	835	C≡N	890
C—H	416	N—N	389
C—F	485	O—H	463
C—Cl	339	F—H	565
C—Br	284	Cl—H	431
C—I	218	Br—H	366
C—N	306	I—H	299

^{a} Enthalpy of atomization of graphite.

ΔH is $-120\,\text{kJ mol}^{-1}$. For benzene, the corresponding value is $-208\,\text{kJ mol}^{-1}$ (Figure 3.17). From (3.111) we might deduce that the enthalpy of hydrogenation of benzene would be -120×3, or $-360\,\text{kJ mol}^{-1}$. Experiment and calculation shows that benzene is more stable (more negative in energy) than a hypothetical cyclohexatriene by $152\,\text{kJ mol}^{-1}$. This calculation is the source of the $150\,\text{kJ mol}^{-1}$ given in Chapter 2 as the approximate resonance energy of benzene.

If we try to determine the enthalpy of formation of benzene from our bond enthalpy data, we obtain $252\,\text{kJ mol}^{-1}$ (Figure 3.18). The experimental value is $83\,\text{kJ mol}^{-1}$: the difference of $169\,\text{kJ mol}^{-1}$ is the delocalization energy of *ca* $150\,\text{kJ mol}^{-1}$ together with errors reflecting the nature of bond enthalpies. We may improve on our data by determining a bond enthalpy for the aromatic carbon–carbon bond (see Problem 3.19).

Another well-known example of delocalization energy is that of buta-1,3-diene:

$$\underset{\underset{0.135 \text{ nm}}{}}{H_2C} \overset{\overset{0.147 \text{ nm}}{}}{=\!=} CH \underset{\underset{0.135 \text{ nm}}{}}{-\!\!-\!\!-CH=\!=CH_2}$$

Figure 3.16. Bond enthalpy cycle for ethane; the experimental value for $\Delta H_f^{\ominus}(C_2H_6(g))$ is $-84\,\text{kJ mol}^{-1}$ (values are in kJ mol^{-1}).

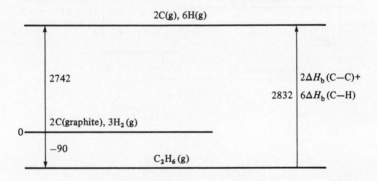

Figure 3.17. Cycle for the hydrogenation of benzene (values are in kJ mol^{-1}).

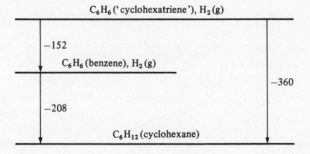

Its enthalpy of hydrogenation is $-239\,\text{kJ mol}^{-1}$, whereas that for two moles of but-1-ene is $-254\,\text{kJ mol}^{-1}$. The smaller value for buta-1,3-diene shows that it is stabilized with respect to two moles of but-1-ene by $-15\,\text{kJ mol}^{-1}$. This value is less per classical double bond than in benzene because the delocalization is less extensive, and the bond length values of the carbon–carbon bonds, shown above, confirm this result. The currently accepted bond lengths for C—C and C=C are 0.154 and 0.133 nm respectively: in benzene, all carbon–carbon bond lengths are 0.140 nm.

3.12 Named enthalpic reactions

The enthalpy of formation is the enthalpy change of a reaction like (3.62). Several 'named' enthalpic reactions are commonly defined, as set out in Table 3.7. The reader should have no difficulty with other named reactions of a similar nature.

Figure 3.18. Determination of the enthalpy of formation of gaseous benzene; the experimental value is $83\,\text{kJ mol}^{-1}$ (values are in kJ mol^{-1}).

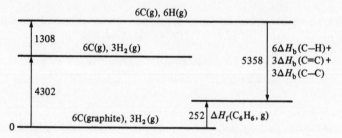

Table 3.7. *Named enthalpic reactions (standard state conditions)*

Name	Initial state	Final state	Typical reaction
Formation	Elements in their normal states at 298.15 K and 1 atm	Compound at 298.15 K and 1 atm	$Hg(l) + \frac{1}{2}O_2(g) \rightarrow HgO(s)$
Combustion	Elements or compounds at 298.15 K and 1 atm	Products of complete combustion in pure O_2 referred to 298.15 K and 1 atm	$C_6H_6(l) + \frac{15}{2}O_2(g) \rightarrow 6CO_2(g) + 3H_2O(l)$
Atomization	Elements in their normal states at 298.15 K and 1 atm	Gaseous atomized elements at 298.15 K and 1 atm	$\frac{1}{2}Br_2(l) \rightarrow Br(g)$
Neutralization	$H^+(aq)$, $OH^-(aq)$ at 298.15 K and at infinite dilution (strong acid and strong base)	$H_2O(l)$ at 298.15 K	$H^+(aq) + OH^-(aq) \rightarrow H_2O(l)$
Dissolution	Solid and solvent at 298.15 K	Solvated solid (or ions) at 298.15 K and infinite dilution	$CaCl_2(s) + aq \rightarrow Ca^{2+}(aq) + 2Cl^-(aq)$

Problems 3

3.1 $^{226}_{88}$Ra undergoes spontaneous disintegration to $^{222}_{86}$Rn with the loss of an α particle ($^4_2$He). Given that the relative atomic masses of the Ra, Rn and He species are 226.0254, 222.0175 and 4.0026 respectively, calculate the molar energy change for this disintegration process. You will need values for the constants u, c and N_A.

3.2 A system of internal energy $-13.3 \, \text{kJ mol}^{-1}$ moves, at constant pressure, to another state in which its internal energy is $-47.8 \, \text{kJ mol}^{-1}$. In the change, the system does work that is equivalent to $56.9 \, \text{kJ mol}^{-1}$. Calculate ΔU and q for the process, and state whether it is exothermic or endothermic.

3.3 Determine the maximum work that can be done by an ideal gas in expanding 10 mole against a constant pressure if the temperature changes by 17 K.

3.4 Six mole of nitrogen gas are compressed isothermally and reversibly from a volume of $1.00 \, \text{m}^3$ to a volume of $0.15 \, \text{m}^3$, at 298 K. Calculate the work done on the gas if it obeys (a) the ideal gas equation, and (b) the van der Waals' equation of state with $a = 0.14 \, \text{N m}^4 \, \text{mol}^{-1}$ and $b = 3.9 \times 10^{-5} \, \text{m}^3 \, \text{mol}^{-1}$.

3.5 For an isothermal process at constant pressure where the only work done is that of expansion against a constant pressure, show from (3.88) that since U is a state function, $(\partial T/\partial V)_S = (\partial P/\partial S)_V$.

3.6 Calculate the difference between ΔU and ΔH when 1 mole of water is boiled at 373.15 K. The water vapour can be assumed to behave ideally, and the volume of the liquid water may be neglected in comparison with the volume of the vapour.

3.7 A gas is compressed adiabatically from 1 atm to 20 atm, the initial temperature being 298 K. The compression is carried out in (a) one stage, and (b) two stages, 1–10 atm and then 10–20 atm, with the gas being allowed to regain its initial temperature between the two stages. Show which of the two processes needs the least energy for the total compression ($C_V(\text{air}) = 20.9 \, \text{J K}^{-1} \, \text{mol}^{-1}$).

3.8 Give and explain the mathematical formulation of Hess's law, and outline a method for its experimental verification. Show how the law may be used to obtain $\Delta H_f^{\ominus}(\text{CuCl,s})$, given that $\Delta H_f^{\ominus}(\text{CuCl}_2\text{,s})$ is $-205 \, \text{kJ mol}^{-1}$ and that $\Delta H^{\ominus}$ for the reaction

$$\text{CuCl(s)} + \tfrac{1}{2}\text{Cl}_2(\text{g}) \rightarrow \text{CuCl}_2(\text{s})$$

is $-71 \, \text{kJ mol}^{-1}$ (construct a cycle to solve this problem).

3.9 In the combustion of naphthalene, C_{10}H_8, in a bomb calorimeter, it was found that $\Delta U = -5166 \, \text{kJ mol}^{-1}$. Construct an equation for the combustion reaction, and calculate the corresponding enthalpy change. Given that $\Delta H_f^{\ominus}(\text{CO}_2\text{,g})$ is $-394 \, \text{kJ mol}^{-1}$ and $\Delta H_f^{\ominus}(\text{H}_2\text{O,l})$ is $-286 \, \text{kJ mol}^{-1}$, evaluate $\Delta H_f^{\ominus}(\text{C}_{10}\text{H}_8\text{,s})$.

3.10 The enthalpy of combustion of hydrogen at 400 K, according to the equation

$$\text{H}_2(\text{g}) + \tfrac{1}{2}\text{O}_2(\text{g}) \rightarrow \text{H}_2\text{O}(\text{g})$$

is $-242 \, \text{kJ mol}^{-1}$. Calculate the enthalpy of combustion at 1000 K, given the following additional data:

$\text{H}_2\text{O(g)}$: $C_P = 30.20 + 9.933 \times 10^{-3}T + 11.17 \times 10^{-7}T^2$

$\tfrac{1}{2}\text{O}_2\text{(g)}$: $C_P = 12.75 + 6.805 \times 10^{-3}T - 21.28 \times 10^{-7}T^2$

$\text{H}_2\text{(g)}$: $C_P = 29.07 - 0.837 \times 10^{-3}T + 20.12 \times 10^{-7}T^2$

3.11 Explain briefly how the heat of solution at infinite dilution may be obtained experimentally for rubidium hydroxide RbOH, which is very hygroscopic. The following data refer to a temperature of 298.15 K and a pressure of $101325 \, N \, m^{-2}$; deduce the standard enthalpy of formation of rubidium sulphate, Rb_2SO_4.

	$\Delta H^{\ominus}/kJ \, mol^{-1}$
$H^+(aq) + OH^-(aq) \rightarrow H_2O(l)$	-56.1
$RbOH(s) + aq \rightarrow Rb^+(aq) + OH^-(aq)$	-62.8
$Rb_2SO_4(s) + aq \rightarrow 2Rb^+(aq) + SO_4^{2-}(aq)$	$+24.3$
$Rb(s) + \frac{1}{2}O_2(g) + \frac{1}{2}H_2(g) \rightarrow RbOH(s)$	-413.8
$H_2(g) + S(s) + 2O_2(g) + aq \rightarrow 2H^+(aq) + SO_4^{2-}(aq)$	-907.5
$H_2(g) + \frac{1}{2}O_2(g) \rightarrow H_2O(l)$	-285.8

As for Problem 3.8, an appropriate thermochemical cycle should be drawn.

3.12 Calculate the entropy change when 1 mole of chloromethane is vaporized at 249 K and 1 atm. The enthalpy change for this process is $21.7 \, kJ \, mol^{-1}$.

3.13 Use the data below to obtain $\Delta G^{\ominus}$ for the reaction

$N_2O_4(g) \rightarrow 2NO_2(g)$

at 298 K:

	$\Delta H_f^{\ominus}/kJ \, mol^{-1}$	$S^{\ominus}/J \, K^{-1} \, mol^{-1}$
$NO_2(g)$	33.9	240
$N_2O_4(g)$	9.70	304

3.14 Utilize the bond enthalpy data (Table 3.7) to determine the enthalpies of formation of propane and propene. Hence, obtain a value for the enthalpy of hydrogenation of the carbon–carbon double bond.

3.15 Give a definition of the enthalpy of sublimation of a substance.

3.16 Using data from Section 3.8.2, determine the change in molar entropy on heating ammonia from 273 K to 773 K.

3.17 In what circumstances do (a) $đq$, and (b) $đw$ become exact differentials?

3.18 Obtain the relationship $đw = -P \, dV = (RT/P) \, dP - R \, dT$, and use it to show that $đw$ is an inexact differential.

3.19 Determine a bond enthalpy for the aromatic carbon–carbon bond from the data in Table 3.7 and the value of $46 \, kJ \, mol^{-1}$ for the enthalpy of formation of benzene, $\Delta H_f(C_6H_6)$. Use the result to evaluate the enthalpy of formation of naphthalene $\Delta H_f(C_{10}H_8)$. If your answer for $\Delta H_f(C_{10}H_8)$ here is different from that obtained in the answer to Problem 3.9, comment on the discrepancy.

3.20 Joule estimated that the temperature of water at the bottom of the Niagara Falls should be 0.1 K higher than at the top. Given that the fall of water is 50 m, the heat capacity of water ($H_2O = 0.018 \, kg \, mol^{-1}$) is $80 \, J \, K^{-1} \, mol^{-1}$ and the gravitational acceleration $9.81 \, m \, s^{-2}$, calculate the rise in temperature to compare with Joule's value.

4

States of matter

Matter exists in three states: gaseous, liquid and solid. It is convenient to discuss the properties of matter under these headings, although certain types of substances, such as vitreous materials and liquid crystals, do not fall clearly into one or other of them.

Gases are characterized by the large volume changes consequent upon variations in their temperature or pressure, and by their ability to flow into and fill space available to them. Gases are miscible in all proportions, and differ markedly from liquids and solids in that many of their properties are independent of their chemical nature and may be described by general gas laws.

Solids have definite volumes and shapes, neither of which changes appreciably with variations in pressure or temperature. A study of the solid state is essentially a study of the crystalline state since almost all solids are crystalline. Many solids that are described as amorphous, such as charcoal or powdered boron, are crystalline although the crystals are of very small dimensions and the material does not appear crystalline to the naked eye. The term amorphous is best restricted to those solids in which the regularity of the structure extends over a range of only a few molecular dimensions, as in glasses, or in which the size has been reduced to these small dimensions by some physical or chemical process, as in the case of colloidal sulphur.

A liquid, like a gas, has no definite form and takes up the shape of any containing vessel into which it is placed, but it occupies a definite volume. It has a boundary surface that restricts its extent and which is responsible for many of the properties associated with the liquid state. The cohesive forces between the molecules of a liquid are, in general, much stronger than those between the molecules of a gas but substantially weaker than those responsible for cohesion in solids. Thus, under normal conditions of temperature and pressure, gases behave as an assemblage of freely moving atoms or molecules whereas solids form rigid structures. The intermediate nature of the forces operative in liquids has made the study of this state of matter difficult, and at the present time no entirely satisfactory theory of the liquid state exists.

The so-called colloidal state is not a state of matter but rather a system composed of very small solid or liquid particles ranging from approximately 10 nm to 100 nm and known as the *disperse phase*, in a suspensory phase, the *dispersion medium*, which may itself be in either the liquid or the gaseous state.

The state of a substance that obtains, at a given temperature, is determined by the result of a competition between the interatomic forces acting between its component atoms and the thermal energy that it contains.

4.1 The gaseous state

The relationship between pressure, volume and temperature can be expressed for most gases in terms of the gas laws and a combination of them. The laws of Boyle and Charles resulted from experimental studies on gases. Boyle's law states that *at constant temperature, the volume of a fixed mass of gas varies inversely as its pressure*. Hence, if two conditions of a gas are represented by P_1, V_1 and P_2, V_2, then

$$P_1V_1 = P_2V_2 \tag{4.1}$$

or

$$(PV)_T = \text{constant} \tag{4.2}$$

the subscript T indicating that the temperature is constant. Charles's law, sometimes referred to as Gay-Lussac's law, states that *at constant pressure, the volume of a fixed mass of gas varies as its temperature*. Thus, if two conditions of a gas are represented by V_1, T_1 and V_2, T_2, we have

$$V_1/T_1 = V_2/T_2 \tag{4.3}$$

or

$$(V/T)_P = \text{constant} \tag{4.4}$$

Alternatively, we may say that at constant volume, the pressure of a fixed mass of gas varies as the temperature, that is

$$(P/T)_V = \text{constant} \tag{4.5}$$

The temperature dependence of the pressure of a gas leads to a temperature scale. A constant-volume gas thermometer (Figure 4.1) compares the pressure P of the volume of gas in the thermometer bulb in contact with an object of temperature T with the pressure P_0 at the triple point of water, the temperature T_0 of 273.15 K at which ice, liquid water and water vapour are in equilibrium. Hence,

$$T \approx T_0 P/P_0 \tag{4.6}$$

Equation (4.6) becomes exact when the gas is behaving ideally. In practice, several readings are taken with successively smaller amounts of gas in the thermometer and the results extrapolated to $P=0$.

It follows from (4.3) that gases that are behaving ideally all expand or contract equally with a change in temperature. From (4.3), we may write

$$V_2 - V_1 = V_1 T_2/T_1 - V_1 = V_1(T_2/T_1 - 1) \tag{4.7}$$

or

$$\Delta V/V_1 = \Delta T/T_1 \tag{4.8}$$

that is, the change in volume per unit volume is solely a function of temperature.

4.1.1 *The gas constant*

Consider a fixed mass of gas under the conditions P_1, V_1, T_1 being changed in two stages, the first an isothermal change to P_2, V, T_1, and the second an isopiestic (constant pressure) change to P_2, V_2, T_2; it can be imagined that the gas is enclosed in a cylinder by a piston (Figure 4.2). From (4.1)

$$(P_1 V_1)_{T_1} = (P_2 V)_{T_1} \tag{4.9}$$

and from (4.3)

$$\left(\frac{V}{T_1}\right)_{P_2} = \left(\frac{V_2}{T_2}\right)_{P_2} \tag{4.10}$$

Eliminating V between (4.9) and (4.10)

$$\frac{P_1 V_1}{T_1} = \frac{P_2 V_2}{T_2} \tag{4.11}$$

Figure 4.1. Constant-volume gas thermometer: the volume of gas in the bulb B is kept constant by adjusting the mercury levels in the reservoir R so that the pointer P just rests on the surface of the mercury. The pressure on the gas in B can be measured in terms of the height h.

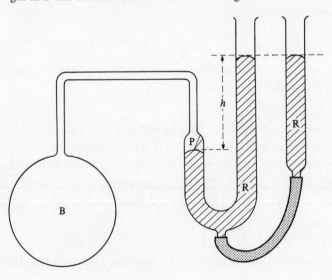

or

$$\frac{PV}{T} = \text{constant} \tag{4.12}$$

This constant is, per mole of gas, the gas constant R. For n mole of gas

$$PV = nRT \tag{4.13}$$

Equation (4.13) is an equation of state for a conceptual ideal gas which obeys exactly the laws of Boyle and Charles; all real gases depart from the equation under most conditions but at low pressures and high temperatures they approximate very closely to the ideal model, and the equation of state is a useful concept with which to describe their behaviour.

Example

0.003 kg of nitrogen gas is confined at a pressure of 708 mmHg and a temperature of 630 K. Assuming that the gas behaves ideally, what volume does it occupy? $R = 8.3160$ J K^{-1} mol^{-1} and the atomic weight of nitrogen is 14.007.

Since 760 mmHg is equivalent to 101 325 N m^{-2}, the volume occupied by 1 mol of nitrogen under the given conditions is $8.3160 \times 630 \times 760/(101\,325 \times 708)$ m^3, which is 0.0555 m^3. Hence, the volume occupied by 0.003 kg of the gas is $0.0555 \times 0.003/0.028014$ m^3, which is 0.00594 m^3, or 5.94 l. It may be noted at this point that the fact that R is a constant to within ± 0.0002 J K^{-1} mol^{-1} is in accord with Avogadro's law, which states that *equal volumes of all gases under the same conditions of temperature and pressure contain equal numbers of molecules.*

A further experimentally derived gas law is Dalton's law of partial pressures. This may be stated: *if two or more gases that do not interact are mixed at constant temperature, then each gas exerts the same pressure as it would if it alone occupied the whole of the given containing volume.* This pressure is its *partial pressure*, the total pressure exerted by the mixture of gases being equal to the sum of the partial pressures of its N component gases:

$$P = p_1 + p_2 + \cdots + p_N \tag{4.14}$$

Figure 4.2. Change of conditions for a gas: $P_2 > P_1$, $T_2 > T_1$.

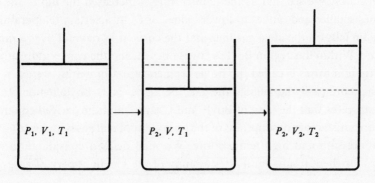

Examples

(a) Let $0.25\,m^3$ of oxygen at 1 atm expand to fill $2\,m^3$ at a constant temperature of 273 K. The new pressure of the oxygen gas is given by Boyle's law: $p(O_2) = 1 \times 0.25/2 = 0.125\,atm$

(b) Let $0.5\,m^3$ of hydrogen expand under the same conditions as in (i). Then $p(H_2) = 1 \times 0.5/2 = 0.25\,atm$

(c) Let $0.25\,m^3$ of oxygen and $0.5\,m^3$ of hydrogen, each at 1 atm pressure and 273 K be mixed and allowed to expand under the same conditions as in (i). From Dalton's law, Equation (4.14), the total pressure P is given by

$$P = p(O_2) + p(H_2) = 0.375\,atm$$

From (4.13), for a mixture of N gases,

$$PV = (n_1 + n_2 + \cdots + n_N)RT \tag{4.15}$$

where P and V are the total pressure and total volume respectively, and $n_1, n_2, \ldots, n_N$ are the numbers of moles of the various species present in the gas mixture.

4.1.2 *Deviations from the gas laws*

An ideal gas obeys the gas laws exactly, but to what extent do real gases depart from ideality and to what causes can we ascribe such behaviour? The gas laws were combined into an equation of state (4.13), and for 1 mole of gas we should expect PV/RT, which is often called the compression factor Z, to have the constant value of unity. Figure 4.3 shows the variation of Z with P for hydrogen and nitrogen at 273 K and for carbon dioxide at 313 K (carbon dioxide liquefies at or below 304 K under a pressure of 75 atm or more). The graph shows that at moderate pressures, say 1–10 atm, Z is unity within 5%, except for the easily liquefiable carbon dioxide. At very high pressures Z is greater than unity for all gases.

Figure 4.4 shows the variation of PV/RT with P for 1 mole of nitrogen at four temperatures. We see that as the temperature is increased the dip in the curve becomes smaller and moves to lower values of P. At a certain temperature the minimum falls on the ideal gas line, and the curve is horizontal over a range of pressure. Within this region Boyle's law is obeyed, and the temperature at which this situation arises is called the Boyle temperature. Comparative studies show that the more easily liquefiable the gas, the higher is its Boyle temperature.

It transpires that the laws of Boyle and Charles, and the derived equation of state, are satisfactory descriptions of the behaviour of real gases under conditions of low pressure and high temperature. We shall delay a consideration of the reasons for these results until after a study of the kinetic theory of gases.

Figure 4.3. Variation of the experimental compression factor Z with pressure P.

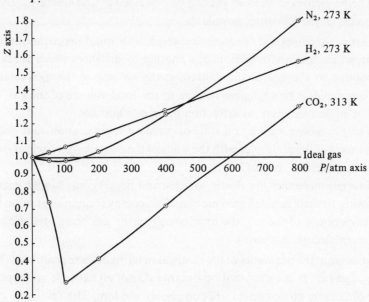

Figure 4.4. Variation of the calculated value of PV/RT with P for nitrogen.

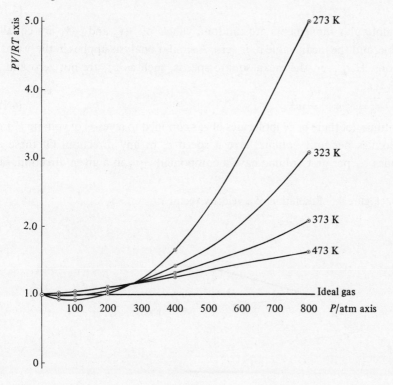

4.1.3 *The kinetic theory of gases*

The properties of an ideal gas can be described by the kinetic theory of gases, in terms of the following postulates.

(*a*) Gases are composed of molecules which, at normal temperatures and pressures, are separated from one another by distances which are large relative to the size of the molecules; the volume of the molecules is considered to be negligible relative to the total volume of the gas.

(*b*) The molecules exert no attraction upon one another.

(*c*) The gas molecules are in a state of constant rapid motion, and elastic impacts of the molecules with the walls of the containing vessel produce the pressure of the gas.

(*d*) The gas molecules are elastic spheres, and possess only kinetic energy owing to their translational motion. This energy is dependent upon the temperature of the gas, the total energy of the gas being unaltered by intermolecular collisions.

We next develop the properties of the ideal gas in terms of these postulates. The molecules of gas are in constant motion but they do not all have the same speed, and their directions of movement are completely random. The velocity $\mathbf{v}$ of a molecule can be represented by a vector having components $\mathbf{v}_x$, $\mathbf{v}_y$ and $\mathbf{v}_z$ parallel to three mutually perpendicular axes x, y and z (Figure 4.5). Thus

$$\mathbf{v} = \mathbf{v}_x + \mathbf{v}_y + \mathbf{v}_z \tag{4.16}$$

Since molecular movements are random, values of $+\mathbf{v}_x$ and $-\mathbf{v}_x$ are equally probable, and the mean value $\overline{v}_x$ is zero. A similar analysis applies in the y and z directions. However, the mean square speeds, such as $\overline{v_x^2}$, are not zero and it follows that

$$\overline{v_x^2} = \overline{v_y^2} = \overline{v_z^2} = \tfrac{1}{3}\overline{v^2} \tag{4.18}$$

To continue, let there be N molecules of gas confined to a vessel of volume V. Let n_a molecules per unit volume have a speed v_a in any direction. Of these n_a molecules, n_1 per unit volume have a component $\pm v_1$ in a given direction, say

Figure 4.5. Resolution of a velocity vector $\mathbf{v}$.

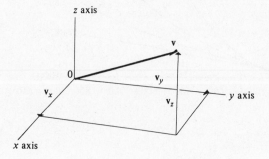

along the x axis, n_2 have a component $\pm v_2$ in the same direction, and so on. In a small interval of time Δt a fraction of the n_1 molecules, those in a volume element $v_1 \Delta t$, strike unit area of the wall of the vessel. The number of molecules in this elemental volume moving towards the wall is $\frac{1}{2} n_1 v_1 \Delta t$, and they hit the wall at the rate of $\frac{1}{2} n_1 v_1$. Since the collisions are postulated to be elastic, no energy is lost and the speeds before and after impact are the same. The momentum of each of the n_1 molecules is $m v_1$, where m is the mass of a molecule. Hence, the rate of change of momentum perpendicular to the wall is $2 m v_1$ for each impact, and $m n_1 v_1^2$ for all of the n_1 molecules. For all positive (towards the wall) components along the x axis, the total rate of change of momentum is given by

$$\Delta p_x = m(n_1 v_1^2 + n_2 v_2^2 + \cdots) \tag{4.19}$$

The mean square speed along the x axis is given by

$$\overline{v_x^2} = \sum_i n_i v_i^2 / \sum_i n_i \tag{4.20}$$

Since $\sum_i n_i = n_a$ for all molecules of speed v_a, and using (4.18) and (4.20), we have

$$\Delta p_x = m n_a \overline{v_x^2} = \tfrac{1}{3} m n_a \overline{v_a^2} \tag{4.21}$$

For all possible speeds $v_a, v_b, v_c, \ldots$ the total change in momentum Δp is given by

$$\Delta p = \tfrac{1}{3} m n_a \overline{v_a^2} + \tfrac{1}{3} m n_b \overline{v_b^2} + \cdots \tag{4.22}$$

If $\overline{v^2}$ is the mean square speed of the molecules, defined by

$$\overline{v^2} = \frac{n_a \overline{v_a^2} + n_b \overline{v_b^2} + \cdots}{n_a + n_b + \cdots} \tag{4.23}$$

and n is the total number of molecules per unit volume given by

$$n = n_a + n_b + \cdots = \frac{N}{V} \tag{4.24}$$

then

$$\Delta p = \tfrac{1}{3} m n \overline{v^2} \tag{4.25}$$

or

$$\Delta p = \frac{1}{3} \frac{m N \overline{v^2}}{V} \tag{4.26}$$

From Newton's second law, the force acting on a body is equal to the rate of change of momentum, and pressure is the force per unit area exerted on the walls of the vessel by the molecules, whence

$$P = \frac{1}{3} \frac{m N \overline{v^2}}{V} \tag{4.27}$$

since Δp, from (4.26), represents the total change of momentum per second resulting from impacts on unit area of wall. Thus

$$PV = \tfrac{1}{3} m N \overline{v^2} \tag{4.28}$$

This is the fundamental equation of the kinetic theory, from which the gas laws may be derived as follows.

4.1.4 *Boyle's law*

In an ideal gas the energy of the molecules arises from their translational motion, and the kinetic energy E_K of all N molecules in a mole of gas is $\frac{1}{2}mN\overline{v^2}$. At constant temperature, the kinetic energy is constant, and from (4.28)

$$(PV)_T = \frac{2}{3}(\frac{1}{2}mN\overline{v^2}) = \frac{2}{3}(E_K) = \text{constant} \tag{4.29}$$

which is equivalent to Boyle's law. From (4.11) and (4.28)

$$\frac{1}{3}mN\overline{v^2} = \frac{2}{3}E_K = RT \tag{4.30}$$

It follows that the internal energy per mole is independent of the volume, which is a thermodynamic requirement of an ideal gas.

4.1.5 *Charles's law*

We may deduce from (4.30) that molecules have zero kinetic energy at the absolute zero of temperature, and that kinetic energy is proportional to the absolute temperature. Hence, at constant pressure

$$V \propto E_K \propto T \tag{4.31}$$

which is equivalent to Charles's law.

4.1.6 *Avogadro's law*

If equal volumes of two ideal gases are confined at the same pressure and temperature

$$P_1 V_1 = P_2 V_2 \tag{4.32}$$

or

$$\frac{1}{3}N_1 m_1 \overline{v_1^2} = \frac{1}{3}N_2 m_2 \overline{v_2^2} \tag{4.33}$$

Since $T_1 = T_2$,

$$E_{K_1} = E_{K_2} \tag{4.34}$$

and

$$\frac{1}{2}m_1 \overline{v_1^2} = \frac{1}{2}m_2 \overline{v_1^2} \tag{4.35}$$

Hence from (4.33)

$$N_1 = N_2 \tag{4.36}$$

which is equivalent to Avogadro's law.

From (4.28), the root mean square (rms) speed is given by

$$(\overline{v^2})^{\frac{1}{2}} = \left(\frac{3PV}{mN}\right)^{\frac{1}{2}} \tag{4.37}$$

If V is the molar volume then mN is the molar mass M of the gas, and from (4.13)

$$\overline{(v^2)}^{\frac{1}{2}} = (3RT/M)^{\frac{1}{2}} \tag{4.38}$$

Note that the rms speed $\overline{(v^2)}^{\frac{1}{2}}$ is not equal to the mean speed $|\bar{v}|$.

Example

For nitrogen gas at 273.15 K the rms speed is

$$\overline{(v^2)}^{\frac{1}{2}} = \left(3 \times 8.316 \times \frac{273.15}{0.028014} \right)^{\frac{1}{2}}$$

$$= 493 \text{ m s}^{-1}$$

The mean speed $|\bar{v}|$ is actually $[(8/3\pi)\overline{v^2}]^{\frac{1}{2}}$ (see for example, E. A. Moelwyn-Hughes, *Physical Chemistry*, 2nd rev. edn, Pergamon Press, Oxford, 1961, Equation (41), p. 39) which is equal to 454 m s^{-1} for nitrogen gas. The rms speed of the molecules of an ideal gas increases with temperature but not with pressure: Equation (4.37) shows that $\overline{(v^2)}^{\frac{1}{2}}$ is proportional to $(PV)^{\frac{1}{2}}$ which, for an ideal gas, is independent of P (Figure 4.3).

4.1.7 *Gaseous diffusion*

The density of a gas is given by

$$d = M/V \tag{4.39}$$

where M and V are respectively the molar mass and molar volume. Hence, from (4.37) and (4.38)

$$\overline{(v^2)}^{\frac{1}{2}} = (3P/d)^{\frac{1}{2}} \tag{4.40}$$

and at constant pressure the rms speed of the molecules in a gas is inversely proportional to the square root of the density, which provides a theoretical basis for Graham's law of gaseous diffusion, namely that the rate of diffusion of a gas is proportional to the rms speed of its molecules.

4.1.8 *Brownian motion and the Avogadro constant*

In 1827 the botanist Brown noticed that microscopic pollen grains suspended in water exhibited continuous random motion. This erratic movement is now called Brownian motion and supports the kinetic theory very strikingly, it has been observed with all kinds of small particles and suspensory media. It occurs because water molecules are in constant motion in every direction with different speeds, and a solid particle is bombarded on all sides by these moving water molecules. At any instant the forces on the particle do not balance and a net impulse in a given direction ensues, but at the next moment the same particle may receive a net impulse in a different direction and with a different magnitude, so that its motion is changing continuously. Smaller particles move more rapidly than larger particles because the larger the particle, the more water molecules bombard it, and the more likely it is that the net impulse on it will be zero.

Perrin studied Brownian movement with suspensions of gamboge and mastic in water and it is from his results that the *Avogadro constant* was evaluated. This is usually given the symbol N_A and represents the number of entities, such as molecules, ions or electrons, in 1 mole of the substance. Perrin's results for N_A ranged between 6.5×10^{23} mol^{-1} and 7.2×10^{23} mol^{-1}, the currently accepted value being 6.0221×10^{23} mol^{-1}.

4.1.9 *Distribution of molecular speeds*

We have noted that not all gas molecules have the same speed, and the distribution of these varying speeds was deduced by Maxwell in 1860. The probability that the speeds of molecules in a gas of molar mass M lie between v and $(v + dv)$ is given, without proof here, by

$$P(v)\,dv = 4\pi \left(\frac{M}{2\pi RT}\right)^{\frac{3}{2}} \exp\left(\frac{-Mv^2}{2RT}\right)v^2\,dv \tag{4.41}$$

In Figure 3.13 the significance of (4.41) is illustrated. The height of an ordinate corresponds to the probability that molecules possess the corresponding speed v, the maximum ordinate indicating the most probable speed.

If the temperature of the gas is increased, the distribution of speeds becomes wider and the most probable speed becomes larger, while the number of molecules with high speeds is, at the same time, increased considerably.

If E_K is the kinetic energy per mole of molecules of speed v

$$E_K = \tfrac{1}{2}Mv^2 \tag{4.42}$$

Hence, in (4.41)

$$M^{\frac{3}{2}}v^2\,dv = \sqrt{(2E_K)}\,dE_K \tag{4.43}$$

From (4.41) and (4.43)

$$P(E_K)\,dE_K = \frac{2\pi}{(\pi RT)^{\frac{3}{2}}} \exp\left(-\frac{E_K}{RT}\right)\sqrt{(E_K)}\,dE_K \tag{4.44}$$

which expresses the probability that molecules possess kinetic energy within the range from E_K to $E_K + dE_K$.

In studying reaction rates we are concerned with the proportion of molecules having energy in excess of a given value. The probability $P(E_K)\,dE_K$ may be equated with dn_{E_K}/n, the fraction of the total molecules having energy in the range E_K to $E_K + dE_K$. It can be shown that (see Appendix A7)

$$n_{E_K} = n \exp(-E_K/RT) \tag{4.45}$$

where n_{E_K} is the number of molecules, in the total number n, that possess energy in excess of the value E_K per mole. For example, if E_K is 40 kJ mol^{-1} at 298 K, n_{E_K}/n is $\exp(-16.14)$, or 9.78×10^{-8}.

4.1.10 *The kinetic theory and real gases*

In the treatment of the ideal gas we assumed *inter alia* that molecules exert no force of attraction upon one another, and that the volume of the molecules is negligible in comparison with the total volume of gas.

Attractive forces

It is evident that particles of solids or liquids attract one another, and since all gases liquefy under suitable conditions a similar attraction, albeit of smaller magnitude, must exist between the molecules of a real gas. The existence of such attractive forces was proved by the porous plug experiment of Joule and Thomson (Figure 4.6). A gas passed through a porous plug of silk, asbestos or unglazed earthenware was generally found to become cooler (the Joule–Thomson effect); a similar effect is observed if the gas expands into a vacuum. Hydrogen and helium were found to become warmer in this experiment, an anomaly that will be discussed later (p. 98).

The apparatus is insulated from the surroundings, so that the process is adiabatic ($q=0$). The gas on the entry side of the plug is at conditions P_1, V_1, T_1 and is compressed isothermally through the plug by means of a piston so that the volume on the entry side becomes zero. The work done on the gas is $P_1 V_1$. The gas on the exit side expands isothermally against another piston increasing the exit volume from zero to V_2, so that the work done on the gas is now $-P_2 V_2$. The total work done on the gas is $P_1 V_1 - P_2 V_2$, and is equal to the change in internal energy $U_2 - U_1$. Hence the expansion is isenthalpic ($dH=0$). The property observed is $\Delta T/\Delta P$, which becomes, in the limit as $P \rightarrow 0$, $(\partial T/\partial P)_H$ the Joule–Thomson coefficient, often called μ_{JT}. We shall consider the thermodynamic implications of this parameter next.

From (3.97) and Section 3.7.2, we have for infinitesimal changes in each property

$$dG = dU + P\,dV + V\,dP - T\,dS - S\,dT \qquad (4.46)$$

Assuming reversibility in a closed system of constant composition, dU may be replaced by $T\,dS - P\,dV$, so that

$$dG = V\,dP - S\,dT \qquad (4.47)$$

Figure 4.6. Joule–Thomson experiment; in general, $T_1 > T_2$.

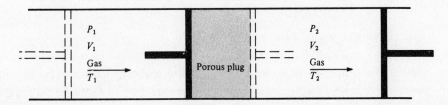

From (3.17)–(3.19)

$$(\partial V/\partial T)_P = -(\partial S/\partial P)_T \tag{4.48}$$

In a similar way, we can show that

$$dH = T\,dS + V\,dP \tag{4.49}$$

Specifying constant temperature, we can write

$$(\partial H/\partial P)_T = T(\partial S/\partial P)_T + V \tag{4.50}$$

which, with (4.48), leads to

$$(\partial H/\partial P)_T = V - T(\partial V/\partial T)_P \tag{4.51}$$

For the experiment, we can write, following (3.15)

$$dH = (\partial H/\partial P)_T\,dP + (\partial H/\partial T)_P\,dT = 0 \tag{4.52}$$

Now, using (3.29), (4.51) and (4.52), we have

$$(\partial T/\partial P)_H = -(\partial H/\partial P)_T/(\partial H/\partial T)_P = [T(\partial V/\partial T)_P - V]/C_P \tag{4.53}$$

which is an expression for μ_{JT} (see also Section 4.1.12). For an ideal gas, $T(\partial V/\partial T)_P = V$, so that μ_{JT} is zero and there is no Joule–Thomson effect. With a real gas the cooling must be attributed to the effect of intermolecular attractive forces being overcome in streaming the gas through the porous plug.

Volume of molecules

When two molecules approach each other, the attraction between them at a small distance of separation is small compared with the repulsion due to the proximity of their electrons. Even if a molecule was negligibly small compared with the volume occupied by the gas, it would have an effective *collision diameter* σ. This parameter may be taken as the distance of closest approach of two molecules in a gas. Molecules must therefore have a finite volume: if the diameter of a molecule is taken to be the collision diameter, then the volume v_m of a molecule is given by

$$v_m = \frac{4\pi}{3}\left(\frac{\sigma}{2}\right)^3 = \frac{1}{6}\pi\sigma^3 \tag{4.54}$$

The collision diameter of a gas molecule is on average about 0.4 nm, so that v_m is 0.034 nm^3. At normal pressures the volume of the molecules is about 0.1% of the total gas volume, but at a pressure between 100 atm and 200 atm the volume of the molecules is about 10% of the total volume and no longer negligible.

By subjecting a gas to cooling and compression, it can be first liquefied and then solidified, showing that molecules have a limiting volume. It follows that the space available for movement in a gas is less than the occupied volume.

4.1.11 *Van der Waals' equation of state*

A modification of the ideal gas equation of state (4.13) was proposed by van der Waals in 1873. As a molecule is about to strike the wall of its containing vessel, the forces on it are unbalanced and a net attraction towards the bulk of the gas exists. The effect is to decrease the impact of the molecule on the wall and so reduce the pressure of the gas. The force exerted on such a molecule is proportional to the density, and the number of molecules striking the wall at any instant is also proportional to the density so that the total attractive force is proportional to (density)^2, or to $1/V^2$. Thus the corrected pressure is $[P+(a/V^2)]$ where a is a constant; a/V^2 is termed the internal cohesive pressure.

The finite size of the molecules reduces the effective volume, so the corrected volume may be represented as $(V-b)$ in which b, the covolume, may be shown to equal $4v_m$ (see Appendix A6). Hence

$$\left[P+\left(\frac{n^2a}{V^2}\right)\right](V-nb)=nRT \tag{4.55}$$

which is the van der Waals' equation of state for n mole of a real gas; Table 4.1 lists the values of a and b for several gases.

At low pressures V is large; a/V^2 is therefore small, and b is small with respect to V. Hence (4.55) approximates to (4.13) and the real gas is behaving ideally. At slightly higher pressures b is still small with respect to V,

$$PV=nRT-n^2a/V \tag{4.56}$$

Under these conditions, and for 1 mole of gas, PV is less than RT and decreases with increase in P, which explains the dip in the curves for N_2 and CO_2 (Figure 4.3) in terms of the internal cohesive pressure.

At fairly high pressure b is significant with respect to V, but a/V^2 is still small compared to P if a is small. Hence (4.55) becomes, for 1 mole of gas

$$PV=RT+Pb \tag{4.57}$$

PV is now greater than RT and rises with increase in P. Hydrogen has a small value of a and so Equation (4.57) predominates at ordinary temperatures (Figure 4.3).

Table 4.1. *Van der Waals' constants*[a] *for some gases*

Gas	$a/\text{dm}^6\,\text{atm mol}^{-2}$	$10^2b/\text{dm}^3\,\text{mol}^{-1}$
He	0.034	2.37
H_2	0.24	2.66
O_2	1.36	3.18
N_2	1.39	3.91
CO_2	3.59	4.27
NH_3	4.17	3.71

[a] $R=8.3160\,\text{J K}^{-1}\,\text{mol}^{-1}=0.082073\,\text{dm}^3\,\text{atm K}^{-1}\,\text{mol}^{-1}$.

4.1.12 *Joule–Thomson coefficient*

The van der Waals' equation may be used to predict the sign and magnitude of the Joule–Thomson coefficient. Expanding (4.55) for $n = 1$

$$PV = RT + Pb - (a/V) + (ab/V^2) \tag{4.58}$$

ab/V^2 is of small magnitude, and V in the small term a/V may be replaced by RT/P. Hence we obtain

$$PV = RT + Pb - aP/RT \tag{4.59}$$

Dividing throughout by P and differentiating Equation (4.59) with respect to T, at constant pressure

$$\left(\frac{\partial V}{\partial T}\right)_P = \frac{R}{P} + \frac{a}{RT^2} \tag{4.60}$$

Substituting for RT from (4.59) leads to

$$\frac{R}{P} = \frac{(V-b)}{T} + \frac{a}{RT^2} \tag{4.61}$$

and using (4.60)

$$T\left(\frac{\partial V}{\partial T}\right)_P - V = \frac{2a}{RT} - b \tag{4.62}$$

whence, using (4.53)

$$\mu_{JT} = \left(\frac{2a}{RT} - b\right) \Big/ c_P \tag{4.63}$$

Table 4.2 illustrates the applicability of (4.63) to the calculation of the Joule–Thomson coefficient for the gases hydrogen, nitrogen and carbon dioxide at 273 K. The negative value of μ_{JT} for hydrogen indicates that at 273 K this gas is warmer after passing through a porous plug from a pressure P_1 to a slightly lower pressure P_2. Evidently the attractive forces in hydrogen are very small: Table 4.1 shows that a has a very low value whereas b is about the same as for other gases, and the term $[(2a/RT) - b]$ is negative for hydrogen at 273 K. The Joule–Thomson inversion temperature T_i is obtained by putting $\mu_{JT} = 0$ in (4.63)

$$T_i = 2a/bR \tag{4.64}$$

Table 4.2. *Calculation of Joule–Thomson coefficients at 273 K*

Gas	$C_P/dm^3\,atm\,K^{-1}\,mol^{-1}$	$\mu_{JT}/K\,atm^{-1}$ Calculated	Observed	T_i/K
H_2	0.285	−0.02	−0.03	220
N_2	0.287	+0.30	+0.32	866
CO_2	0.366	+0.76	+1.3	2049

Evidently for $T < T_i$ μ_{JT} is positive, and for $T > T_i$ μ_{JT} is negative. Calculated values of T_i for the three gases considered are also listed in Table 4.2.

4.1.13 *Virial equation of state*

A more generalized equation of state expresses the departure of a real gas from ideality by means of a power series. For 1 mole of gas, it may be written

$$PV = RT + BP + CP^2 + \cdots \tag{4.65}$$

and is known as the virial equation of state. The terms $B, C, \ldots$ are the first, second and subsequent virial coefficients; their values for nitrogen gas are listed in Table 4.3. These data were used in the construction of the curve in Figure 4.4. The virial coefficients can be evaluated by the methods of statistical thermodynamics, but this topic is mathematically involved and outside the scope of this book.

4.1.14 *Molar heat capacity of a gas*

The heat capacity of a gas has been considered in the chapter on thermodynamics, Sections 3.7.2 and 3.7.3. It can be defined as the heat imparted to a body to cause a rise in temperature of 1 K in that body. Assuming the ideal gas equation of state, a rise in temperature must be accompanied by an increase in P or V, or in both. If V increases, work is done by the gas against the constant external pressure and the energy required for this work is supplied as heat, less heat being needed if the volume of the gas is kept constant. Two important molar heat capacities have been recognized.

At constant volume, when no external work is done, the heat supplied to the gas increases the energies of the molecules. According to the kinetic theory, gas molecules possess translational energy only, in the amount $\frac{3}{2}RT$ per mole. Hence, from (3.23)

$$C_V = 3R/2 \tag{4.66}$$

which is approximately $12.5 \, \text{J K}^{-1} \, \text{mol}^{-1}$. This argument applies only to monatomic gases. If the gas is diatomic, then the molecules possess additional

Table 4.3. *Second and third virial coefficients[a] for nitrogen*

T/K	$10^2 B/\text{dm}^3 \, \text{mol}^{-1}$	$10^5 C/\text{dm}^3 \, \text{atm}^{-1} \, \text{mol}^{-1}$
223	−2.8790	14.980
273	−1.0512	8.626
373	0.6662	4.411
473	1.4763	2.775

[a] See footnote to Table 4.1.

rotational and vibrational energies which are dependent upon temperature. The molar heat capacity is given approximately by the relationship

$$C_V = (3R/2) + p \qquad (4.67)$$

where $p = 0$ for monatomic gases; at 298 K, $p = 8 \, \text{J K}^{-1} \, \text{mol}^{-1}$ for diatomic gases and $p = 17–21 \, \text{J K}^{-1} \, \text{mol}^{-1}$ for polyatomic gases.

At constant pressure, heat absorbed by a gas increases the energies of translation, rotation and vibration, and does work of expansion against a constant external pressure. We have shown in Section 3.5 that the work of expansion is $P \, \Delta V$. For an ideal gas, $P \, \Delta V = R \, \Delta T$, but since $\Delta T = 1$, the work of expansion is equal to R. Hence

$$C_P = (5R/2) + p \qquad (4.68)$$

and

$$C_P - C_V = R \qquad (4.69)$$

per mole.

4.2 The solid state

Most well-defined solid substances are crystalline. The word crystal comes from the Greek *krustallos* meaning ice, first applied to quartz, which was thought to be water permanently congealed by intense cold. The term *crystal* now covers a multitude of solid substances, including minerals, rocks, soil, sand, snow, diamonds, rubies and other precious stones, salt, sugar, penicillin, the vitamins and nearly all other solid chemical substances. Still crystalline, although not so completely, are hair, silk, cotton, nylon and stretched rubber, but a number of solids such as glass and certain synthetic plastics are not crystalline. We shall see that this term implies a high degree of regularity in most solids, whereas in gases and liquids order exists in regions of only a few molecular dimensions in size. The very shape of crystals seems unnatural, since curved or rounded outlines are more common in nature, the sharp edges and plane faces associated with crystals occurring only infrequently.

An early observation by Stensen (1669) revealed a fundamental property of crystals. Stensen cut transverse and longitudinal sections of natural crystals such as quartz and traced their outlines on paper (Figure 4.7). He noted that whatever the shape or size of the section, the tracings showed that the angles between corresponding faces in crystals of the same substance always had a constant value. This finding is embodied in the law of constant interfacial angles, which has subsequently been extended and fully explained in terms of the structure of crystals. Sodium chloride crystallizes from water in cubes, but when it is crystallized from a 5–10% solution of urea, other faces (octahedra) are also present (Figure 4.8) and the sodium chloride is said to have changed its *habit*.

Increasing the percentage of urea increases the development of the octahedral faces until finally the shape of the sodium chloride crystal is an octahedron, but the angles between the corresponding faces are constant whatever the habit. Figure 4.9 shows two habits of potassium sulphate; it may be shown by goniometric measurement that the angles between faces such as ab, ac and ad have constant values in each crystal and are characteristic of the substance. Figures 4.7–4.9 suggest symmetrical arrangements of various sets of faces.

Figure 4.7. Stensen's figures: (a) idealized quartz crystal, (b) vertical sections of actual quartz crystals, (c) transverse sections of actual quartz crystals.

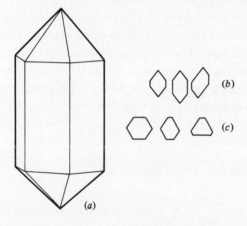

Figure 4.8. Variations in the habit of sodium chloride crystals.

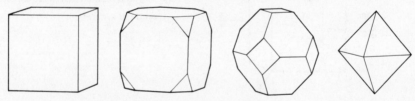

Figure 4.9. Two habits of crystalline potassium sulphate.

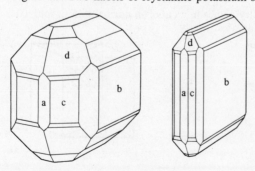

Crystal symmetry is an important subject but we cannot discuss it in detail in this book (for a more detailed treatment, M. F. C. Ladd & R. A. Palmer, *Structure Determination by X-Ray Crystallography*, 2nd edn, Plenum Press, NY, 1985, is recommended). The drawings in Figures 4.10 and 4.11 illustrate some of the symmetry elements of a cube. Rotation of a crystal about an *n*-fold axis of symmetry brings the crystal into self-coincidence *n* times during a rotation of 360°. Planes of symmetry (mirror planes) divide a crystal into two halves related to each other as an object is to its mirror image. In all there are 23 symmetry elements associated with a cube, but there are no symmetry planes normal to the three-fold rotation axes. Only shapes based on the rotation symmetries of degree (*n*) 1, 2, 3, 4, 6, or combinations of these, are found in crystals because only with regular figures of such symmetry can space be filled (Figure 4.12.).

The symmetrical arrangement of crystal faces is an external manifestation of the internal structure of the crystal. The underlying geometrical feature of a crystal is its lattice, which may be defined as an infinite, regular, three-dimensional arrangement of points in space such that each point has the same environment as every other point. Figure 4.13 illustrates the most general space lattice. A unit cell

Figure 4.10. Typical symmetry axes in a cube.

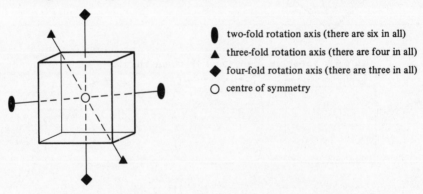

two-fold rotation axis (there are six in all)

three-fold rotation axis (there are four in all)

four-fold rotation axis (there are three in all)

O centre of symmetry

Figure 4.11. Symmetry planes in a cube: (*a*) normal to four-fold axis, (*b*) normal to two-fold axis.

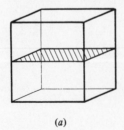

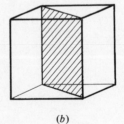

(*a*)   (*b*)

Figure 4.12. Patterns based on (*a*) two-fold, (*b*) three-fold, (*c*) four-fold, (*d*) six-fold, (*e*) five-fold, (*f*) eight-fold symmetry; the five-fold and eight-fold patterns do not fill space completely. In (*e*) and (*f*) the v-marks represent voids in the pattern.

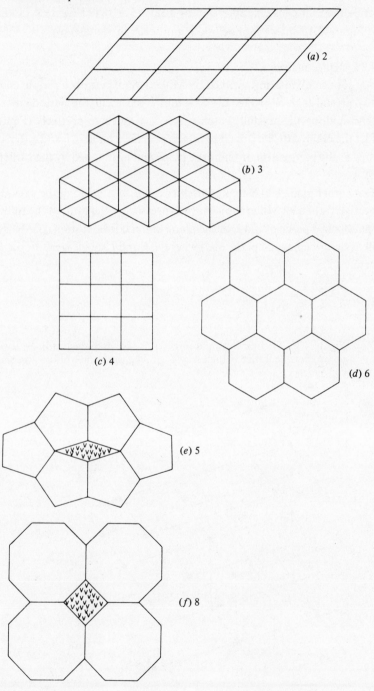

(*a*) 2

(*b*) 3

(*c*) 4

(*d*) 6

(*e*) 5

(*f*) 8

is shown in dashed lines, and may be defined by three non-coplanar vectors **a**, **b** and **c** or alternatively, by the three scalars a, b and c and the angles α, β and γ. Special relationships between these axial lengths and angles can be used to describe the unit cells characterized by Bravais in 1848 (Figure 4.14 and Table 4.4). Figure 4.15 shows monoclinic P unit cells in stereoview.

4.2.1 *Miller indices*

A useful notation evolved by Miller (1839) permits a simple description of any crystal face or plane. Let a, b and c represent the periodicities (repeat distances) along the crystallographic axes x, y and z respectively (Figure 4.16), and let the plane ABC be drawn to intercept these axes at a, b and c. This plane is defined as the parametral or reference plane and is designed by the Miller indices (111).

If any other plane LMN makes intercepts a/h, b/k, c/l along the axes x, y and z respectively, then its Miller indices are expressed by the ratio of the intercepts of the parametral plane of those of the plane under consideration, LMN. Thus this plane is described as the plane (hkl) where h, k and l are integers. In the example shown,

$$\frac{a}{h}=\frac{a}{2} \quad \frac{b}{k}=\frac{b}{4} \quad \frac{c}{l}=\frac{c}{3}$$

Figure 4.13. Triclinic lattice; a unit cell is shown in dashed lines. It should be noted that the lattice is the array of points; the framework is a geometrical convenience.

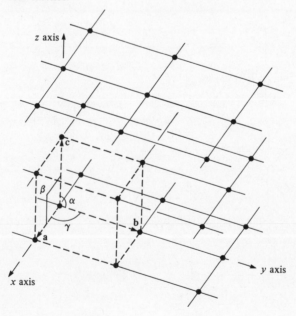

hence LMN is the plane (243). If the plane had intercepted the y axis at $-b/4$, the other intercepts remaining the same, then the Miller indices would have been $(2\bar{4}3)$.

Figure 4.14. Unit cells of the 14 Bravais lattices; interaxial angles are 90° unless indicated otherwise by a numerical value or symbol: (a) triclinic P, (b) monoclinic P, (c) monoclinic C, (d) orthorhombic P, (e) orthorhombic C, (f) orthorhombic I, (g) orthorhombic F, (h) tetragonal P, (i) tetragonal I, (j) cubic P, (k) cubic I, (l) cubic F, (m) hexagonal P, (n) trigonal R. (Key: P, primitive; C, C-face centred; I, body centred; F, all-face centred; R, rhombohedral (primitive).)

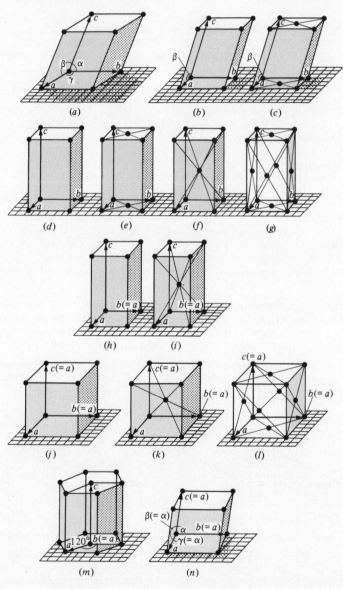

A plane such as ABDE is parallel to the z axis and we may say that it intercepts this axis at infinity, its Miller indices being (110) since it intercepts the x and the y axes at a and b respectively. Again, the plane BDGF intercepts the y axis at b and is parallel to both the x and z axes, so that its indices are (010). We therefore have a geometric description of any crystal plane in terms of the parametral plane (111). In terms of the unit cells (Figure 4.14) O is the origin of the unit cell, whence h, k and l are simply the reciprocals of the fractional intercepts of the plane on the unit cell edges a, b and c respectively.

Table 4.4. *Characteristics of the seven crystal systems*

System	Axial relationships	Examples
Triclinic	$a \neq b \neq c$ $\alpha \neq \beta \neq \gamma \neq 90°, 120°$	$CuSO_4 \cdot 5H_2O$ $K_2Cr_2O_7$
Monoclinic	$a \neq b \neq c$ $\alpha = \gamma = 90°; \beta \neq 90°, 120°$	$CaSO_4 \cdot 2H_2O$ $C_{12}H_{22}O_{11}$
Orthorhombic	$a \neq b \neq c$ $\alpha = \beta = \gamma = 90°$	$BaSO_4$ KNO_3
Tetragonal	$a = b \neq c$ $\alpha = \beta = \gamma = 90°$	KH_2PO_4 SnO_2
Cubic	$a = b = c$ $\alpha = b = \gamma = 90°$	NaCl C (diamond)
Hexagonal	$a = b \neq c$ $\alpha = \beta = 90°; \gamma = 120°$	C (graphite) H_2O (ice)
Trigonal	$a = b = c$ $\alpha = \beta = \gamma \neq 90°, < 120°$	$CaCO_3$ $NaNO_3$

Figure 4.15. Stereoscopic drawings of eight adjacent P unit cells in a monoclinic lattice. The sharing of the corner points can be seen readily by focussing attention on the central lattice point in the drawings. Thus, the corner points are shared equally by eight adjacent unit cells.

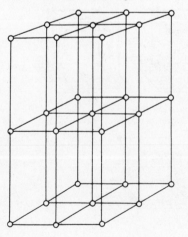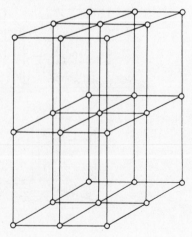

Experience has shown that the Miller indices of the actual faces of a crystal are small integers and rarely does such an index exceed four (the law of rational indices). Figure 4.17 represents the projection of an array of lattice points on to the xy plane, with the traces of the (100), (110) and ($2\bar{3}0$) families of planes outlined. We see immediately that as the Miller indices increase in value numerically, the density of lattice points (reticular density) on the corresponding planes decreases. If the lattice under discussion is cubic P, the reticular densities are in the ratio $1:1/\sqrt{2}:1/\sqrt{13}$ for the planes (100), (110) and ($2\bar{3}0$) respectively.

In simple crystals, the component atoms or ions often occupy the positions of lattice points, so that the population of atoms on any given plane may be related directly to the corresponding reticular density. The faces of a crystal represent the terminations of families of planes, and the crystal grows in such a way that the external faces are planes of the highest reticular density – a more stable system

Figure 4.16. Miller indices of crystal planes.

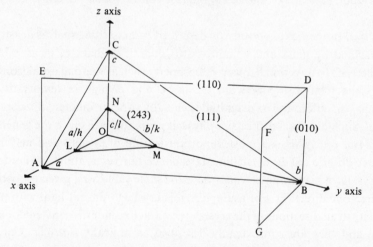

Figure 4.17. Families of parallel planes: (100), (110), ($2\bar{3}0$).

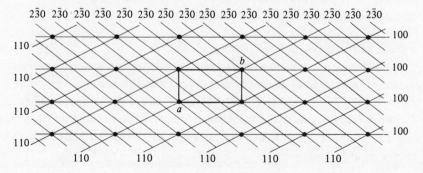

than that of a crystal with surfaces exhibiting, on the atomic scale, relatively large holes. As we have already shown that the planes of highest reticular density are also those of lower Miller indices, the law of rational indices now follows.

A family of planes is a set of parallel equidistant planes in a lattice, and its Miller indices can always be determined by examining the intercepts of the plane closest to, but not passing through, the origin of the lattice. For example, if we select the family $(2\bar{3}0)$ and the origin of the lattice is taken as the intersection of the vectors **a** and **b** (Figure 4.17), the nearest plane to the origin intercepts the x axis at $a/2$ and the y axis at $-b/3$. The intercept along the z axis is c/∞ as the Miller index l is zero. We might have chosen another plane of this family, the same distance from the origin, and discovered that it intercepts the x axis at $-a/2$ and the y axis at $b/3$ so that its Miller indices are $(\bar{2}30)$. It is a consequence of the geometry of lattices that the two Miller symbols $(2\bar{3}0)$ and $(\bar{2}30)$ identify the same family of planes. The *different* planes (hkl) which are related by the *symmetry* of the lattice constitute a crystal *form*, written $\{hkl\}$; in this example, the form $\{230\}$ comprises the planes (230), $(\bar{2}\bar{3}0)$, $(2\bar{3}0)$ and $(\bar{2}30)$, the last two of which are contained in Figure 4.17.

The physical properties of crystals are described by quantities, such as density, which is defined by mass and volume – two quantities that are measurable without reference to direction. However, a property such as thermal expansion is defined by the relationship between extension and change in temperature. Extension has direction, and its magnitude generally varies in different directions in a crystal. Such directional effects in physical properties are given the general name of *anisotropy*. This is a most important feature of crystals and may be regarded as a criterion of the crystalline state, but certain reservations are needed because crystals of high symmetry may have the same value for a given property in more than one direction. For example, cubic crystals transmit light with the same velocity in all directions in the crystal (the refractive index is constant for all directions) and they are consequently described as optically *isotropic*. Cubic crystals are isotropic for many but not all properties; they are anisotropic for elasticity and photoelasticity. Cubic crystals and indeed all crystals must be regarded as potentially anisotropic, but their symmetry may lead to special relationships for certain physical properties.

4.2.2 *Isomorphism, polymorphism and allotropy*

During a study of the crystalline phosphates and arsenates of the alkali metals, Mitscherlich (1819) noted that salts of similar chemical composition, such as $Na_2HPO_4 \cdot H_2O$ and $Na_2HAsO_4 \cdot H_2O$, exhibited the same crystalline form. The word *isomorphous* was introduced to describe the replaceable element, the term being later extended to include the relationship between compounds. Many examples are known, such as K_2SO_4 and K_2SeO_4, $CuSO_4 \cdot 5H_2O$ and

$CuSeO_4 \cdot 5H_2O$, and $K_2SO_4 \cdot Cr_2(SO_4)_3 \cdot 24H_2O$ and $K_2SO_4 \cdot Al_2(SO_4)_3 \cdot 24H_2O$.

Subsequent investigations into the structure of solids showed that an even wider definition of isomorphism is desirable. Consider the following pairs of substances:

$NaNO_3$ $CaCO_3$

$BaSO_4$ KBF_4

KIO_4 $CaWO_4$

These pairs of compounds exhibit similar crystal structures despite their chemical dissimilarity. The important conditions for structural isomorphism in its wider sense are firstly that the components of each pair of substances must be similar in size and shape, and secondly that an electrical charge balance must be maintained; for example, in the first pair, the radii of the spherical sodium and calcium ions are 0.095 nm and 0.099 nm respectively, while the planar nitrate and carbonate ions also have similar size and shape:

However, sodium nitrate and potassium nitrate, although chemically similar, are not isomorphous ($r(K^+) = 0.133$ nm). The radius of the barium ion is 0.135 nm, and the ions sulphate and tetrafluoroborate both form tetrahedral groupings in which the bond distances S–O and B–F are 0.144 nm and 0.140 nm respectively; consequently barium sulphate and potassium tetrafluoroborate are isomorphous.

Numerous solids exist in more than one crystalline modification (polymorphism). A particular crystal structure depends for its formation upon its environment during the growth process. Physical factors such as temperature, pressure and concentration as well as the presence of foreign substances all help to modify crystal growth. Interconversion of polymorphs takes place at a given transition temperature for a specified pressure, but not all transitions are reversible:

$$\text{zinc blende (ZnS)} \xrightarrow{1300K} \text{wurtzite (ZnS)}$$

$$\text{quartz (SiO}_2) \xrightarrow{1145K} \text{tridymite (SiO}_2) \underset{}{\overset{1745K}{\rightleftharpoons}} \text{cristobalite (SiO}_2)$$

Polymorphism in elements is often called *allotropy*, the different modifications being called *allotropes*. Only one of the modifications is stable under given conditions, but it does not follow that the other forms revert spontaneously to the stable phase. The term stable is meaningful only in reference to a given set of conditions; the reference state or *standard state* for elements refers to a temperature of 298 K and a simultaneous pressure of 1 atm, the phrase 'normal conditions' often meaning conditions of temperature and pressure closely similar to the standard conditions. The standard state of elemental carbon is graphite. Diamond, however, shows no tendency to transform to graphite except at very high temperatures (in an inert atmosphere). Considerable energy is required to disrupt the diamond structure and rearrange it to that of graphite under normal conditions, even though the less stable form is in a higher (more positive) energy state. Polymorphism can be subdivided into two classes, monotropism and enantiotropism.

4.2.3 *Monotropism*

This term describes the condition where one crystal modification is stable over the whole range of its existence, the other form (or forms) invariably being metastable. Since the vapour pressure curve for the β-phase in Figure 4.18 lies everywhere above that for the α-phase, the β-phase tends to revert to the α-phase. In no circumstances can the solid α-form be converted directly to the β-form, and to obtain the metastable form the α-phase must be melted or vaporized and cooled rapidly. Other examples of monotropic systems are oxygen(O_2)–ozone (O_3), calcite($CaCO_3$)–aragonite($CaCO_3$), iodine monochloride and benzophenone.

Figure 4.18. Phase diagram showing the monotropic relationship between red phosphorus (α) and white phosphorus (β): AB is the vapour pressure–temperature curve for the α-form, and DE that for β; BC is the liquid–vapour boundary, and B is a triple point at which α, liquid and vapour are in equilibrium; it is the melting point of the α-form; E is the melting point of the β-phase under its own vapour pressure, and F is the transition temperature between the α- and β-phases; it is hypothetical because it lies above the melting points of both solids.

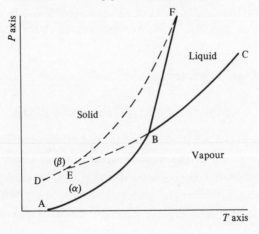

4.2.4 *Enantiotropism*

Two polymorphs are enantiotropic if each has a definite range of stability and changes into the other at a definite transition temperature in either direction. If the α-phase is heated rapidly, transformation may be suspended and so BF (Figure 4.19) represents a range in which the α-phase is metastable. CE and BG represent the supercooling curves of the liquid and β-phases respectively. The diagram shows that an enantiotropic substance can melt at different temperatures, according to the rate of heating. Other examples of enantiotropic systems are those of elemental tin, ammonium chloride, mercuric iodide and hexachloroethane.

Some substances show both monotropic and enantiotropic transitions; for example, silica (SiO_2):

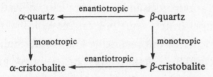

Figure 4.19. Phase diagram showing the enantiotropic relationship between orthorhombic sulphur (α) and monoclinic sulphur (β): AB is the vapour pressure–temperature curve for the α-form, and BC that for β; B is the triple point for α, β and vapour; it is the transition temperature (369.7 K) for $\alpha \rightarrow \beta$; C is the melting point of monoclinic sulphur (387.7 K), and CD is the liquid–vapour curve; BF and CF show the effect of external pressure on the transitions $\alpha \rightarrow \beta$ and $\beta \rightarrow$ liquid respectively; E is the melting point of the α-form under its own pressure, and EF shows the effect of external pressure on the transition supercooled liquid $\rightarrow \alpha$-sulphur.

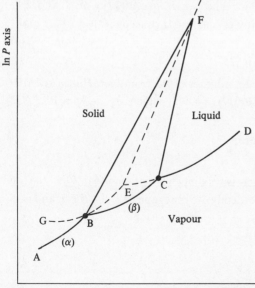

4.2.5 *Thermodynamics of change of state*

Since U and S are state functions, it follows that the combination $U - TS$ is also a state function; it is called the Helmholtz free energy A:

$$A = U - TS \tag{4.70}$$

Differentiating (4.70) completely gives

$$dA = dU - T\,dS - S\,dT \tag{4.71}$$

For a reversible process in which the only work done is that of expansion (see Section 4.1.10)

$$dA = -P\,dV - S\,dT \tag{4.72}$$

Following (3.17)–(3.19)

$$(\partial P/\partial T)_V = (\partial S/\partial V)_T \tag{4.73}$$

Consider a pure liquid and its vapour in equilibrium, in a closed system. The pressure is dependent on the temperature T but not on the volume V. If the transition from liquid to vapour is carried out reversibly and isothermally at the transition temperature T, the enthalpy change may be equated at $T\,\Delta S$. Furthermore, if ΔV is the volume change accompanying vaporization, then $\Delta S/\Delta V$ has a constant value. Utilizing all this information (4.73) leads to

$$dP/dT = \Delta S/\Delta V = \Delta H/T\,\Delta V \tag{4.74}$$

which is the Clapeyron equation, the term on the extreme right-hand side being an extension due to Clausius. It may be applied generally to changes of state of pure substances.

Example

We shall calculate the effect of applied pressure on the melting point of ice. At 273.15 K the densities of ice and water are $1000\ \mathrm{kg\,m^{-3}}$ and $917\ \mathrm{kg\,m^{-3}}$ respectively, and the enthalpy of fusion ΔH of ice is $6009.5\ \mathrm{J\,K^{-1}\,mol^{-1}}$. From (4.74) we have

$$dT/dP = T(V_L - V_S)/\Delta H$$

where V_L and V_S are the molar volumes of water and ice. Hence, $dT/dP = 273.15$ $[(0.018016/1000) - (0.018016/917)]/6009.5$ which is $-7.4 \times 10^{-8}\ \mathrm{K\,N^{-1}}$, or $-0.0075\ \mathrm{K\,atm^{-1}}$

For the reversible transition

$$\text{solid} \rightleftharpoons \text{gas} \tag{4.75}$$

the molar volume V_G of the gas is very much larger than that, V_S, of the solid, and if the gas is assumed to behave ideally, we may replace V_G by RT/P and neglect V_S in relation to V_G. Hence

$$dP/dT = \Delta H\,P/RT^2 \tag{4.76}$$

or

$$d(\ln P)/dT = \Delta H/RT^2 \tag{4.77}$$

where ΔH is the molar enthalpy of sublimation: this approximate form of Clapeyron's equation was deduced by Clausius.

The equilibrium constant (see Chapter 5) for the process in (4.75) is given by

$$K_p = p_G \tag{4.78}$$

since the partial pressure of the solid phase is constant. Hence

$$\mathrm{d}\ln K_p / \mathrm{d}T = \Delta H / RT^2 \tag{4.79}$$

which equation was deduced by van't Hoff, and is of general applicability. We may note here that (4.79) provides an explanation for Le Chatelier's principle. Integrating (4.79) leads to

$$\ln K_p = -\Delta H / RT + \text{constant} \tag{4.80}$$

If we have a closed system at equilibrium and the enthalpy change for the reaction is positive (endothermic), it follows immediately from (4.80) that an increase in temperature leads to an increase in K_p, which is equivalent to enhancing the amounts on the products side of the equation, always assuming that ΔH does not change significantly over the temperature range considered.

4.3 Determination of crystal structures

Until 1912, information on the internal structures of crystals was essentially speculative, although some of the ideas put forward were very close to the truth. In that year von Laue showed that X-rays would behave towards crystals rather like light passing through a diffraction grating. The positions and intensities of the diffracted beams are governed by the geometry of the crystal and by the relative positions of its component atoms. It was soon shown that diffraction data could be used to provide direct evidence on the structures of crystals, and we shall discuss this subject after the manner given first by Bragg.

4.3.1 *Bragg equation*

Bragg's treatment of X-ray diffraction from crystals considers it as a 'reflexion' of X-rays from planes of atoms, or electron density, in the crystal. This view was based upon the experimental observation that if a crystal in a diffracting position was rotated about an axis, through an angle ϕ, to the next diffracting position, the direction of the diffracted beam was found to have rotated through 2ϕ. A similar situation is experienced with the reflexion of light from a plane mirror (Figure 4.20), as in the optical lever.

To derive the Bragg equation, let X-rays be incident to an angle θ on a family of parallel, equidistant planes, and be 'reflected' (diffracted) from them at an equal angle (Figure 4.21). We need to determine the condition that X-rays reflected from successive planes interfere constructively, or reinforce one another, in the given direction of reflexion.

In Figure 4.21, AB, CD and EF represent typical parallel rays in an X-ray wave incident on the (hkl) family of interplanar spacing d_{hkl} at an angle θ_{hkl}. The rays are in phase both along the normal BQ to the incident wavefront and, after reflexion, along the normal BS to the reflected wavefront. For the first two planes, the excess path δ of CDH over ABG is given by

$$\delta = PD + DR \tag{4.81}$$

From the construction

$$PD = DR = d_{hkl} \sin \theta_{hkl} \tag{4.82}$$

Thus

$$\delta = 2d_{hkl} \sin \theta_{hkl} \tag{4.83}$$

For constructive interference between reflected rays, the path difference must be equal to the wavelength λ of the X-radiation. Hence

$$2d_{hkl} \sin \theta_{hkl} = \lambda \tag{4.84}$$

which is Bragg's equation. A simple extension of the construction shows that if

Figure 4.20. Reflexion of light from a plane mirror, initially at M_1 and after rotation through an angle ϕ to M_2; the reflected beam has turned through 2ϕ.

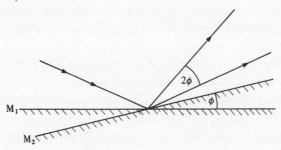

Figure 4.21. Bragg reflexion of X-rays from crystal planes; three planes of an (hkl) family are shown.

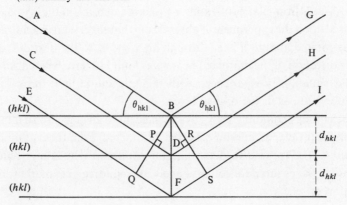

(4.84) is satisfied for the first and second planes in a family, it will be satisfied also for the second and third, third and fourth, and so on for the whole family.

Sometimes the Bragg equation is written as

$$2d \sin \theta_n = n\lambda \tag{4.85}$$

However, in X-ray crystallography all families of planes are defined uniquely by their Miller indices, including multiples such as (nh, nk, nl). Thus d in (4.85) referring to the fundamental spacing $(n = 1)$ in a family, is equal to nd_{hkl} where d_{hkl} is defined in (4.84). For example, $d_{200} = d_{100}/2$, $d_{336} = d_{112}/3$, and so on. The reader should sketch the families (100), (110), (120), (200), (220), (2$\bar{4}$0) and ($\bar{3}$60) for a crystal unit cell, in projection on the xy plane, given that $a = 0.6$ nm, $b = 0.9$ nm and the interaxial angle $\gamma = 110°$.

4.3.2 *Interplanar spacing*

It is convenient to relate d_{hkl} to the values of h, k and l. Consider Figure 4.22; for simplicity, the axes will be assumed to be mutually perpendicular, and the plane ABC is the first plane from the origin O in the family (hkl). In the right-angled triangle ONA

$$ON = OA \cos \alpha \tag{4.86}$$

Figure 4.22. Relationship between d_{hkl} and hkl.

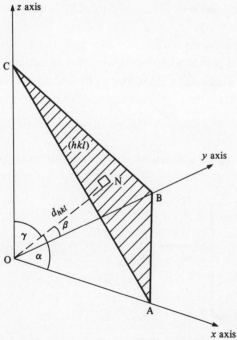

Since, by definition of Miller indices, $OA = a/h$, where a is the distance from the origin to the intercept of the parametral plane on the x axis. Hence

$$d_{hkl} = (a/h)\cos\alpha \tag{4.87}$$

or

$$(h^2/a^2)d_{hkl}^2 = \cos^2\alpha \tag{4.88}$$

Similarly, from triangles ONB and ONC

$$(k^2/b^2)d_{hkl}^2 = \cos^2\beta \tag{4.89}$$

and

$$(l^2/c^2)d_{hkl}^2 = \cos^2\gamma \tag{4.90}$$

Summing (4.89)–(4.90), remembering that the sum of the squares of the direction cosines of a line is unity,

$$d_{hkl}^2 = (a^2/h^2) + (b^2/k^2) + (c^2/l^2) \tag{4.91}$$

In a cubic unit cell, $a = b = c$. Hence,

$$1/d_{hkl}^2 = (h^2 + k^2 + l^2)/a^2 \tag{4.92}$$

and using (4.81) and (4.92)

$$\sin\theta_{hkl} = (\lambda/2a)(h^2 + k^2 + l^2)^{\frac{1}{2}} \tag{4.93}$$

4.3.3 *Structure analyses of potassium and sodium chlorides*

The first crystal structure analyses, those of potassium and sodium chlorides, were reported by von Laue in 1912 and confirmed by Bragg in 1913; we shall consider briefly the latter work.

Potassium and sodium chlorides crystallize as cubes. In order to study the reflexion of X-rays from crystal planes with the apparatus then available, faces other than those of the cube itself had to be obtained by cutting the crystals. Bragg's work was based on data for the families of planes of the types $(h00)$, $(hh0)$

Figure 4.23. Cube, showing the orientations of the (100), (110) and (111) planes.

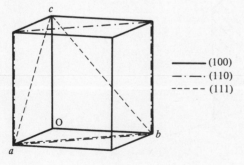

and (*hhh*), parallel to (100), (110) and (1̄11) respectively. Figure 4.23 shows the traces of these planes on a cube.

X-rays were allowed to fall on these planes in turn, and the intensity I of the reflected (diffracted) radiation was measured as a function of the Bragg angle θ. The six sets of results are presented in the form of bar graphs in Figure 4.24.

It is convenient to use the Bragg equation in the form (4.85). All of the sets of spectra obey the Bragg equation. Consider the *h*00 spectra for potassium chloride. The three reflexions occur at θ values at 5.36, 10.76 and 16.26°. The ratios of the corresponding values of $\sin \theta$ are 1:2.00:3.00, which are integral within experimental error; similar results are obtained with (*hh*0) and (*hhh*). Consider next the *first* reflexion for potassium chloride in the *h*00, *hh*0 and *hhh* sets, their θ values being 5.36, 7.59 and 4.65°. Since $d \propto 1/\sin \theta$, the ratios of these d values are $1:1/\sqrt{2}:2/\sqrt{3}$. In the original work, the very weak *hhh* reflexion at $\theta = 4.65°$ was not recorded: this fact led to the d ratios being taken as $1:1/\sqrt{2}:1/\sqrt{3}$, which indicated a primitive cubic arrangement of atoms. Such a structure could be based on the cubic unit cell in Figure 4.14(*j*); we can show easily from the figure that the d ratios for the (100), (110) and (111) planes are $1:1/\sqrt{2}:1/\sqrt{3}$.

Figure 4.24. Bar graphs of $(I)^{\frac{1}{2}}$ for potassium chloride and sodium chloride up to $\theta \approx 26°$. For each compound the ordinates are normalized to the value of $(I)^{\frac{1}{2}}$ for the first *h*00 reflexion.

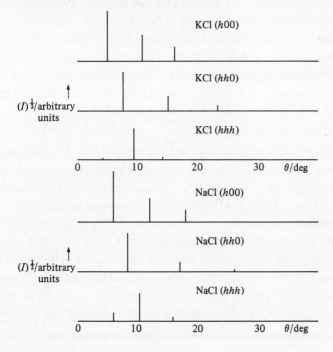

In the sodium chloride results, the true d ratios are clearly established, and these ratios are consistent with a face-centred (see Figure 4.14(l)) arrangement of atoms (Figure 1.2). We can now explain the intensity data for both sodium and potassium chlorides more fully in terms of the F unit cell. The $h00$ and $hh0$ series of spectra must be labelled 200, 400, 600 and 220, 440, 660, respectively ($n = 2, 4, 6$). Reflexions such as 100 or 110 ($n = 1$) are not possible, because successive planes in these families are interleaved by planes of equivalent scattering power, at exactly $d/2$ apart, and interference occurs that causes complete cancellation of such reflexions.

The (hhh) planes, however, are populated by either all cations or all anions, and can be labelled, in order, (111), (222) and (333). Consider the (111) family of planes. They are interleaved by planes at a spacing $d_{111}/2$: from the Bragg equation, it is clear that when successive (111) planes are in the reflexion position, the interleaving planes scatter exactly out of phase, because the path distance between the planes of anions and cations is, from (4.83), $2(d_{111}/2) \sin \theta$. Eliminating $d_{111} \sin \theta$, the path difference becomes $n\lambda/2$, because d in (4.85) is d_{111}. Thus when n is an odd number, the reflexion intensity will be weaker than when n is even, and this is indicated clearly in Figure 4.24. We can see further how 111_{KCl} is much weaker in intensity than 111_{NaCl}. X-rays are scattered by electrons, and to a first approximation the intensity of scattering will be governed by the atomic number of a species. Now potassium and chloride ions are isoelectronic, with 18 electrons each, while the sodium ion has 10 electrons: hence the out-of-phase scattering has a lesser effect on the intensity in sodium chloride than in potassium chloride, which is again completely in accord with experiment.

It is easy to show that the unit cell of the sodium chloride structure type (Figure 1.2) contains four cations and four anions. The density of sodium chloride is 2165 kg m^{-3} and the relative atomic masses of sodium and chlorine are 22.990 and 35.453 respectively. Using the relationship density $=$ mass/volume in terms of the unit cell, and the atomic mass unit u, we may write

$$2165 = 4 \times 58.443 \times 1.660 \times 10^{-27}/a^3$$

where a is the length of the unit cell edge. Hence $a = 5.64 \times 10^{-10}$ m, or 0.564 nm.

We have not used the wavelength of the X-radiation involved in the experimental measurements of potassium and sodium chlorides. Indeed, X-ray wavelengths were unknown when these measurements were made. Taking the θ-value for the 200 reflexion from sodium chloride as 5.99° we have, from (4.85) with $n = 2$ and d, the fundamental $h00$ spacing, set equal to a, $\lambda = 0.0589$ nm. This wavelength corresponds to palladium $K\alpha$ X-radiation, used by Bragg in 1913.

It should be noted that crystal structures are not analyzed today by this simple method, although several of its principles have their counterparts in modern X-ray crystallography. We shall illustrate a simple approach by the powder method.

4.3.4 *Powder method*

This technique employs finely powdered, crystalline material made into a cylindrical specimen of about 0.3 mm diameter in a borosilicate glass capillary tube. It is irradiated with monochromatic X-rays, and the diffraction pattern may be recorded on a photographic film.

Since λ is constant, and for a given family of planes d is fixed, the Bragg equation (4.84) must be satisfied by providing for the correct angle θ. It is achieved through the polycrystalline nature of the specimen. In Figure 4.25(*a*), the Bragg equation is satisfied, for a given interplanar spacing, by the generators of a cone of semi-vertical angle 2θ, coaxial with the incident X-ray beam. Thus, a complete cone of diffracted X-rays is produced which intersects a photographic film, placed normally, in a circular trace (Figure 4.25(*b*)).

The measurable parameters of this trace are its diameter and its intensity; they can be obtained equally well from the strip of film ABCD. Consequently, a powder camera usually consists of a cylindrical enclosure for the specimen and the film, shown schematically in Figure 4.26(*a*). The specimen can be made to rotate on its own axis: although not essential to the method, it enhances the number of crystallites that come into the reflecting position for any given d-value.

From Figure 4.26(*a*), it is clear that the arc T becomes the linear distance T on the strip of film (Figure 4.26(*b*)). The geometry of the apparatus shows that

$$T = 4R\theta \qquad (4.94)$$

where R is the radius of the film. The camera is designed so that the radius is equal to $180/\pi$ mm. Hence, if T is measured in mm

$$\theta/\text{deg} = T/4 \qquad (4.95)$$

Figure 4.25. Development of a powder diffraction line: (*a*) geometry of diffraction – X is the X-ray source, S is the specimen, $2\theta_{hkl}$ is twice the Bragg angle for the (*hkl*) family of planes, giving rise to reflexion of intensity I_{hkl}, (*b*) photographic film – ABCD is the strip of film normally used.

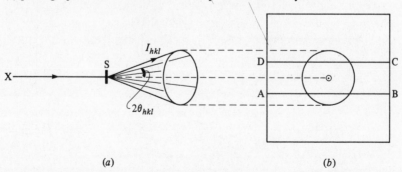

(*a*)   (*b*)

4.3.5 *Powder method for cubic crystals*

Cubic crystals (see Table 4.4) are particularly straightforward under powder photographic analysis. From (4.93)

$$\sin^2 \theta_{hkl} = (\lambda^2/4a^2)(h^2 + k^2 + l^2) = (\lambda^2/4a^2)N \qquad (4.96)$$

where N must be integral, and will increase as the Miller indices become numerically larger. It follows that the values of $\sin^2 \theta$ for a sequence of lines on a powder photograph must have a simple integral ratio to one another, within experimental error.

Certain values of N cannot be achieved in any circumstances. They are those that cannot be expressed as the sum of the squares of three integers: for example 7, 15, 23 and, in general, $m^2(8n-1)$ where m and n are integers. We may call these values mathematically forbidden.

Then there are values that are not permitted because of the unit-cell type: we mentioned this effect in Section 4.3.3. More fully, we may note that for the three cubic lattices P, I and F we can set up Table 4.5. The table shows that there are no particular restrictions on the hkl values for P unit cells, but for I $h+k+l$ is even, and for F h, k and l are either all even or all odd integers.

It should be noted that the first six lines on a powder photograph of a cubic substance which is either P or I can produce the same ratios of $\sin^2 \theta$ values, so that it is essential to record at least seven lines, preferably more.

Example

The following values of $\sin^2 \theta$ were obtained from a powder photograph of a

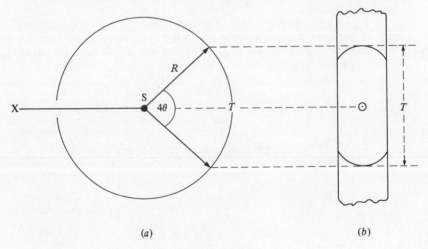

Figure 4.26. Geometry of the powder camera: (*a*) schematic diagram of a powder camera of radius R with specimen at S; T is the arc intercepted by the cone of X-rays, (*b*) strip film, showing the *arc* T now as a linear measure.

(*a*)   (*b*)

cubic oxide, using copper $K\alpha$ X-radiation ($\bar{\lambda} = 0.15418$ nm).

   0.1028   0.2735   0.3756   0.4102   0.5467   0.8205   0.9227

We are required to determine the unit-cell type, the indices of the reflexions and the unit-cell dimension a.

We begin by searching for a common factor for the $\sin^2 \theta$ values: usually it is equal to the first value, one-half of it, one-third of it, or so on, and we begin to set up a table of results (Table 4.6). It is not too difficult to show that, in this example, a common factor is about 0.0343 (column 2), whence the values of the integer N can be determined (column 3).

From Table 4.5, it is clear that we are dealing with an F unit cell, so that we can now complete the indexing (column 4). Using (4.95), values for a may be calculated from the lines, the average of which is 0.4169 nm. Certain values of N are missing from the sequence of values for an F unit cell listed in Table 4.5. They are too weak to be recorded, and are a manifestation of the particular

Table 4.5. *Values of $h^2 + k^2 + l^2(N)$ up to 30*

N	P hkl	I hkl	F hkl
1	100		
2	110	110	
3	111		111
4	200	200	200
5	210		
6	211	211	
8	220	220	220
9	300,221		
10	310	310	
11	311		311
12	222	222	222
13	320		
14	321	321	
16	400	400	400
17	410,322		
18	411,330	411,330	
19	331		331
20	420	420	420
21	421		
22	332	332	
24	422	422	422
25	500,430		
26	510,431	510,431	
27	511,333		511,333
29	520,432		
30	521	521	

arrangement of atoms in the crystal. We must always be prepared for this situation, which is of not infrequent occurrence.

Although we should not expect to solve complex structures by powder methods, the technique is useful in many ways: identification, measurement of physical properties, such as thermal expansion coefficients or particle sizes, study of solid state reactions, determination of phase diagrams (see Figures 4.39 and 4.42).

4.4 Crystal chemistry

In Chapter 2 we discussed the nature and characteristics of interatomic forces, and noted that chemical structures could be classified according to the predominant interatomic force. We conclude our study of the solid state of matter by amplifying this classification and by describing the structures of some compounds.

Crystal chemistry relates the physical and chemical properties of a substance to its internal structure, endeavouring both to interpret properties in terms of known structural features and to associate certain structural characteristics with measured properties. The subject originated in 1920, following Bragg's analyses of the structures of the crystalline alkali metal halides, which apart from caesium chloride, caesium bromide and caesium iodide, crystallize with the sodium chloride structure type (Figure 1.2). The equilibrium distance r_e between the cation M^+ and the anion X^- is one-half of the repeat distance a in the cubic unit cell for the sodium chloride structure type, and $a\sqrt{3}/2$ for the caesium chloride structure type (caesium chloride, caesium bromide, caesium iodide). The r_e values are listed in Table 4.7. We can trace the progressive increase in r_e in, for example, moving vertically from sodium fluoride to sodium iodide or in moving horizontally from lithium bromide to caesium bromide. The increase is regular, the difference $\Delta(X)$ between r_e for two halides of a given cation being almost independent of the cation, while the difference $\Delta(M)$ between r_e for two alkali

Table 4.6. *Powder data for the cubic oxide*

$\sin^2 \theta$	$\sin^2 \theta/0.0343$	N	hkl	a/nm
0.1028	3.00	3	111	0.4164
0.2735	7.97	8	220	0.4169
0.3756	10.95	11	311	0.4172
0.4102	11.96	12	222	0.4170
0.5467	15.94	16	400	0.4170
0.8204	23.92	24	422	0.4170
0.9227	26.90	27	511,333	0.4170

halides with a given anion is almost independent of the anion. This result suggests that each ion possesses its own characteristic radius:

$$r_e(NaCl) = r(Na^+) + r(Cl^-)$$
$$r_e(KCl) = r(K^+) + r(Cl^-)$$

whence

$$r_e(KCl) - r_e(NaCl) = r(K^+) - r(Na^+) \tag{4.97}$$

which is the same for all pairs of sodium and potassium halides.

It must be remembered that the radius of a given species depends upon its environment. For example, $r(Na^+)$ in sodium chloride is 0.095 nm but in sodium metal, $r(Na) = 0.186$ nm. Evidently the radius of any given species is strongly dependent upon the nature of the interatomic forces acting upon it.

4.4.1 *Structures of elements*

The metals of the groups IA, IB and IIA of the periodic table and the transition-type metals form relatively simple structures: the cubic close-packed arrangement A1, the body-centred cubic arrangement A2, and the hexagonal close-packed arrangement A3. The types A1 and A3 represent two ways of close packing to the same degree spheres of equal size (Figure 4.27). Any three successive close-packed layers have two of these layers in an identical spatial arrangement while the third layer can be superimposed upon the other two layers in one of two ways. Thus we obtain the successions

A1: 1 2 3 1 2 3 1 2...

A3: 1 2 1 2 1 2 1 2...

In the structure types A1 and A3 each sphere is in contact with twelve other spheres, so the coordination number is twelve. The radius of the metal atom can be taken as one-half the distance of closest approach of any two spherical atoms.

If the spheres have radius r, we can show that the volume of space occupied per sphere is $5.66r^3$; the volume of each sphere is $4.19r^3$. In Figure 4.27(a) let the

Table 4.7. *Equilibrium interionic distances r_e/nm in the alkali metal halides*

	ΔLi	Δ(M)	Na	Δ(M)	K	Δ(M)	ΔRb	Δ(M)	Cs
F	0.201	0.030	0.231	0.036	0.267	0.015	0.282	0.018	0.300
Δ(X)	0.056		0.050		0.047		0.046		0.056
Cl	0.257	0.024	0.281	0.033	0.314	0.014	0.328	0.028	0.356
Δ(X)	0.018		0.017		0.015		0.015		0.016
Br	0.275	0.023	0.298	0.031	0.329	0.014	0.343	0.028	0.371
Δ(X)	0.025		0.025		0.024		0.023		0.024
I	0.300	0.023	0.323	0.030	0.353	0.013	0.366	0.029	0.395

side length of the cubic unit cell be a. Then the volume of the cube is a^3, but $a\sqrt{2}$ (a face diagonal) is equal to $4r$, because the spheres are close packed and in contact along this diagonal. Hence, the volume of the cube may be written as $(4r/\sqrt{2})^3$. Since four atoms are associated with a face-centred unit cell, the volume occupied per sphere is $(4r/\sqrt{2})^3/4$, or $5.66r^3$, and the fraction of space occupied is 0.74. A similar calculation for the A3 structure shows the same degree of close packing. Figure 4.28 shows the A1 and A3 structures in projection on to the close-packed planes.

The body-centred cubic structure A2 represents a slightly less close-packed array of equal spheres (Figure 4.29); the volume occupied per sphere is $6.16r^3$ in this arrangement, and the coordination number is eight.

Figure 4.27. Stereoviews of the close packing of spheres of equal size: (*a*) close-packed cubic A1 – close-packed planes are {111}, (*b*) close-packed hexagonal A3 – close-packed planes are {0001}.

(*a*)

(*b*)

Figure 4.28. Close packing in projection on to the close-packed planes: (*a*) a first layer 1 – full lines, (*b*) a second layer 2 – dashed lines, (*c*) close-packed cubic structure – 12312..., (*d*) close-packed hexagonal structure – 12121.... In (*c*) and (*d*) the third layers are shown, for clarity, just larger than the first-layer circles although they are, of course, exactly superimposed in this view.

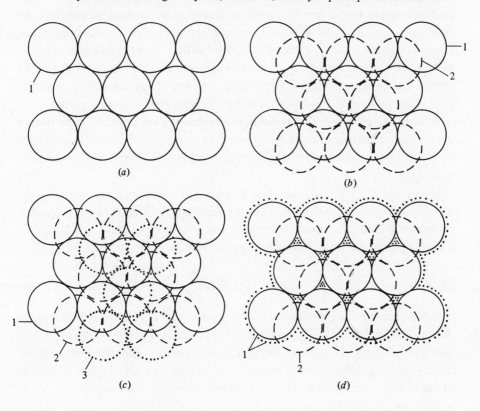

(*a*)　　(*b*)

(*c*)　　(*d*)

Figure 4.29. Stereoview of the body-centred cubic A2 packing of spheres of equal size.

The high coordination numbers eight and twelve are characteristic of the metallic bond. Important mechanical properties of metals arise through deformation of their structures by gliding along close-packed planes, there being four close-packed planes in A1, parallel to $\{111\}$, one in A3, normal to the vertical c axis of the hexagonal unit cell, while in the eight-coordinated A2 structure the packing is less close. We find that the malleable metals such as copper, silver, nickel and γ-iron have the cubic close-packed structures, whereas the harder and more brittle metals such as chromium, tungsten and α-iron have the A3 or A2 structures. The metallurgical importance of iron is related to its ability to crystallize with either the A1 or A2 structure according to heat treatment.

A study of elements in the B-groups of the periodic table enables us to trace a continuous transition of bond type from the true metals of group IB, which we have discussed already, to the molecular compounds formed by the halogen molecules.

Iodine (Figure 4.30) and the other halogens form diatomic molecules of the type X_2. The bond between the atoms is mainly covalent, and cohesion between molecules in the crystal arises through the much weaker van der Waals' forces. As a consequence, the melting points of the crystalline halogens are low:

	F_2	Cl_2	Br_2	I_2
m p/K	50	170	266	387

The distance between molecules in the crystal varies from about 0.25 nm in fluorine to about 0.35 nm in iodine; the corresponding X–X interatomic distances vary from about 0.14 nm in fluorine to 0.27 nm in iodine. Selenium forms infinite helical chains parallel to the c axis of a hexagonal unit cell (Figure 4.31). The atoms in each chain are linked by covalent bonds and the chains are linked by van der Waals' forces. The Se–Se–Se bond angle is about 105°, which value is indicative of directional, or covalent, bond character. Selenium also forms a structure which contains Se_8 molecules in the form of puckered rings. The orthorhombic form of sulphur is similar (Figure 4.32), with an S–S–S angle of approximately 107°.

Figure 4.30. Stereoview of the structure of iodine.

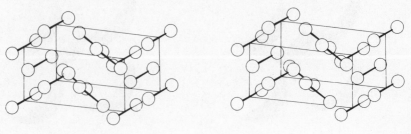

The structures of arsenic (Figure 4.33) and bismuth (group VB) may be considered as superimposed sheets of atoms with bond angles of about 100°, leading to non-planarity of the sheets. Each atom is coordinated by six other

Figure 4.31. Stereoview of the structure of β-selenium.

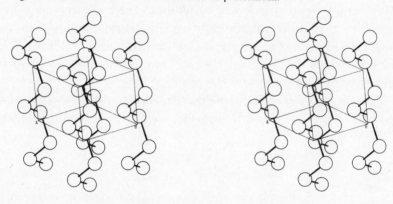

Figure 4.32. Stereoview of the sulphur S_8 molecule.

Figure 4.33. Stereoview of the structure of arsenic.

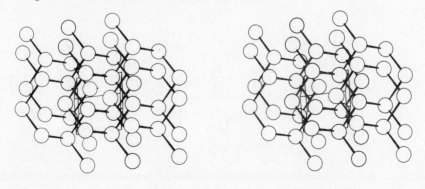

atoms, but three of these neighbours are appreciably closer than the other three. The high melting points and high electrical conductivity of antimony and bismuth indicate a degree of metallic character.

The structures of germanium and grey (α) tin resemble that of diamond (Figure 4.34) but both exhibit some metallic character, unlike diamond, which is more evident still in white (β) tin (Figure 4.35). In elemental zinc and cadmium (group IIB) metallic bond type is well established. These elements have hexagonal structures that are very nearly close packed, the axial ratio, c/a, in zinc being about 1.9 while for the hexagonal close-packed structure (Figure 4.27(b)) this ratio is 1.63. Of the twelve neighbours of any zinc atom six are closer than the others, because of the relative elongation of the c axis.

Thallium and lead are B-group elements, but they have close-packed structures possibly owing to stable electron configurations representing atoms that are not

Figure 4.34. Stereoview of the structure of diamond – also representative of germanium and grey (α) tin.

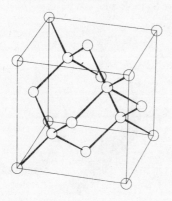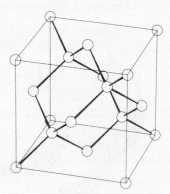

Figure 4.35. Stereoview of the structure of white (β) tin.

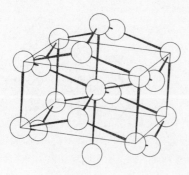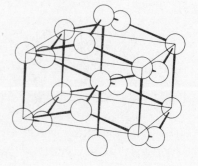

fully ionized:

Tl$^+$ [Kr] $4d^{10} 4f^{14} 5s^2 5p^6 5d^{10} 6s^2$

Pb^{2+} [Kr] $4d^{10} 4f^{14} 5s^2 5p^6 5d^{10} 6s^2$

The states of ionization expected from the position of these elements in the periodic table are Tl^{3+} and Pb^{4+}. These ions are known but they are unstable in aqueous solution with respect to the species Tl$^+$ and Pb^{2+}, this feature of the elements thallium and lead being called the 'inert pair' (in this case 6s^2) effect.

The continuous change in bond type exemplified by the elements in groups IIB–VIIB is reflected in their properties. Brittleness increases along the periodic group and 'open' or low coordination structures are formed which produce a volume contraction on melting to a relatively close-packed liquid structure:

Contraction/m^3 Mg^{-1}

Zn	-0.0108
Bi	$+0.0034$
H$_2$O	$+0.083$

The so-called $(8-n)$ rule can be observed to apply in this series of elements: an element in periodic group nB will generally form $(8-n)$ covalent bonds, as with zinc to bromine for example, but there are exceptions.

The inert gases crystallize with cubic close-packed structures. The atoms bond through very weak van der Waals' forces, and the melting points of the crystals are correspondingly low:

He 1 K   Rn 202 K

Certain diatomic molecules such as carbon monoxide, nitrogen, hydrogen chloride and oxygen have cubic close-packed structures because they are in free rotation in the solid state, a phenomenon noticed by Hendricks in 1930 during a study of long-chain alkyl ammonium halides and discussed theoretically by Pauling in the same year.

The combination of two elements produces a range of widely differing binary compounds including ionic structures such as potassium fluoride and magnesium oxide, metallic alloys such as the copper–gold systems, and molecular compounds such as mercuric chloride and carbon dioxide.

4.4.2 *Binary metal systems*

The pure metals copper and gold have the close-packed cubic structure already described (A1). They form a complete range of solid solutions which, like liquid solutions, are homogeneous within the given range. Initially the addition of copper to gold leads to a structure in which the atoms are distributed in a completely random manner on the same sites that are occupied in the pure metals (Figure 4.36), the unit cell dimension decreasing in proportion to the concentration of copper added; the atomic radii are $r(Cu) = 0.128$ nm and $r(Au) =$ 0.144 nm. At a composition corresponding to CuAu a similar random replacement occurs if the molten alloy is quenched rapidly, but if the same alloy is carefully annealed the copper and gold atoms separate into layers; because of their different atomic sizes the ratio c/a is equal to 0.93 and the structure is tetragonal (Figure 4.37). Another ordered phase exists at the composition Cu_3Au, in this case a cubic structure in which the lattice sites are occupied in a regular manner (Figure 4.38); these ordered structures are termed superlattice structures. The rearrangement of the atoms into an ordered phase evidently lowers the free energy of the system with respect to the random solid solution, although the entropy is decreased in the same process. Silver and gold also form a complete range of solid solutions: no superlattice structures are obtained because the radii of the silver and gold atoms are approximately equal, owing to the lanthanide contraction (see p. 138), and under these conditions there is no free energy

Figure 4.36. Stereoview of the disordered (random) structure of copper–gold.

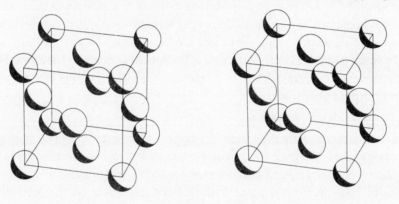

Figure 4.37. Stereoview of the ordered, tetragonal CuAu structure; circles in decreasing order of size represent gold and copper.

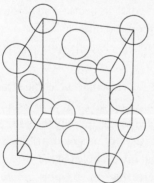
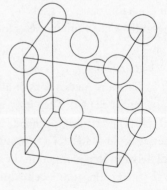

lowering for a superlattice structure. It may be noted that both of the ordered phases are described by *P* unit cells.

The electrical conductivity of quenched Cu–Au alloys shows a smooth variation with composition, but with the annealed specimens maxima occur at the compositions CuAu and Cu_3Au. In general it is not possible to characterize

Figure 4.38. Stereoview of the ordered, cubic Cu_3Au structure; circles in decreasing order of size represent gold and copper.

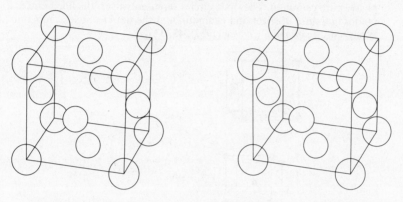

Figure 4.39. Equilibrium phase diagram for the silver–cadmium system. (A. Westgren, *Angewandte Chemie* (1932) **45**, 33.)

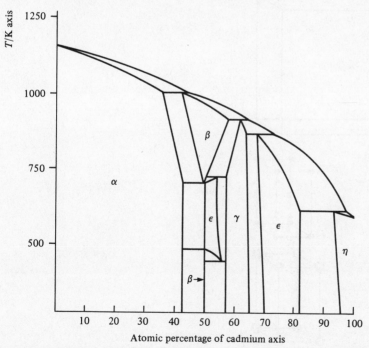

an alloy by a chemical analysis, and more sophisticated methods such as X-ray diffraction become very important.

The silver–cadmium binary system (Figure 4.39) is more complex than the one just discussed. The α-phase is pure silver and can form solid solutions with up to

Figure 4.40. Structures of phases in the silver–cadmium system: (*a*) α-phase, pure silver, (*b*) β-phase, represented ideally by the composition AgCd, (*c*) γ-phase, a complex cubic structure represented ideally by the composition Ag_5Cd_8, (*d*) ε-phase, represented ideally by the composition $AgCd_3$, (*e*) η-phase, pure cadmium. The open circles represent silver, the filled circles a random mixture of silver and cadmium and the double circles cadmium. (A. Westgren, *Angewandte Chemie* (1932), **45**, 33.)

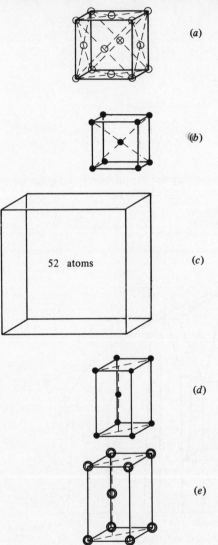

(*a*)

(*b*)

52   atoms   (*c*)

(*d*)

(*e*)

about 42% of cadmium, while a β-phase which has a structure type similar to caesium chloride appears at 50% cadmium. A comparison of Figure 4.40, showing the structure of some of the phases present in the silver–cadmium system, and Figure 4.41, representing the caesium chloride structure type, reveals that the β-AgCd is statistically body-centred whereas caesium chloride has a primitive unit cell. The γ-phase is a complex cubic unit cell containing 52 atoms, and is mechanically very hard and brittle. The ε-phase represents an approximately close-packed hexagonal structure, the lattice sites being occupied by both silver and cadmium atoms in a completely random manner. Pure cadmium is represented by the η-phase which is also a nearly close-packed hexagonal structure, and it can be seen from the phase diagram that cadmium takes up only about 4% of silver into solid solution. Figure 4.42 shows X-ray powder photographs of various alloys of silver and cadmium. The diffraction lines characteristic of each phase can be readily identified, and this data was used to derive the phase diagram. The regions in which two phases coexist can be seen and compared with the phase diagram. No superlattice structures are formed in this system.

Hume Rothery's rule

The factors that govern the appearance of different phases in the alloys of a true metal and a B sub-group element were discussed in 1926 by Hume Rothery, and subsequently given a theoretical basis in the wave-mechanical theory of the metallic bond.

The frequent occurrence of the so-called β-, γ- and ε-phases, among chemically dissimilar binary systems and at widely differing chemical compositions, is determined by the ratio of the number of valence electrons to the number of atoms (see Table 4.8).

Figure 4.41. Stereoview of the caesium chloride structure type, circles in decreasing order of size represent chlorine and caesium.

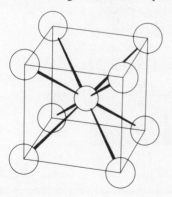 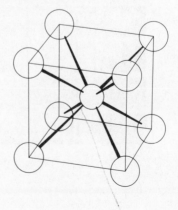

Figure 4.42. X-ray powder photographs of silver–cadmium alloys. The percentage compositions are determined by measuring the intensities of selected diffraction lines and comparing the results with those from known compositions. (A. Westgren & G. Phragén, *Metallwirtshaft* (1928), **7**, 700.)

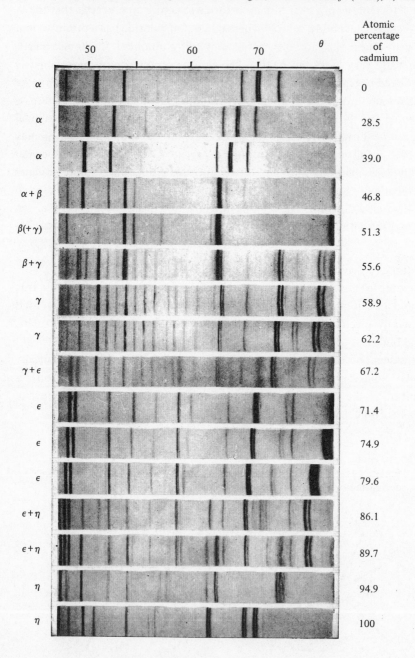

Figure 4.43 shows the structure of sodium thallide NaTl. Each atom is coordinated by eight atoms, four of each type, and the radius of the sodium atom is 0.162 nm. This value lies between the value found in sodium metal (0.186 nm) and that found in sodium chloride (0.095 nm), which indicates that the bonding in sodium thallide is mainly metallic but with some degree of ionic character, and its electrical conductivity and the relatively high coordination of the structure support this view.

The ideal covalent compound is diamond (Figure 4.34) in which a three-dimensional structure (giant molecule) is formed by tetrahedral sp^3 bonds of length 0.1545 nm. The hardness of this substance is well known and its melting point is greater than 3775 K, and it is worth emphasizing that these are properties of covalent compounds; the low melting points and softness of organic compounds, often attributed incorrectly to covalent bonding, arise from the fact that covalent molecules are linked in the solid state by van der Waals' forces, and it is these weaker bonds that are broken when the crystal is melted or mechanically ruptured.

The closely related structures of zinc sulphide, blende and wurtzite, are shown in Figures 4.44–4.46. If these structures were mainly ionic, the ions would be

Table 4.8. *Electron:atom ratios in some binary alloys*

Phase	Composition	Electrons	Atoms	Ratio
β	AgCd	1+2	2	3/2
β	Cu_3Al	3+3	4	3/2
γ	Ag_5Cd_8	5+16	13	21/13
γ	Cu_9Al_4	9+12	13	21/13
ε	$AgCd_3$	1+6	4	7/4
ε	Ag_5Al_3	5+9	8	7/4

Figure 4.43. Stereoview of the structure of sodium thallide, NaTl; circles in decreasing order of size represent thallium and sodium.

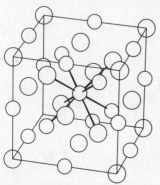
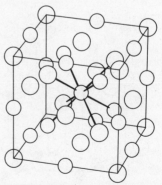

Figure 4.44. Stereoview of the structure of blende (β-ZnS); circles in decreasing order of size represent sulphur and zinc.

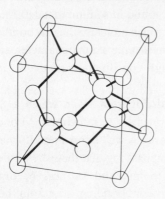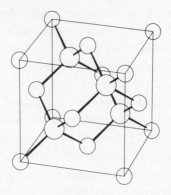

Figure 4.45. Stereoview of the structure of wurtzite (α-ZnS); circles in decreasing order of size represent sulphur and zinc.

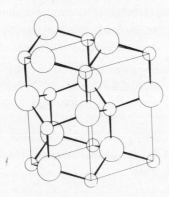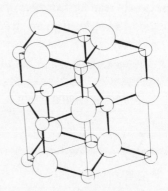

Figure 4.46. Illustration of (*a*) blende and (*b*) wurtzite, showing the close similarity of the two structures.

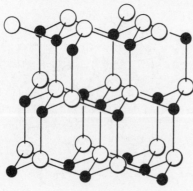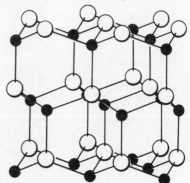

represented as Zn^{2+} and S^{2-}. Covalency requires overlap of the atomic orbitals of the ions, but the true state is intermediate between these extremes and so the charges on the ions are numerically smaller than those representing the ionic state. Several of these compounds with the wurtzite structure are *semiconductors*, their resistivities lying between those of a typical conductor (copper) and a typical insulator (diamond):

	Cu	GaAs	GaSb	InSb	C (diamond)
$\rho/\Omega\,m$ (at 298 K)	1.7×10^{-8}	0.40	1.00	10^{-4}	10^{10}

We have emphasized the absence of pure bond type in the structures of compounds, and it must be accepted that a given compound will contain some degree of each of the principal types of interatomic forces. Often one of these will predominate: for example, we think of sodium chloride as an ionic compound, but accurate calculations of the lattice energy of sodium chloride reveal that there are small contributions to the energy from covalent and van der Waals' forces. We cannot describe this intermediate, resonance structure pictorially and its evaluation can be made only in terms of wave mechanics.

4.4.3 *Ionic compounds, MX and MX$_2$*

Since the ionic bond is non-directional, the fundamental feature of the MX and MX_2 structure types is the geometric disposition of a set of ions of characteristic radius in a manner which is electrically neutral. The principal MX structure types are caesium chloride, sodium chloride and zinc sulphide. The early study of the alkali halides by Bragg led to the interionic distances (Table 4.7) and also to the idea of ions of approximately constant radii. The following observations by Landé (1920) led to an assignment of individual radii for ions (Table 4.9). From the similarities in the interionic distances, it was considered that in the two sulphides and, more especially, the two selenides the anions were in a close-packed array with the cations occupying the interstices (Figure 4.47). The replacement of magnesium by manganese does not alter the anion contact appreciably, except in the two oxides, and we can deduce from the figure:

$$2r_0\sqrt{2} = 4r(X^{2-}) \tag{4.98}$$

Table 4.9. *Sodium chloride type crystals*

	r_e/nm		r_e/nm
MgO	0.2104	MnO	0.2218
MgS	0.2595	MnS	0.2605
MgSe	0.2725	MnSe	0.2725

whence $r(Se^{2-})=0.193$ nm and $r(S^{2-})=0.184$ nm. A table of ionic radii was drawn up in 1926 by Goldschmidt, a modern version of which is reproduced in Table 4.10.

Large differences exist between the radii of a given element in different states of combination:

	S^{2-}	S	S (in SO_2)
r/pm	184	112	94

In general, radii decrease with increasing positive charge since the electrons move in a progressively stronger field and are bound more strongly to the central portion of the atom. In the case of the lanthanons, where the occupancy of the inner 4f electron level is increased with a rise in atomic number, the effective nuclear charge is not exactly balanced by the added electron without a decrease in volume. Thus the ionic radius decreases from 118 pm to 99 pm along the series Ce^{3+} to Lu^{3+} although the atomic number increases from 58 to 71; this effect is

Figure 4.47. Close-packed arrangements of ions, as seen in projection on to the face of the cubic unit cell; note the separation of the anions in magnesium oxide and in manganese oxide.

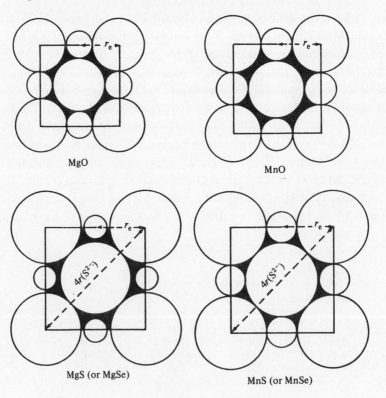

Table 4.10. *Ionic radii/pm; values correspond to six-coordination*

	I	II	III	IV	V	VI	VII	VIII
1							H^- 154	
2	Li^+ 60	Be^{2+} 31	B	C —	N —	O^{2-} 140	F^- 136	
3	Na^+ 95	Mg^{2+} 65	Al^{3+} 50	Si^{4+} 41	P^{3-} 200	S^{2-} 184	Cl^- 181	
4	K^+ 133	Ca^{2+} 99	Sc^{3+} 81	Ti^{3+} 76, Ti^{4+} 68	V^{2+} 88, V^{3+} 74, V^{4+} 60	Cr^{2+} 84, Cr^{3+} 63, Cr^{4+} 56	Mn^{2+} 80, Mn^{3+} 66, Mn^{4+} 54	Fe^{2+} 80, Fe^{3+} 64; Co^{2+} 72, Co^{3+} 63; Ni^{2+} 69, Ni^{3+} 62
	Cu^+ 96	Zn^{2+} 74	Ga^{3+} 62	Ge^{4+} 62	As	Se^{2-} 198	Br^- 195	
5	Rb^+ 148	Sr^{2+} 113	Y^{3+} 93	Zr^{4+} 80	Nb	Mo	Tc	Ru^{4+} 63; Rh^{3+} 68; Pd^{2+} 80, Pd^{4+} 65
	Ag^+ 126	Cd^{2+} 97	In^{3+} 81	Sn^{4+} 71	Sb	Te^{2-} 221	I^- 216	
6	Cs^+ 169	Ba^{2+} 135	La^{3+} 115	Hf^{4+} 78	Ta	W^{4+} 66	Re^{4+} 72	Os^{4+} 65; Ir^{4+} 64; Pt^{4+} 65
	Au^+ 137	Hg^{2+} 110	Tl^+ 144, Tl^{3+} 95	Pb^{2+} 121, Pb^{4+} 84	Bi	Po	At	
7	Fr^+ 176	Ra^{2+} 143	Ac^{3+} 111					

Lanthanide elements

Ce^{3+} 102	Pr^{3+} 100	Nd^{3+} 99	Pm^{3+} 98	Sm^{3+} 97	Eu^{3+} 97	Gd^{3+} 97
Tb^{3+} 100	Dy^{3+} 99	Ho^{3+} 97	Er^{3+} 96	Tm^{3+} 95	Yb^{3+} 93	Lu^{3+} 93

Actinide elements

Th^{3+} 108	Pa^{3+} 106	U^{3+} 104	Np^{3+} 102	Pu^{3+} 101	Am^{3+} 100
Th^{4+} 95	Pa^{4+} 91	U^{4+} 89	Np^{4+} 88	Pu^{4+} 86	Am^{4+} 85

Other ions

NH_4^+ 148 OH^- 153

called the lanthanide contraction. A series of isoelectronic positive ions shows the expected decrease in radius with increase in positive charge:

	Na^+	Mg^{2+}	Al^{3+}
r/pm	95	65	50

but a series of isoelectronic negative ions shows very little change in radius with change in ionic charge

	P^{3-}	S^{2-}	Cl^-
r/pm	200	184	181

because the addition of electrons does not alter the principal quantum number of the occupied orbitals.

MX structures

The highest coordination exhibited in MX structures is that of 8:8 in the caesium chloride type. (By $n:m$ coordination we mean that the first species in the formula is coordinated by m of the second species, and vice versa.) This is the structure of caesium chloride, caesium bromide, caesium iodide, ammonium chloride, ammonium bromide and ammonium iodide, to quote just a few. The structural arrangement has been illustrated in Figure 4.41. When anions of radius, say, 0.1 nm are in contact with each other, the radius of the 'hole' at the centre of any eight anions is 0.0732 nm. This result may be deduced from the figure. The length of the cube diagonal is $2r(+) + 2r(-)$, and the length of the cube edge is $2r(-)$. But $2r(-)\sqrt{3} = 2r(+) + 2r(-)$, so that $r(+)/r(-) = 0.732$. The ratio $r(+)/r(-)$ is called the *radius ratio*, and it gives some idea of the coordination to be expected for given ions, although its predictions are not always accurate.

If we consider reducing the value of $r(+)$ while keeping $r(-)$ constant, it is clear that for $r(+)/r(-)$ less than 0.732 no closer approach of the anions can be obtained. A reduction in the cohesive energy of the crystal (lattice energy) with further decrease in $r(+)$ can be produced, however, if the sodium chloride structure with 6:6 coordination is adopted (Figure 1.1); this structure has a lower limit of the radius ratio at 0.414.

The lattice energy of an ionic crystal can be represented closely by the equation

$$U(r_e) = (-N_A|z(+)||z(-)|Ae^2/4\pi\varepsilon_0 r_e)(1 - \rho/r_e) \qquad (4.99)$$

where $|z(+)|$ and $|z(-)|$ are the numerical values of the cationic and anionic charges respectively. For the alkali metal halides ρ/r_e is approximately 0.1, and substituting the values of the constants in (4.99) we obtain, with r_e in nm,

$$U(r_e) = -125A/r_e \text{ kJ mol}^{-1} \qquad (4.100)$$

The Madelung constant A is a geometrical factor, determined by the structural arrangement of the ions present. It has the following values for MX structures:

MX	CsCl	NaCl	α-ZnS	β-ZnS
A	1.7627	1.7476	1.6407	1.6381

It may be thought at first that the caesium chloride structure type will always be more stable (more negative value of the lattice energy) than those of sodium chloride and zinc sulphide whatever the value of r_e. However, if in the caesium chloride structure type $r(+)$ is decreased such that $r(+)/r(-)$ is less than 0.732, no change in the effective value of r_e occurs because the anions are in contact; hence from (4.100) there is no change in the lattice energy. If we assume a constant anion radius, say 0.181 nm (Cl^-), we can plot $U(r_e)$ from (4.100) as a function of $r(+)/r(-)$ taking $r_e = r(+) + r(-)$. This has been done for the caesium chloride, sodium chloride and β-zinc sulphide structures (Figure 4.48). Below the value of 0.732 for the radius ratio, lower lattice energies should evidently be obtained for the sodium chloride structure type. Assuming that the same degree of ionic

Figure 4.48. Variation of the lattice energy $U(r_e)$ with radius ratio $r(+)/r(-)$ for MX structure types; unit charges apply throughout.

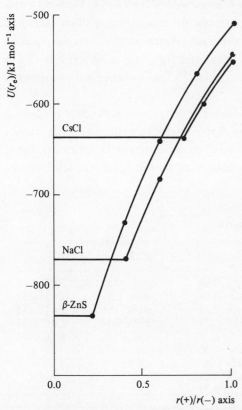

character is retained below the sodium chloride structure lower limiting radius ratio of 0.414, the β-zinc sulphide structure should have the greater stability, down to its limiting value of 0.225 for $r(+)/r(-)$. In Table 4.11 the radius ratios of some MX structures are listed under the structure types actually observed at normal temperatures and pressures. The radius ratio rule is evidently not strictly obeyed. The sodium chloride structure type is often preferred outside its radius ratio limits mainly because this parameter takes the structures as purely ionic, and this rule is not completely true even for the alkali metal halides and less so for the blende and wurtzite structures.

A complete treatment requires a more quantitative lattice theory. We shall discuss one effect qualitatively by using the idea of *polarization*, which can be envisaged as a distortion of the electron distribution of an ion by the presence of neighbouring ions. The polarizability P of an ion measures the distortion which can be produced in this way and is associated with easily deformable electron clouds, a property usually associated with anions although large cations also have appreciable polarizabilities (Table 4.12).

The effectiveness of an ion of radius r in deforming the electron distribution of its neighbours, its polarizing power p, can be measured approximately by the strength of its electric field, $|z|e/(4\pi\varepsilon_0 r^2)$; some values are listed in Table 4.13. It is known that the non-inert-gas type cations are anomalous in their ability to deform the electron clouds of anions to a much greater extent than is suggested by their size. For example, the ions Cu^+, Ag^+ and Hg^{2+} are comparable in size to Na^+, K^+ and Sr^{2+} respectively, although the physical and chemical properties of their halides are quite different.

The effects of polarization are two-fold. The distortions of the electron density of the ions lead to electrostatic dipole–dipole attractions although, because they are symmetrical, there is no permanent dipole moment in ionic structures. Secondly, the deformation of the orbitals of the ions increases the tendency for

Table 4.11. *Radius ratios for MX structures*

CsCl type		NaCl type			ZnS type	
CsCl	0.93	LiF	0.44	MgO 0.46	AgI	0.58	ZnS 0.40
CsBr	0.87	LiI	0.28	BaO 0.96	BeS	0.17	ZnSe 0.37
CsI	0.78	NaF	0.44	MgS 0.35	BeSe 0.16	ZnTe 0.33
NH_4Cl 0.82		KF	0.98	BaS 0.73	BeTe 0.14	CuF 0.71
NH_4Br 0.76		KI	0.62	MgSe 0.33	CuCl 0.53	CuI 0.44
NH_4I	0.69	RbF	1.09	BaSe 0.68		
TlCl	0.80	RbI	0.69	CaTe 0.45		
TlI	0.67	CsF	1.24	BaTe 0.61		
		AgF	0.93	AgCl 0.70		

overlap, or covalent character. Both of these effects produce a more stable structure than would be expected on the basis of the simple ionic picture.

Table 4.14 lists experimentally determined r_e values for the potassium and silver halides and the sum of the corresponding ionic radii. It may be seen that the departure from additivity of ionic radii is more marked in the silver halides, where the polarization is more severe. For the effect on physical properties, one may consider the great differences in the solubilities of the potassium and silver halides (Table 4.15). A clue to the behaviour of the Cu^+ and Ag^+ ions may be obtained through the values of the screening constants (see Appendix A21).

Table 4.12. *Polarizability* $(10^{40}P/F\,m^2)$ *for selected ions*

Li^+ 0.03	Be^{2+} 0.01	O^{2-} 2.7	F^- 1.0
Na^+ 0.30	Mg^{2+} 0.10	S^{2-} 6.1	Cl^- 3.4
K^+ 1.3	Ba^{2+} 2.8	Se^{2-} 7.8	Br^- 4.8
Rb^+ 2.0		Te^{2-} 10.0	I^- 7.3
Cs^+ 3.4			

Table 4.13. *Polarizing power* $(10^{18}p/V\,m^{-1})$ *for selected ions*

Li^+ 3.4	Be^{2+} 13.2	O^{2-} 3.5	F^- 1.7
Na^+ 2.3	Mg^{2+} 5.7	S^{2-} 2.5	Cl^- 1.2
K^+ 1.7	Ba^{2+} 3.4	Se^{2-} 2.4	Br^- 1.2
Rb^+ 1.5		Te^{2-} 2.2	I^- 1.1
Cs^+ 1.3			

Table 4.14. *Interionic distances in the potassium and silver halides*

Halide	r_e/nm	$\sum r$/nm
KF	0.267	0.269
KCl	0.315	0.314
KBr	0.330	0.328
KI	0.353	0.349
AgF	0.246	0.262
AgCl	0.278	0.307
AgBr	0.289	0.321
AgI	0.280	0.324[a]

[a] The ionic radii have been corrected to the 4:4 coordination found in silver iodide. The radii for eight coordination and four coordination are 3% greater and 5% less respectively than the corresponding six coordination radii which are the standard values listed in Table 4.10.

MX₂ structures

The MX_2 structures are very numerous and of a more diverse character than those of the MX type. The maximum coordination of 8:4 is observed in the calcium fluoride or fluorite structure type (Figure 4.49). Each calcium ion is coordinated by eight fluoride ions making up the corners of a cube, and each fluoride ion is coordinated tetrahedrally by four calcium ions; the positions of anions and cations can be interchanged, as in sodium oxide Na_2O for example, this arrangement being termed the antifluorite structure. As with the caesium chloride arrangement, the radius ratio for the fluorite structure type is $r(+)/r(-) > 0.732$.

The 6:3 coordination is typified by rutile, titanium dioxide (Figure 4.50) in which each titanium atom is coordinated by six oxygen atoms making up the corners of a nearly regular octahedron, and each oxygen atom is coordinated by three titanium atoms to form an isosceles triangle. The radius ratio limits for this structure type are 0.732 to 0.414, as with the sodium chloride structure type.

The third simple MX_2 structure type which we shall describe is the 4:2 coordinated structure of cristobalite (Figure 4.51). This may be considered as derived from blende (Figure 4.44), because the cristobalite structure is obtained if silicon atoms are placed on both the zinc and sulphur sites with oxygen atoms mid-way between them. Hence, each silicon atom is coordinated tetrahedrally by four oxygen atoms, and each oxygen atom is coordinated linearly by two silicon

Table 4.15. *Solubility* (s/mol dm⁻³) *for potassium and silver halides*

	F	Cl	Br	I
K	17	4.8	5.3	8.7
Ag	14	1.1×10^{-5}	7.5×10^{-7}	1.0×10^{-8}

Figure 4.49. Stereoview of the structure of calcium fluoride; circles in decreasing order of size represent fluorine and calcium.

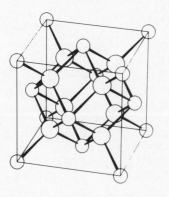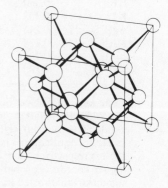

atoms. The linearity of the Si—O—Si linkage indicates a high degree of ionic character; in quartz, another modification of silicon dioxide, the Si—O—Si bond angle at room temperature is less than 180°. Table 4.16 lists the radius ratios of several MX_2 structures and the structure type in which each occurs under normal

Table 4.16. *Radius ratios of some MX_2 structures*

Fluorite				Rutile				Cristobalite	
CaF_2	0.73	CdF_2	0.71	MgF_2	0.48	MnO_2	0.57	BeF_2	0.23
SrF_2	0.83	HgF_2	0.81	MnF_2	0.59	SnO_2	0.51	SiO_2	0.29
BaF_2	0.99			FeF_2	0.59	TiO_2	0.49	GeO_2	0.38
Li_2O	0.43	Li_2S	0.17	CoF_2	0.53	GeO_2	0.38		
Na_2O	0.68	Rb_2S	0.80	NiF_2	0.51				
K_2O	0.95			ZnF_2	0.54				
Rb_2O	1.06								
Li_2Se	0.16	Li_2Te	0.14						
Rb_2Se	0.75	K_2Te	0.60						

Figure 4.50. Stereoview of the structure of rutile (TiO_2); circles in decreasing order of size represent oxygen and titanium.

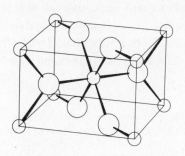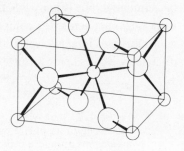

Figure 4.51. Stereoview of the structure of cristobalite (SiO_2); circles in decreasing order of size represent oxygen and silicon.

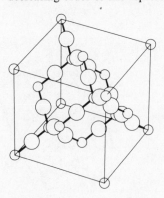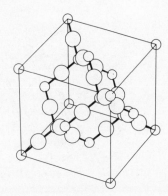

conditions: it is noteworthy that many of the MX and MX_2 structures undergo polymorphic transformations under different conditions of temperature and pressure. For example, under high pressure sodium chloride adopts the caesium chloride structure.

In general, the MX_2 structures are in better agreement with the radius ratio predictions than are the MX structures, which may at first sight seem surprising as the larger polarization in the MX_2 structures would be expected to lead to greater departures from ionic character. However, remembering the curves of Figure 4.48 and considering the Madelung constants of these structures, it is not unreasonable to conclude that the lattice energies of MX_2 structures exhibit greater numerical differences than do those of the caesium chloride and sodium chloride types:

	Fluorite	Rutile	Cristobalite
A	5.04	9.64	8.24

Thus the small energy terms which are important in deciding between the caesium chloride and sodium chloride structures are of less structural effect in the MX_2 series.

Strong polarization leads to the development of layer structures such as those of cadmium iodide (Figure 4.52). The asymmetry of the coordination is evident; weak forces act across the composite layers (I—Cd—I), and the crystals exhibit good cleavage along planes parallel to these layers. Further polarization leads to discrete molecules being formed which are held together in the solid state by van

Figure 4.52. Stereoview of the structure of cadmium iodide; circles in decreasing order of size represent iodine and cadmium. The coordination of iodide ions around a cadmium ion is very nearly regular octahedral, but that of cadmium ions around an iodide ion is asymmetric: there are three near cadmium neighbours on one side of the iodide ion with respect to three much more distant cadmium neighbours on the opposite side of the iodide ion.

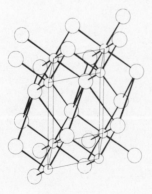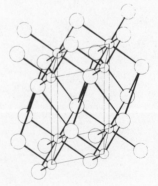

der Waals' forces: the structures have the properties associated with these forces, such as low melting and boiling points and low mechanical strength. Two examples are mercury(II) chloride and carbon dioxide: the atomic orbitals in both mercury(II) chloride and carbon dioxide overlap to a large extent and covalent bonds are formed. We might consider that extreme polarization in ionic structures leads to covalency, but this picture is only qualitative.

Ionic crystals are formed also between cations and complex anions such as sulphate, carbonate and nitrate ions, while complex cations such as $\{Co(NH_3)_6\}^{3+}$ and $\{Cu(H_2O)_4\}^{2+}$ may also be involved in ionic structures. The complex ions have shapes and sizes (Figure 4.53) which remain approximately constant throughout a wide range of structures. One of these compounds is illustrated in Figure 1.3, where the tetrahedral sulphate groups can be seen.

Figure 4.53. Shapes and sizes of some polyatomic anions (see also Figure 2.16).

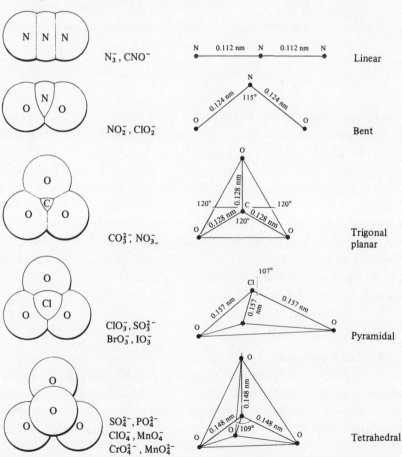

Cohesion in this structure is obtained partly by the strong electrostatic linkages between the calcium ion and six oxygen atoms in sulphate groups and the two oxygen atoms in water molecules. Hydrogen bonding is also important in this compound.

Hydrogen-bonded ionic structures

The structural effects of the hydrogen bond are far reaching. We have referred to the hydrogen bonding already, in the first two chapters. It is responsible for the high degree of structure in liquid water and ice, the openness of the ice structure leading to a volume contraction on melting to form a more closely packed liquid (see also p. 2).

Ammonium fluoride and ammonium hydrogen difluoride exhibit strong hydrogen bonding, unlike the other ammonium halides of coordination number eight. The hydrogen fluoride ion HF_2^- in NH_4HF_2 is hydrogen bonded in addition to the ionic forces expected for such structures. Since the ammonium ion has a tetrahedral configuration, this is imposed upon the whole structure, both ammonium fluoride and ammonium hydrogen difluoride having tetrahedrally coordinated arrangements. In potassium hydrogen fluoride only the hydrogen fluoride ion is hydrogen bonded and this ion is linked electrostatically to the potassium ions to give a higher coordination.

Hydrogen bonding is present in most hydrates. We refer to the structure of gypsum once again (Figure 1.3), to note that the cohesion in one direction is due entirely to hydrogen bonds. The gypsum crystal exhibits a perfect cleavage, and we find that the cleavage plane ruptures the weaker hydrogen bonds in the structure.

Only a few crystal structures have been discussed in order to amplify the earlier description of interatomic forces in chemical compounds, and to illustrate the classification of these structures based upon the types of bonds present. A wealth of structural information is now available and for further study the reader is directed to the works of Evans and of Bragg and Claringbull (R. C. Evans, *Crystal Chemistry*, 2nd edn, Cambridge University Press, Cambridge, 1964: W. L. Bragg & G. F. Claringbull, *Crystal Structures of Minerals* (*The Crystalline State, vol. IV*), Bell, London, 1965).

4.5 The liquid state

The theory of the liquid state is much less well developed than that of the gaseous state with its molecular chaos, or of the solid state with its characteristic regularity. Models and mathematical functions can often be envisaged for extreme ranges of certain properties but not for their intermediate values, and approximations made in the extreme cases may not be sufficiently accurate for the intermediate case. Liquids likewise represent an intermediate state, in this instance between order and disorder.

Similarity to the solid state is apparent in the way that molecules are packed in a liquid. Cohesive forces strong enough to lead to a condensed state of matter exist, but the appreciable translational energy of the molecules inhibits long-range order. In the solid state by comparison, the kinetic energy of the molecules is negligible because although atoms or ions vibrate about their mean positions, free movement is normally prevented by strong and well-defined interatomic forces. In a liquid, molecules cluster together to form localized, partially-ordered groups that continually disperse and re-form. Certain distances of approach, those within the cluster, are more probable than others despite the fact that the range of distances within a given volume is very variable.

Liquids represent a state of higher internal energy and of lower degree of order than in crystalline solids. The cohesion in a liquid may be due to ionic forces (molten electrolytes), to metallic forces (molten metals), to hydrogen bonding (water, for example) or to van der Waals' forces (organic liquids); in some cases more than one of these forces may operate.

4.5.1 *Disorder in liquids*

All substances that do not decompose on heating form liquids, provided that the external pressure is sufficiently great. A crystal at its melting point and the liquid at the same temperature are in chemical equilibrium, but energetically the crystalline state is more favourable than that of the liquid. To melt, the enthalpy (latent heat) of fusion must be taken up by the system. The equilibrium situation is determined by the free energy change for the process, and the increased disorder on melting causes the entropic term $T \Delta S$ to overcome the enthalpy change, and the crystal melts when the following relationship has been established

$$T[S(l) - S(s)] \geqslant H(l) - H(s) \tag{4.101}$$

The sharpness of a melting point indicates a discontinuity between the solid state and the liquid state of a pure substance. In a gas the energy distribution of the molecules is given by the Boltzmann (classical) equation (4.45), but in a solid energy changes are quantized and when melting takes place enough energy must be supplied for the whole crystal to melt, otherwise some of it remains as a solid in equilibrium with the liquid at the melting point.

The transition between liquid and gas is not so marked as that between solid and liquid. Owing to its relatively high vapour pressure a liquid may be evaporated isothermally, vapour pressure being governed by temperature according to the empirical equation

$$\ln p = a + b \ln T + (c/T) \tag{4.102}$$

If the vapour obeys the gas laws, the molar heat of vaporization is given by

$$\Delta H_v = \alpha + \beta RT \tag{4.103}$$

The approximate values of β are 1.5, 2.5, 3.5 and 4.5 for liquids composed of monatomic, diatomic, triatomic and tetratomic molecules respectively. The term α is the molar heat of vaporization extrapolated to $T = 273$ K so that $\Delta H_v / RT$ is a constant (approximately 11); this relationship is known as Trouton's rule and is obeyed moderately well. Not surprisingly, liquids seem to obey either Boltzmann or quantized conditions in differing circumstances, but the approximate constancy of Trouton's rule indicates that a definite amount of energy must be expended to evaporate one mole of a liquid even though it evaporates continuously.

4.5.2 *Compressibility*

Liquids are generally more compressible than solids but much less compressible than gases. The compressibility coefficient β measures the relative decrease in volume with increase in applied external pressure, and varies with temperature according to the approximate equation

$$\beta = \beta_0 \exp(bT) \tag{4.104}$$

For pentane and mercury, b has the values 7.97×10^{-3} K^{-1} and 1.37×10^{-3} K^{-1} respectively. Although a solid and its melt are in chemical equilibrium, the discontinuity between these two states of matter is emphasized by a marked difference in their compressibilities.

The variation of the compressibility of a liquid with pressure is indicated in Figure 4.54. The decrease is rapid at low pressures, and at 1000 atm the compressibility has fallen to about one-half of its value at 1 atm.

Figure 4.54. Variation of compressibility coefficient β with pressure.

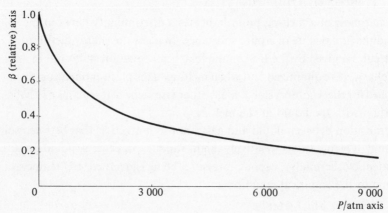

4.5.3 *Liquid flow*

The fact that liquids flow under applied stress does not imply the absence of interatomic forces, but rather it demonstrates their inability to maintain a fixed shape, unlike a solid. The viscosity η of a liquid may be obtained from studies on the rate of flow through capillary tubes:

$$\eta = \pi R^4 P / 8 l V_t \tag{4.105}$$

where R is the radius of a capillary tube of length l, P is the pressure difference across the ends of the capillary and V_t is the volume flow per second. The variation of viscosity with temperature is given approximately by

$$\eta = A \exp\left(\frac{-\xi}{RT}\right) \tag{4.106}$$

where ξ is an energy barrier which must be overcome before flow takes place, and A is a constant, the limiting value of η as T tends to infinity. The relationship is demonstrated well by both ethanol and benzene (Figure 4.55), but it fails for water because the hydrogen-bonded structure changes with a change in temperature.

4.5.4 *Evidence for structure in liquids*

X-ray diffraction studies on liquids indicate a degree of ordering of the atomic arrangement. A typical diffraction pattern is shown in Figure 4.56 and is interpretable in terms of a radial distribution function, $\rho(r)$, such that the probability of finding another atom distant between r and $(r + dr)$ from any one atom is given by $4\pi r^2 \rho(r)\,dr$ (Figure 4.57). Evidently there is an order in liquids, and a principal difference between liquids and solids emerges from this study: in a solid, the structural unit, an array of atoms, is repeated periodically in three dimensions by the translations of a Bravais lattice, but in a liquid there is no

Figure 4.55. Temperature dependence of viscosity η.

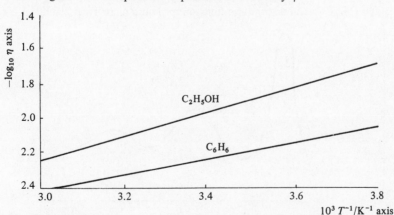

regular repetition of this nature. These features are implied in the terms long-range order and short-range order.

4.5.5 *Structure models*

The cohesive forces operating in the liquid state are of prime importance in deciding its structure, and an estimate of their magnitude is given by the internal pressure P_i. For an ideal gas P_i is zero, since by definition the internal energy of a gas is independent of its volume. For a real gas, however, forces of attraction exist between the molecules, and an internal cohesive pressure results for which the term a/V^2 in the van der Waals' equation of state for a gas attempts to account. In Table 4.17 the magnitudes of the internal pressures for a number of gases and liquids are compared. The internal pressure of a liquid varies much more rapidly with external pressure than does that of a gas. Thus, whereas a

Figure 4.56. Intensity of X-ray diffraction as a function ϕ of the scattering angle 2θ for liquid mercury at 298 K.

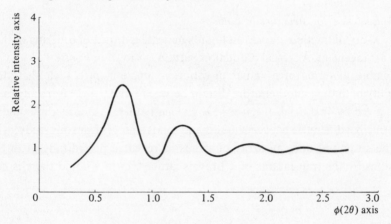

Figure 4.57. Radial distribution function for liquid mercury at 298 K.

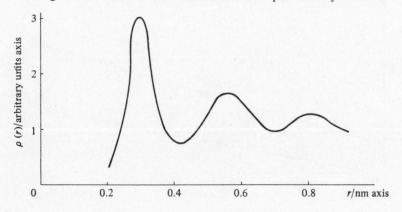

moderately simple extension of the equation of state for an ideal gas explains many of the properties of real gases no such simple equation holds for a liquid.

A theory of the liquid state has to account for many diverse factors; they include the general decrease in enthalpy, entropy and volume in forming a solid, the viscosity and its variation with temperature, the compressibility, isothermal evaporation, and the X-ray evidence of localized structure. No single theory has emerged yet to account satisfactorily for all features of the liquid state.

Eyring considered a free space theory. The free space in a liquid is the volume which is not occupied by the molecules themselves. At ordinary temperatures about 3% of free space exists. This figure may be deduced from the compressibility studies already discussed. Between 1 atm and 1000 atm the large decrease in compressibility corresponds to a reduction in volume by 3%. The high initial fall in the compressibility coefficient β may be interpreted as a removal of the free space in the liquid; after this has occurred, the liquid is much less compressible.

Eyring's liquid model is illustrated in Figure 4.58. The vapour above the liquid contains relatively few molecules, moving about completely at random. In the liquid itself, the space is mostly filled and the molecules move about at random, singly or in clusters. If the temperature is increased, molecules pass into the vapour, and holes in the liquid increase in number. At the critical temperature, the density of the liquid should equal that of the vapour; hence, the average density should remain constant. In practice a small decrease in the average density $\bar{D}$ with temperature is observed, so that

$$\bar{D} = D_0 - \alpha T \tag{4.107}$$

D_0 is a limiting density at $T = 273$ K and α is $d\bar{D}/dT$. This relationship was discovered in 1886 by Cailletet and Mathias and is known as the law of rectilinear diameters. For pentane, $D_0 = 0.3231$ and $\alpha = 0.00046$, leading to Figure 4.59: X is the average density at 400 K, and XY represents the small decrease in the average density with increase in temperature.

The holes in a liquid may be of molecular size. A molecule adjacent to a vacant space would have gas-like properties, whereas a molecule adjacent to other

Table 4.17. *Internal pressures at 298 K and 1 atm external pressure*

Gas	P_i/atm	Liquid	P_i/atm
Ideal	0	C_7H_{14}	2510
H_2	0.0005	CCl_4	3310
O_2	0.0027	C_6H_6	3640
CO_2	0.0072	CS_2	3670
H_2O	0.0109	H_2O	13200
Hg	0.0162	Hg	20000

molecules would experience forces similar to those in a solid. Eyring and Ree considered the equation of state for a liquid in terms of gas-like and solid-like functions.

Bernal considered the structure of a liquid in terms of the packing of molecules. X-ray evidence had shown the existence of five-fold coordination in liquids.

Figure 4.58. Eyring's model for the liquid state. (H. Eyring in *Physical Chemistry* by W. J. Moore, 3rd edn, Prentice-Hall, NJ, 1962.)

Figure 4.59. Density change at the vapour–liquid transition point for pentane.

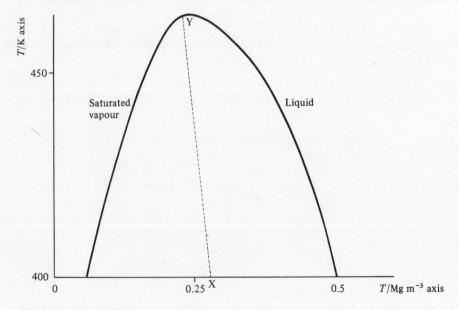

Bernal arranged five spheres around one sphere and then added other spheres in an ordered array. A single point of abnormal coordination in several hundred spheres produced long-range disorder in the model of the liquid. Bernal concluded that the particles in a liquid are in a close-packed array or *heap* (Figure 4.60). Each molecule is in contact with many others but not in a regular manner. The coordination numbers are distributed between five and twelve with an average at about 8.5. This distribution of coordination is another fundamental structural difference between liquids and solids.

In a close-packed solid, the arrangement is called a *pile* (Figure 4.61). In contradistinction to the heap, a pile is a regular arrangement, and it exhibits the single coordination number of twelve as in the metal structures A1 and A3. The density of closest random packing is about 64% of the total volume occupied. For loose packing it is reduced to 60% and for close packing in a crystal it is 74%.

A characteristic feature of liquids is their ability to dissolve substances to form an extensive range of homogeneous mixtures. This can be understood by

Figure 4.60. Random, close-packed *heap* of spheres. (After J. D. Bernal in *Liquids: Structure, Properties, Solid Interactions*, ed. T. J. Hughel, Elsevier, Amsterdam, 1965.)

Figure 4.61. Regular *pile* of spheres. (After J. D. Bernal in *Liquids: Structure, Properties, Solid Interactions, ed. T. J. Hughel, Elsevier, Amsterdam, 1965.*)

considering a liquid structure in terms of equal spheres. If a sphere of a different size were introduced into the pile, a dislocation would result which would extend throughout the structure. In the heap, however, the effect of a similar impurity is likely to be less structure-breaking since the heap already contains many such dislocations.

Problems 4

4.1 To what temperature must 1 dm³ of an ideal gas at 1 atm be cooled from room temperature (298 K) in order to reduce its volume to 10^5 mm³? If the lowest temperature attainable was twice that just calculated, what pressure would then be needed to obtain the same final volume?

4.2 Gases conform more and more closely to the ideal gas equation as the temperature increases, or as the density decreases. Obtain an equation between pressure and density D for an ideal gas. The following data have been measured for dimethyl ether at 298 K.

P/atm	0.1207	0.2486	0.3649	0.5958	0.8411	1.0000
$10^3 D/\text{Mg m}^{-3}$	0.2276	0.4695	0.6898	1.129	1.589	1.903

Draw a graph to show how closely the gas approaches ideal behaviour, considering (4.65) as far as the second virial coefficient. Calculate the relative molar mass M_r of diethyl ether.

4.3 In a small-scale process for synthesizing ammonia from nitrogen and hydrogen, the containing vessel has a volume of 50 dm³ and contained 4 mol hydrogen and 2 mol nitrogen at 304.6 K.

(a) What are the partial pressures of each gas in the mixture, and what is the total pressure?

(b) If all of the hydrogen were converted to ammonia, what would be the partial pressures of hydrogen, nitrogen and ammonia?

4.4 The general expression for the mean $\bar{X}$ of a property X distributed according to a probability function $P(X)$ is given by $\bar{X} = \int XP(X)\,dX / \int P(X)\,dX$, where the integration is taken over the range of the variable. Use (4.38) and (4.41) to show that the ratio of the mean speed $|\bar{v}|$ of molecules to their rms speed $(\overline{v^2})^{\frac{1}{2}}$ is $(8/3\pi)^{\frac{1}{2}}$.

4.5 Use the van der Waals' equation with data from Tables 4.1 and 4.2 to calculate the following:

(a) μ_{JT} at 350 K and T_i for ammonia ($C_P = 0.287\,\text{dm}^3\,\text{atm}\,\text{K}^{-1}\,\text{mol}^{-1}$).

(b) The Boyle temperature for ammonia.

4.6 Calculate the compression coefficient Z for nitrogen gas at 273 K and 100 atm.

4.7 At 298 K the ratio C_P/C_V for water vapour is 1.32. What would this ratio become at higher temperatures, when the molecular vibrations are fully developed?

4.8 Two identical cubes are exactly superimposed. What is the crystal system of the resulting figure? Draw a sketch to show the symmetry elements of the figure.

4.9 Show by means of a diagram, or otherwise, that the structure of sodium chloride can be represented by a rhombohedral unit cell. Calculate the length of the side of the rhombohedral unit cell, in terms of that of the conventional face-centred cubic unit cell, and the interaxial angle.

4.10 What are the Miller indices of planes that make the following intercepts on the x, y and z axes respectively?

(a) $a/2$ b $-c/3$ (b) a $\|y$ $c/4$

(c) $\|x$ $\|y$ $-c$ (d) $2a$ $-b$ $c/2$

(e) $-a/3$ $-b$ $c/2$ (f) $a/2$ $2b$ $-3c/2$

4.11 What are the ratios of the reticular densities of atoms on the planes (100), (110), (111) and (230) in a body-centred cubic unit cell with side a and having identical atoms at the lattice points?

4.12 The vapour pressure of nitric acid depends on temperature as follows:

T/K	273	293	313	323	343	353	363	373
P/atm	0.019	0.063	0.175	0.274	0.614	0.882	1.233	1.687

(a) Calculate the molar enthalpy of vaporization of nitric acid.

(b) What is the boiling point of the acid?

4.13 The unit cell dimension of potassium chloride is 0.628 nm. Calculate the interplanar spacing d_{123} and the corresponding Bragg angle for copper $K\alpha$ X-radiation ($\lambda = 0.15418$ nm).

4.14 The following values of $\sin^2\theta$ were obtained for the first ten lines on a powder photograph of a cubic substance, taken with cobalt $K\alpha$ X-radiation ($\lambda = 0.17902$ nm):

0.0376   0.0556   0.0744   0.0926   0.112   0.167   0.186   0.205   0.261   0.316

Find the unit-cell type, index the lines and determine a mean value for the unit-cell dimension a and the standard deviation of that mean value.

4.15 In the close-packed cubic metal structure there are voids, or holes. The tetrahedral holes are surrounded by four other atoms, and the octahedral holes are surrounded by six. How many holes of each type belong to one and the same unit cell, and what are their coordinates (take the cubic unit-cell side as unity)?

4.16 Show that the volume occupied per sphere of radius 1 nm in the close-packed hexagonal structure is 5.66 nm^3.

4.17 What phase would be expected for each of the binary alloys of compositions AuZn, Au$_5$Al$_3$, Cu$_{31}$Si$_8$ and Ag$_5$Al$_3$?

4.18 Describe the unit-cell type, coordination numbers and patterns, unit-cell contents and coordinates of ions belonging to one and the same unit cell of the fluorite (calcium fluoride) structure.

4.19 Use Figure 4.41 and Table 4.10, including the rider to Table 4.14, to determine the density of caesium chloride (Cs = 132.905, Cl = 453; $u = 1.66057 \times 10^{-27}$ kg).

4.20 The enthalpies of vaporization at the boiling point for five substances are as follows:

	Diethyl ether	Benzene	Propyl ethanoate	Water	Mercury
ΔH_v/kJ mol^{-1}	27.1	31.4	34.8	40.6	59.4
T_{bp}/K	307	353	375	373	630

Calculate the values of the Trouton's rule constant and comment upon the results.

5

Chemical equilibrium

'Insoluble' barium sulphate can be converted to barium carbonate by boiling it with sodium carbonate solution. The extent to which the conversion takes place depends upon the concentration of sodium carbonate – it is a mass effect (Rose, 1842); we shall consider such effects in the present chapter.

5.1 Law of mass action

The equation for the precipitation of silver chloride from solution may be written

$$Ag^+(aq) + Cl^-(aq) \rightarrow AgCl(s) \tag{5.1}$$

This reaction is sensibly complete. Not all reactions take place to completion, however. The esterification of ethanol is represented by

$$CH_3CO_2H(l) + C_2H_5OH(l) \rightleftharpoons CH_3CO_2C_2H_5(l) + H_2O(l) \tag{5.2}$$

Starting with one mole each of ethanol and ethanoic acid, at 298 K, a position of equilibrium is reached, where

$$[CH_3CO_2H] = [C_2H_5OH] = 0.33 \text{ mole}$$

$$[CH_3CO_2C_2H_5] = [H_2O] = 0.66 \text{ mole}$$

where [ ] specifies an amount of substance or concentration. At this composition the rate of conversion in each direction is the same, so that no further change in composition occurs with time; this is what is meant by the establishment of equilibrium. The double arrow in (5.2) implies that this reaction is thermodynamically reversible. Strictly, (5.1) should be written in the same way, ie

$$Ag^+(aq) + Cl^-(aq) \rightleftharpoons AgCl(s) \tag{5.3}$$

but the concentrations of silver and chloride ions at equilibrium are very small (unless there is an excess of either) – about $10^{-5} \text{ mol dm}^{-3}$ – so that the reaction may be regarded as complete for all practical purposes.

Considering again (5.2), the rate of the forward (L → R) reaction is found by experiment to depend initially upon the concentrations of ethanol and ethanoic acid in the reaction mixture, ie

$$\text{rate } (L \rightarrow R) \propto [CH_3CO_2H][C_2H_5OH] \tag{5.4}$$

$$= k_1[CH_3CO_2H][C_2H_5OH] \tag{5.5}$$

where k_1 is the *rate constant* of the forward reaction. If we studied the reverse reaction ($R \rightarrow L$ in (5.2)), we should find the initial rate to depend upon the concentration of the reactants ethyl ethanoate and water

$$\text{rate } (R \rightarrow L) = k_{-1}[CH_3CO_2C_2H_5][H_2O] \tag{5.6}$$

The factors determining the rate of a chemical reaction were enunciated by Guldberg and Wage (1864–67); in modern terminology: the rate of reaction is proportional to the product of the activities of the reactants in the rate-determining step of the reaction, in this case (5.4) or (5.6), at the given temperature. The distinction between activity and concentration is very important in the case of electrolyte solutions; in the present case we are concerned with reactions between molecules, and may replace activities by concentrations without introducing appreciable error. Activity is a thermodynamic concept; it is an 'ideal' concentration, in which the actual concentration has been corrected for interactions between the solute species in solution, and between solute and solvent, so that the activity tends to equal the concentration c as $c \rightarrow 0$. Further consideration of activities is given in Chapter 7, and of the rates of chemical reactions in Chapter 8.

This dynamic approach to equilibrium is useful in that it introduces the idea of a mobile process, in which both forward and reverse reactions continue at equal rates. During the course of (5.2) the concentrations of acid and alcohol decrease, and the rate of the forward reaction, (5.5), decreases correspondingly (Figure 5.1). As reaction proceeds, the concentrations of ester and water increase, so that the rate of the reverse reaction increases. When equilibrium is reached, the two rates are equal; thus, we can write

$$\frac{k_1}{k_{-1}} = \frac{[CH_3CO_2C_2H_5][H_2O]}{[CH_3CO_2H][C_2H_5OH]} = K \tag{5.7}$$

where K is the equilibrium constant, at the particular temperature.

To apply the dynamic approach it is necessary to decide which is the rate-determining step of the reaction being considered, and this must always be settled by recourse to experiment; however, the result (5.7) is valid whatever the mechanism of reaction.

Considering the general reaction

$$aA + bB \rightleftharpoons pP + qQ \tag{5.8}$$

and assuming that we may replace activities by concentrations, the equilibrium constant for this reaction is given by

$$K = \frac{[P]^p[Q]^q}{[A]^a[B]^b}$$ (5.9)

This same expression may be derived by a purely thermodynamic argument, without consideration of the kinetics of the reactions.

5.2 Criterion for equilibrium

A system is in equilibrium when the temperature and pressure remain constant, and there is no resultant change in the distribution of reactants and products with the passage of time. Van't Hoff (1883) proposed that the tendency for a reaction to occur should be measured by the magnitude of the quantity we now call the free energy decrease accompanying the reaction.

At constant temperature and pressure, a process can take place in an isolated system (ie one which cannot exchange matter or energy with an adjacent system) only in the direction of a decrease in free energy G, ie if ΔG is negative. Whether or no a reaction which has a negative ΔG will occur at a measurable rate depends

Figure 5.1. Variation of rates of (5.2) with time.

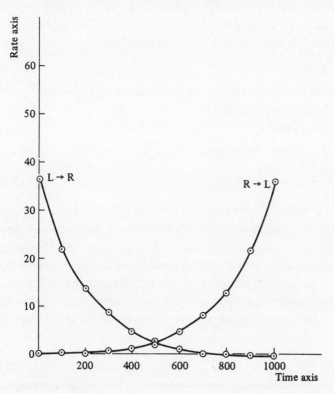

upon how readily it can acquire the necessary *activation energy* (Section 8.3). The following example illustrates this point. The standard free energy decrease for the formation of sulphur dioxide from sulphur and oxygen is

$$\Delta G_f^{\ominus} = -301.2 \text{ kJ mol}^{-1}$$

so that the reaction should be spontaneous. However, for it to occur the temperature must be raised; sulphur 'does not react' (or reacts immeasurably slowly) with oxygen at 298 K.

The free energy of a system at equilibrium, at a given temperature and pressure, is a minimum; thus one thermodynamic condition for equilibrium is

$$(\Delta G)_{T,P} = 0 \tag{5.10}$$

The two conditions, that $(\Delta G)_{T,P}$ is negative for spontaneous reaction to occur, and that $(\Delta G)_{T,P}$ is zero at equilibrium, are of fundamental importance in the study of physical chemistry.

For the further study of equilibria in chemical systems, it will be convenient to make the subdivision (i) homogeneous equilibria and (ii) heterogeneous equilibria.

5.3 Homogeneous equilibria

Reactions in homogeneous (single-phase) systems take place between molecules or ions; examples have been given in (5.1) and (5.2). The application of the law of mass action to such systems is next considered.

5.3.1 *Gaseous systems – no change in the total number of molecules*
In the hydrogen–iodine reaction

$$H_2(g) + I_2(g) \rightleftharpoons 2HI(g) \tag{5.11}$$

the equilibrium constant may be written, from (5.9)

$$K = \frac{[HI]^2}{[H_2][I_2]} \tag{5.12}$$

Let a mixture of a moles of hydrogen and b moles of iodine be heated until equilibrium is attained, at a given temperature. If y moles of hydrogen have been used in forming hydrogen iodide, then at equilibrium $(a - y)$ moles of hydrogen remain; similarly, $(b - y)$ moles of iodine remain, and $2y$ moles of hydrogen iodide are produced. The total number of moles present initially and at equilibrium is the same, $(a + b)$. In considering reactions in the gas phase we may express the concentrations by the corresponding partial pressures. Thus (5.12) may be reformulated

$$K_p = \frac{p^2(HI)}{p(H_2)p(I_2)} \tag{5.13}$$

If P is the total pressure, then the partial pressures of the components, at equilibrium, are as follows:

$$p(HI) = \frac{2y}{(a+b)}P \quad p(H_2) = \frac{(a-y)}{(a+b)}P \quad p(I_2) = \frac{(b-y)}{(a+b)}P$$

Note that $\sum_i p_i = P$, for summation over the i components of the system. Thus

$$K_p = \frac{\left\{\frac{2y}{(a+b)}\right\}^2 P^2}{\frac{(a-y)}{(a+b)}P\frac{(b-y)}{(a+b)}P}$$

$$= \frac{4y^2}{(a-y)(b-y)} \tag{5.14}$$

Equation (5.14) contains no term referring to the total pressure or to the volume of the system. The equilibrium composition of a homogeneous gas system in which the total number of molecules is constant is independent of the volume and of the total pressure. Consequently, the relative amount of hydrogen iodide formed in (5.11) is independent of the total pressure. The data of Table 5.1 relate to this system at 730 K. The values of K_p at the given temperature are sensibly constant, and independent of the total pressure, from whichever side of (5.11) we approach equilibrium.

By integrating the van't Hoff equation

$$d(\ln K_p)/dt = \Delta H^\ominus/RT^2 \tag{5.14a}$$

we obtain

$$\ln\left\{\frac{(K_p)_{T_2}}{(K_p)_{T_1}}\right\} = \frac{-\Delta H}{R}\left(\frac{1}{T_2} - \frac{1}{T_1}\right) \tag{5.15}$$

For the system (5.11), $K_p = 48.8$ at 730 K and 60.8 at 394 K; thus $\Delta H = -15.83\,\text{kJ mol}^{-1}$. Note that K_p decreases with increasing temperature, corresponding to a decrease in enthalpy, ie to an exothermic reaction. This is in accordance with Le Chatelier's principle (Chapter 1).

Table 5.1. *The hydrogen–iodine equilibrium*

$10^3 H_2/\text{mol dm}^{-3}$	$10^3 I_2/\text{mol dm}^{-3}$	$10^3 HI/\text{mol dm}^{-3}$	P/atm	K_p
5.617	0.5936	12.70	1.135	48.4[a]
4.580	0.9733	14.86	1.223	48.8[a]
3.841	1.524	16.87	1.334	48.6[a]
1.433	1.433	10.00	0.771	48.7
1.696	1.696	11.81	0.910	48.5
4.123	4.213	29.43	2.271	48.8

[a] Equilibrium approached from $H_2 + I_2$ in these cases; in the remainder, equilibrium was approached from HI.

The reaction between nitrogen and oxygen to form nitric oxide (nitrogen monoxide) is under suitable conditions an endothermic process. The equilibrium constant is given by

$$K_p = \frac{p^2(NO)}{p(N_2)p(O_2)} \tag{5.16}$$

and its variation with temperature has been found experimentally to be represented approximately by

$$\ln K_p = \frac{22\,000}{T} + 1.09 \tag{5.17}$$

Application of (5.15) shows that the formation of nitric oxide is endothermic. The variation of $\ln K_p$ with $1/T$ is linear and the slope of the line is $-\Delta H/R$. Figure 5.2 shows the plot of (5.17) for nitric oxide; ΔH is $182.9\,\text{kJ mol}^{-1}$.

5.3.2 *Gaseous systems – change in the total number of molecules*

Consider the gas phase reaction between nitrogen and hydrogen:

$$N_2 + 3H_2 \rightleftharpoons 2NH_3 \tag{5.18}$$

This reaction forms the basis of the industrial synthesis of ammonia. Let there be initially a moles of nitrogen and b moles of hydrogen in the system. At equilibrium, let y moles of nitrogen be converted to ammonia. The number of moles of each species is shown in Table 5.2.

Figure 5.2. Variation of K_p with T for the nitric oxide equilibrium.

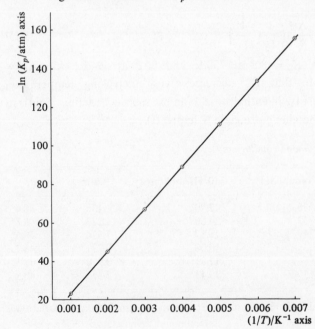

The total number of moles at equilibrium is $(a+b-2y)$; K_p is given by

$$K_p = \frac{p^2(\mathrm{NH}_3)}{p(\mathrm{N}_2)p^3(\mathrm{H}_2)} \tag{5.19}$$

In contrast to the hydrogen–iodine reaction, the equilibrium composition for the nitrogen–hydrogen reaction depends upon the total pressure, P. Substituting the equilibrium concentrations into (5.19), K_p may be written as

$$K_p = \frac{4y^2(a+b-2y)^2}{P^2(a-y)(b-3y)^3} \tag{5.20}$$

The data in Table 5.3 relate to the nitrogen–hydrogen reaction.

Clearly, an increase in pressure increases the yield of ammonia, but an increase in temperature decreases the yield of ammonia. The formation of ammonia is exothermic, unlike that of nitric oxide.

5.4 Principle of mobile equilibrium

This principle, stated in Chapter 1, is a consequence of the second law of thermodynamics and may be applied to all systems at equilibrium. We have illustrated its application with (5.11) and (5.18), for example. The effects of changes in pressure and in temperature upon the position of equilibrium should not be confused with their effects upon the rate of reaction (or the rate of

Table 5.2. *Nitrogen–hydrogen–ammonia gaseous equilibrium*

	Number of moles of each species	
	Initially	Equilibrium
N_2	a	$(a-y)$
H_2	b	$(b-3y)$
NH_3	0	$2y$

Table 5.3. *The nitrogen–hydrogen equilibrium* $(a/b=\frac{1}{3})$

P/atm	10	50	100	10	50	100
T/K		Mole % NH_3 formed		Equilibrium constant, $10^4 K_p$/atm^2		
673	3.85	15.27	25.12	1.64	1.72	1.90
723	2.0	9.2	16.4	0.43	0.48	0.53
773	1.2	5.6	10.4	0.15	0.15	0.16

attainment of equilibrium). As discussed in the chapter on kinetics, an increase in temperature or pressure almost always increases the rate of a chemical reaction.

We consider again the nitrogen–oxygen and the nitrogen–hydrogen reactions which are re-stated here for convenience,

$$N_2 + O_2 \rightleftharpoons 2NO \qquad \Delta H \text{ positive} \tag{5.21}$$

$$N_2 + 3H_2 \rightleftharpoons 2NH_3 \qquad \Delta H \text{ negative} \tag{5.22}$$

The nitrogen–oxygen reaction is endothermic but there is no change in the total number of molecules. From the Le Chatelier principle, a change in pressure has no effect upon the position of equilibrium. An increase in temperature increases the rates of both the forward and backward reactions, and favours the formation of nitric oxide, the endothermic reaction; decreasing the temperature retards the rates of both reactions and increases the concentrations of nitrogen and oxygen in the equilibrium mixture. These conclusions are in agreement with the variation of K_p, (5.17), with temperature illustrated in Figure 5.2.

In the equilibrium of (5.22), changes both in pressure and in temperature will affect the composition of the equilibrium mixture. Since the reaction is exothermic, an increase in temperature favours the reaction which will absorb heat, that is the decomposition of ammonia into nitrogen and hydrogen; in this case K_p, (5.19), decreases with an increase in temperature. An increase in total pressure favours the reaction which is accompanied by a decrease in volume, that is the formation of ammonia. From (5.20), if the total pressure (P) is increased, then since K_p is constant at constant temperature, y must increase. The effects have been illustrated quantitatively by the data in Table 5.3.

5.5 Liquid systems

The esterification of ethanoic acid by ethanol was discussed at the beginning of this chapter. We consider it further as an example of a homogeneous liquid equilibrium. If the system is ideal, that is Raoult's law (Chapter 6) is applicable, then the activities in (5.7) may be replaced by the corresponding number of moles, n, of each species, leading to

$$K = \frac{n(CH_3CO_2Et)n(H_2O)}{n(CH_3CO_2H)n(EtOH)} \tag{5.23}$$

If there are present initially a moles of ethanoic acid and b moles of ethanol, and if y moles of ester are formed, then the number of moles are as shown in Table 5.4. Hence, from (5.23) the equilibrium constant may be expressed by

$$K = \frac{y^2}{(a-y)(b-y)} \tag{5.24}$$

It should be noted that K may be calculated from (5.23) with the concentration of each species expressed in moles per litre. This leads to the same result for this reaction as there is no change in the total number of molecules. If we consider the system

$$A(l) + B(l) \rightleftharpoons C(l)$$

with concentrations at equilibrium $(a-y)/V$, $(b-y)/V$, and y/V, where V is the volume, then the equilibrium constant in terms of concentration K_c is given by

$$K_c = \frac{yV}{(a-y)(b-y)}$$

and depends upon the volume units. This procedure can be misleading in other types of reactions and should be avoided.

In Table 5.5 data relating to the equilibrium in (5.23) have been listed. The variations in K could be due to experimental error and to departure from ideal behaviour. The equilibrium constant for this reaction has been found to vary little with change in temperature. Hence from the Le Chatelier principle the enthalpy change of the reaction should be small; this has been confirmed experimentally.

5.6 Heterogeneous equilibria

The characteristic feature of a heterogeneous system is the presence of two or more *phases*. A phase is a homogeneous and physically distinct part of a

Table 5.4. *Ethanoic acid–ethanol–ethyl ethanoate–water equilibrium*

	Number of moles of each species	
	Initially	Equilibrium
CH_3CO_2H	a	$a-y$
EtOH	b	$b-y$
CH_3CO_2Et	0	y
H_2O	0	y

Table 5.5. *The esterification of ethanoic acid with ethanol at 373 K*

a mole	b mole	y mole	K
1.00	0.19	0.18	4.0
1.00	0.32	0.29	3.9
1.00	0.48	0.41	4.1
1.00	1.00	0.67	4.1
1.00	2.00	0.85	4.2
1.00	8.00	0.97	4.5
			4.1 (Mean)

system separated from other phases by a definite bounding surface. Ice, liquid water and steam are three phases in the water system. This system has one component; the number of *components* of a system at equilibrium is the smallest number of independently variable constituents required to express the composition of each phase present. Comparing

$$H_2O(s) \rightleftharpoons H_2O(l) \rightleftharpoons H_2O(g) \tag{5.25}$$

and

$$CaCO_3(s) \rightleftharpoons CaO(s) + CO_2(g) \tag{5.26}$$

we note that the water system has one component and three phases but the calcium carbonate system has two components and three phases. For if we select certain masses of calcium carbonate and calcium oxide in (5.26), the mass of carbon dioxide is determined by the equilibrium conditions and is not independently variable. In the reaction between iron and steam,

$$3Fe(s) + 4H_2O(g) \rightleftharpoons Fe_3O_4(s) + 4H_2(g) \tag{5.27}$$

three components must be chosen, iron, water and hydrogen for example. Again if ammonium chloride is vaporized *in vacuo*, the system consists of one component. If ammonia gas or hydrogen chloride gas is admitted to the system, there are two components because the vapour no longer has a composition representative of the solid ammonium chloride in terms of

$$NH_4Cl(s) \rightleftharpoons NH_3(g) + HCl(g) \tag{5.28}$$

A third quantity is needed to describe a heterogeneous system completely, namely the *number of degrees of freedom*. The number of degrees of freedom of a system is the number of variable properties, such as temperature, pressure and composition, which must be specified in order to define completely each phase in a system at equilibrium. In a system of one component, for example the water system, each phase has two degrees of freedom; it is bivariant. Figure 5.3 illustrates the water system. Any general point such as m requires both the temperature and pressure at that point to be stated in order to specify it completely. At points such as m′ and m″, water is in equilibrium with ice, and with steam, respectively and we have two-phase systems. It is then necessary to specify only *T* or *P*; the system along the lines OA, OB, OC and OD is univariant. The three phases coexist at the triple point O, and the system is invariant at this point. Note that the slope of OD, d*P*/d*T*, is negative: the melting point of ice decreases with increasing pressure, as illustrated by the physical phenomenon of regelation.

The relationship between the number of phases (*p*), the number of components (*c*) and the number of degrees of freedom (*f*) is expressed by the phase rule,

$$p + f = c + 2 \tag{5.29}$$

We consider next the equilibrium in (5.26) in more detail. The law of mass action is applicable strictly to homogeneous systems but may be extended to heterogeneous systems by assuming that the activity of each solid phase is constant; if the pure solid is chosen as the standard state, its activity is unity. The equilibrium constant for (5.26) is given by

$$K = \frac{a(CO_2)a(CaO)}{a(CaCO_3)} \tag{5.30}$$

Assuming the standard state for the solids, and that the activity of an ideal gas is equal to its pressure, then at low pressure and high temperature where carbon dioxide may be considered to behave ideally, K may be written

$$K_p = p(CO_2) \tag{5.31}$$

The constancy of the activity of a solid at any given temperature results from its constant vapour pressure; it is independent of the number of moles of solid present. An alternative point of view is to imagine that the heterogeneous reaction takes place homogeneously in the gas phase. Thus, we are led to

$$K = \frac{p(CaO)p(CO_2)}{p(CaCO_3)} \tag{5.32}$$

The vapours of calcium carbonate and of calcium oxide are in equilibrium with the solids calcium carbonate and calcium oxide respectively. Hence $p(CaCO_3)$ and $p(CaO)$ are constant and independent of the amounts of solids present. Again, (5.31) follows. The equilibrium dissociation pressure of carbon dioxide in this system varies from about 1 mmHg at 800 K to about 3000 mmHg at 1300 K. Since ΔH is positive for this reaction, an increase in temperature increases the percentage decomposition of calcium carbonate at equilibrium.

The activity of a solid in contact with a liquid phase saturated with the solid is a constant at a given temperature. Consider the reaction between barium sulphate

Figure 5.3. The ice–water–steam equilibrium.

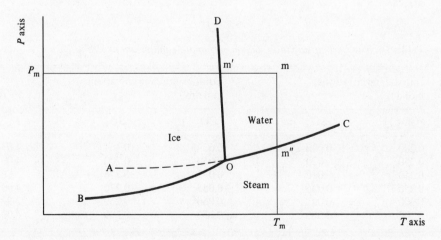

and sodium carbonate solution, with which we began this chapter, given by

$$BaSO_4(s) + CO_3^{2-}(aq) \rightleftharpoons BaCO_3(s) + SO_4^{2-}(aq) \tag{5.33}$$

The terms in the expression for the equilibrium constant of this reaction which involve the solids may be omitted and assuming that molar concentrations may replace activities, we obtain

$$K = \frac{[SO_4^{2-}]}{[CO_3^{2-}]} \tag{5.34}$$

The data in Table 5.6 relate to (5.33); the figures in column 2 refer to the concentrations of added sodium sulphate. The agreement in the values of K for different concentrations is very good bearing in mind the approximations made in the treatment. Closely related systems lead next to a discussion on ionic equilibria.

5.7 Ionic equilibria

Consider again (5.1). Silver chloride is a sparingly soluble electrolyte. The term 'sparingly soluble' cannot be defined exactly; it can be applied to those substances the concentrations of which at saturation are so low that their activities may be replaced by molar concentrations without appreciable error.

Since two or more ionic species must be present to preserve electrical neutrality, the relevant concentration here is the ionic strength I of the solution, defined by

$$I = \frac{1}{2} \sum_i c_i z_i^2 \tag{5.35}$$

where summation is over the i species of ions present, c_i is the concentration in mol dm^{-3}, and z_i the charge, of the ith species. Thus for c mol dm^{-3} potassium chloride, KCl, $I = c$; for c mol dm^{-3} potassium sulphate, K_2SO_4, $I = 3c$; and for a

Table 5.6. *The barium sulphate–sodium carbonate equilibrium at 298 K*

Initial concentrations mol dm^{-3}		Equilibrium concentrations mol dm^{-3}		
$[CO_3^{2-}]$	$[SO_4^{2-}]$	$[CO_3^{2-}]$	$[SO_4^{2-}]$	K
0.200	0.000	0.0395	0.161	4.07
0.250	0.000	0.050	0.200	4.00
0.350	0.000	0.072	0.278	3.86
0.250	0.025	0.055	0.220	4.00
0.300	0.025	0.066	0.259	3.93
0.200	0.050	0.050	0.200	4.00

solution of $c_1 \, \text{mol dm}^{-3}$ potassium chloride and $c_2 \, \text{mol dm}^{-3}$ potassium sulphate, $I = (c_1 + 3c_2)$. The replacement of activities by molar concentrations is usually permissible at low ionic strengths; for $I = 10^{-3} \, \text{mol dm}^{-3}$, the activity of univalent ions is $0.964 \times 10^{-3} \, \text{mol dm}^{-3}$. The activity of a pure solid or liquid, eg AgCl(s) or $\text{Br}_2(\text{l})$, is taken to be unity.

5.7.1 *Solubility product*

If s moles of silver chloride dissolve in $1 \, \text{dm}^3$ of water to form a saturated solution, and dissociate into s moles of each of the species Ag^+ and Cl^-, equilibrium is established

$$\text{AgCl(s)} \rightleftharpoons \text{Ag}^+(\text{aq}) + \text{Cl}^-(\text{aq}) \tag{5.36}$$

The equilibrium constant is given by

$$K = \frac{a(\text{Ag}^+)a(\text{Cl}^-)}{a(\text{AgCl})} \tag{5.37}$$

As explained above, $a(\text{AgCl})$ is defined as unity, and since silver chloride is sparingly soluble, the ionic activities may be replaced by their molar concentrations, leading to

$$K' = S = [\text{Ag}^+][\text{Cl}^-] \tag{5.38}$$

S is the solubility product of silver chloride, ie the product of the concentrations of the ions, in mol dm^{-3}, at saturation and at a given temperature for such sparingly soluble electrolytes. In the general case, for an electrolyte A_xB_y which ionizes

$$\text{A}_x\text{B}_y \rightarrow x\text{A}^{z+} + y\text{B}^{z-} \tag{5.39}$$

the solubility product S is given by

$$S = [\text{A}^{z+}]^x[\text{B}^{z-}]^y \tag{5.40}$$

If the solubility of A_xB_y is $s \, \text{mol dm}^{-3}$, then at saturation $[\text{A}^{z+}] = xs$, and $[\text{B}^{z-}] = ys$, so that

$$S = (x^x y^y)s^{x+y} \tag{5.41}$$

Example

(a) For calcium fluoride, $s = 2.04 \times 10^{-4} \, \text{mol dm}^{-3}$ at 298 K.

$$S = (2^2 \times 1)(2.04 \times 10^{-4})^3 \tag{5.42}$$
$$= 3.40 \times 10^{-11} \, \text{mol}^3 \, \text{dm}^{-9}$$

at this temperature.

(b) The solubility product of lead iodate, $\text{Pb(IO}_3)^2$, is $1.2 \times 10^{-13} \, \text{mol}^3 \, \text{dm}^{-9}$ at 298 K; hence its solubility is given by

$$s = \left(\frac{1.2 \times 10^{-13}}{4}\right)^{\frac{1}{3}} = 3.103 \times 10^{-5} \, \text{mol dm}^{-3}$$
$$= 0.0173 \, \text{g dm}^{-3}$$

Common ion effect

The so-called common ion effect is an application of the solubility product principle and does not really merit a separate title. Several good examples of this effect are afforded by the methods of classical qualitative analysis. Consider the separation of the ions Fe^{3+}, Al^{3+} and Cr^{3+} (analytical group III). The hydroxides of the cations of analytical groups III–V and of magnesium can be precipitated by ammonia:

$$Al^{3+} + 3OH^- \rightleftharpoons Al(OH)_3 \tag{5.43}$$

$$Mg^{2+} + 2OH^- \rightleftharpoons Mg(OH)_2 \tag{5.44}$$

The equilibrium between water and ammonia is expressed by

$$NH_3 + H_2O \rightleftharpoons NH_4^+ + OH^- \tag{5.45}$$

Nmr studies of the rate of isotope exchange for the ammonium ion suggest that the ammonia molecule is hydrogen bonded to three water molecules; there is no evidence in favour of the existence of 'NH_4OH'. The equilibrium constant for the dissociation of the base, K_b, (5.45) is

$$K_b = \frac{[NH_4^+][OH^-]}{[NH_3]} \tag{5.46}$$

Since K_b is constant at a given temperature, an increase in the concentration of ammonium ions (the common ion here) produces a corresponding decrease in the concentration of hydroxide ions. In these circumstances, only the hydroxides of the cations of analytical group III are precipitated, being less soluble than those of analytical groups IV and V and of magnesium – see Appendix A11.

   In considering solubility products, we point out a common error. The application to sodium chloride and similar electrolytes of relatively high solubility is incorrect. The equilibrium

$$Na^+Cl^-(s) \rightleftharpoons Na^+(aq) + Cl^-(aq) \tag{5.47}$$

must be formulated in terms of activities:

$$K = \frac{a(Na^+)a(Cl^-)}{a(Na^+Cl^-)} \tag{5.48}$$

This equation is thermodynamically exact; its application requires a knowledge of the activities of the species present at saturation. The solubility of sodium chloride is about $6 \, mol \, dm^{-3}$ at 298 K, and at this concentration it is not permissible to replace activities by concentrations.

Effect of ionic strength

The solubility of an electrolyte is changed by the addition of 'indifferent' (non-common) ions; these increase the ionic strength (5.35) of the solution, and thus affect the activity coefficients of all the ions (see below, (5.52)). The effect is illustrated in Table 5.7, where the solubility of silver bromate is seen to increase with the addition of potassium nitrate to the solution. Following (5.37) we can write the equilibrium constant for the silver bromate system

$$K = \frac{a(\text{Ag}^+)a(\text{BrO}_3^-)}{a(\text{AgBrO}_3)} = \frac{[\text{Ag}^+][\text{BrO}_3^-]f(\text{Ag}^+)f(\text{BrO}_3^-)}{a(\text{AgBrO}_3)} \tag{5.49}$$

where generally $a(\text{i}) = [\text{i}]f(\text{i})$ where $f(\text{i})$ is the activity coefficient, a dimensionless quantity. Experimentally we can determine only the product $f(+)f(-)$, and the mean ionic activity coefficient is given by

$$f(\pm) = \{f(+)f(-)\}^{\frac{1}{2}} \tag{5.50}$$

We have defined the activity of a pure solid as unity, hence

$$K = [\text{Ag}^+][\text{BrO}_3^-]f^2(\pm) \tag{5.51}$$

The value of $f(\pm)$ depends upon the total ionic strength, as defined by (5.35); it is given approximately, for very dilute solutions, by the Debye–Hückel limiting law for activity coefficients

$$\log f(\pm) = -0.509z(+)z(-)(I)^{\frac{1}{2}}$$

or

$$\ln f(\pm) = -1.172z(+)z(-)(I)^{\frac{1}{2}} \tag{5.52}$$

where $z(+)$ and $z(-)$ are the numerical charges on the cation and anion respectively.

For the silver bromate–potassium nitrate system the relevant values of I are listed in Table 5.7. Using these values in (5.52) the data of Table 5.8 were obtained. Since $f(\pm)$ decreases with increasing I, and K is the thermodynamic constant at a given temperature, the solubility of silver bromate increases with increasing concentration of potassium nitrate. The effect of the ionic strength of the saturated solution of silver bromate is evident at $[\text{KNO}_3] = 0$; for salts of lower solubility, this effect is correspondingly less important. Thus, the solubility of silver chloride is about 1.3×10^{-5} mol dm^{-3} at 298 K, and $f(\pm) = 0.996$ for the saturated solution. This demonstrates that as $c \to 0$ so $f(\pm) \to 1$, as already stated. The values of $f(\pm)$ may be calculated by (5.52) for univalent ions in aqueous solution at 298 K up to about 5×10^{-3} mol dm^{-3}. The calculation may be extended to 0.1 mol dm^{-3} by the empirical equation of C. W. Davies

$$\log f(\pm) = -0.50z(+)z(-)\left\{\frac{I^{\frac{1}{2}}}{(1+I^{\frac{1}{2}})} - 0.30I\right\} \tag{5.53}$$

Table 5.7. *The solubility of silver bromate in the presence of potassium nitrate at 298 K*

$[\text{KNO}_3]$/mol dm^{-3}	$[\text{AgBrO}_3]$/mol dm^{-3}	I/mol^2 dm^{-6}
0.00	0.0083	0.0083
0.026	0.0090	0.0350
0.047	0.0095	0.0565
0.100	0.0103	0.1103
0.139	0.0106	0.1500

We consider next a system involving two sparingly-soluble electrolytes with an ion in common; a practical example is afforded by the Mohr titration, in which the chloride ion concentration is determined by titration with silver ions, using chromate ions as an indicator, in neutral solution.

$$AgCl \rightleftharpoons Ag^+ + Cl^- \qquad S(AgCl) = 1.7 \times 10^{-10} \, mol^2 \, dm^{-6} \qquad (5.54)$$

$$Ag_2CrO_4 \rightleftharpoons 2Ag^+ + CrO_4^{2-} \quad S(Ag_2CrO_4) = 1.9 \times 10^{-12} \, mol^3 \, dm^{-9} \qquad (5.55)$$

Although the solubility product of silver chromate is the lower, silver chloride is precipitated first – its solubility is the lower:

$$s(AgCl) = S^{\frac{1}{2}}(AgCl) = 1.3 \times 10^{-5} \, mol \, dm^{-3}$$

$$= 1.9 \times 10^{-3} \, g \, dm^{-3}$$

$$s(Ag_2CrO_4) = \left\{ \frac{S(Ag_2CrO_4)}{4} \right\}^{\frac{1}{3}} = 7.5 \times 10^{-5} \, mol \, dm^{-3}$$

$$= 2.5 \times 10^{-2} \, g \, dm^{-3}$$

Both of the equilibria in (5.54) and (5.55) are satisfied simultaneously in the analytical system. The concentration of silver ion is given by

$$[Ag^+] = \frac{S(AgCl)}{[Cl^-]} = \left\{ \frac{S(Ag_2CrO_4)}{[CrO_4^{2-}]} \right\}^{\frac{1}{2}} \, mol \, dm^{-3} \qquad (5.56)$$

Thus, no silver chromate will be precipitated until $[Cl^-] = 1.2 \times 10^{-4}[CrO_4^{2-}]^{\frac{1}{2}}$; if, as usual, $[CrO_4^{-2}] = 10^{-2} \, mol \, dm^{-3}$ approximately, ie a few drops of the indicator solution are added, the precipitation of silver chloride is quantitatively complete.

Gross inaccuracy would result if the titration were carried out in the presence of acid. Further equilibria would then be involved:

$$CrO_4^{2-} + H^+ \rightleftharpoons HCrO_4^- \qquad (5.57)$$

$$2HCrO_4^- \rightleftharpoons H_2O + Cr_2O_7^{2-} \qquad (5.58)$$

The formation of the ions $HCrO_4^-$ and $Cr_2O_7^{2-}$ in the presence of hydrogen ions causes more silver chromate to dissociate, since $S(Ag_2CrO_4)$ is constant, at a given temperature. This is the rationale for the statement that 'all dichromates are

Table 5.8. *Variation of mean ionic activity coefficient with ionic strength*

$I/mol^2 \, dm^{-6}$	$f(\pm)$
0.0083	0.899
0.0350	0.803
0.0565	0.757
0.1103	0.678
0.1500	0.635

soluble'. The enhancement of the solubilities of silver ethanoate by nitric acid and of barium sulphate by sulphuric acid are due to the establishment of similar equilibria.

If the solubility of silver chloride in varying concentrations of chloride ion is measured, it will be found that the solubility decreases at first, then passes through a minimum and finally increases, with increasing chloride ion concentration, Figure 5.4. Common-ion action alone, curve (a), predicts that $s \propto 1/[Cl^-]$. The inclusion of the mean ionic activity coefficient, calculated from the ionic strength using equations (5.35) and (5.52), into the calculation of s produces curve (b) with a decrease in slope at the higher $[Cl^-]$; the effect of the common ion cannot be overcome by ionic strength corrections. Neither curve (a) nor curve (b) represents the measured solubility, curve (c). A further explanation could be the formation of complex ions, in which silver ions are taken into solution as an anionic complex,

$$Ag^+ + 2Cl^- \rightleftharpoons (AgCl_2)^- \tag{5.59}$$

There is no reason to suppose that molecules of AgCl are formed in solution.

5.7.2 Acid–base equilibria

Water is ionized to a small but finite extent in accordance with

$$2H_2O \rightleftharpoons H_3O^+ + OH^- \tag{5.60}$$

The hydrogen ion (which is a bare proton) is always hydrated by water, so that H_3O^+ is a more precise designation than H^+. For general purposes, however, the

Figure 5.4. The solubility of silver chloride in varying concentrations of chloride ion.

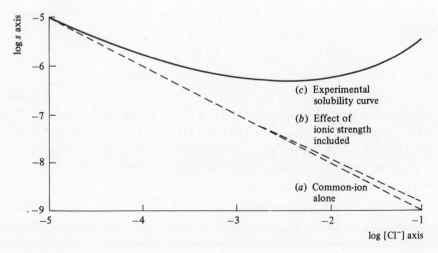

(c) Experimental solubility curve

(b) Effect of ionic strength included

(a) Common-ion alone

simpler designation is acceptable, and the equilibrium constant is given by

$$K = \frac{a(H^+)a(OH^-)}{a(H_2O)} \tag{5.61}$$

In dilute solutions, water is in excess, and its concentration (and activity) does not change appreciably with change of solute concentration; we may write

$$K[H_2O] = [H^+][OH^-] \tag{5.62}$$

in which $[H_2O]$ is approximately $55.6 \, \text{mol dm}^{-3}$, at 298 K. Under these conditions, $K = 1.80 \times 10^{-16} \, \text{mol dm}^{-3}$ and thus $K[H_2O] = 1.00 \times 10^{-14} \, \text{mol}^2$ dm^{-6}. This quantity is known as the *ionic product* for water and is expressed by:

$$K_w(298 \, \text{K}) = [H^+][OH^-] = 1.00 \times 10^{-14} \, \text{mol}^2 \, \text{dm}^{-6} \tag{5.63}$$

From (5.60), $[H^+] = [OH^-]$ and from (5.63), each is equal to $10^{-7} \, \text{mol dm}^{-3}$. Equation (5.60) represents the *autoprotolysis* of a protic solvent, ie a solvent which produces solvated protons. Another example is sulphuric acid:

$$2H_2SO_4 \rightleftharpoons H_3SO_4^+ + HSO_4^- \tag{5.64}$$

By comparison with (5.63)

$$K(H_2SO_4) = [H_3SO_4^+][HSO_4^-] \tag{5.65}$$

At 298 K, $K(H_2SO_4) = 2.4 \times 10^{-4} \, \text{mol}^2 \, \text{kg}^{-2}$; this constant is called the autoprotolysis constant of sulphuric acid. At 298 K, $pK(H_2SO_4)$, ie $-\log\{K(H_2SO_4)/\text{mol}^2 \, \text{kg}^{-2}\} = 3.6$, compared to $pK_w = 14.0$.

It is interesting that nitric acid does not undergo this equilibrium; presumably the solvated proton $H_2NO_3^+$ is unstable, and dissociates

$$2HNO_3 \rightleftharpoons H_2NO_3^+ + NO_3^- \rightarrow NO_2^+ + NO_3^- + H_2O \tag{5.66}$$

For a discussion of equilibria in sulphuric and nitric acids, see W. H. Lee, in *The chemistry of non-aqueous solvents*, Chapters 3 and 4, ed. J. J. Lagowski, Vol. II, Academic Press, NY, 1967.

pH notation – strong acids and strong bases

It is usual to express $[H^+]$ and $[OH^-]$ according to the pH scale: pH may be defined by

$$pH = -\log[H^+] \quad pOH = -\log[OH^-] \tag{5.67}$$

We adopt the notation pK, defined in a similar manner by

$$pK = -\log K \tag{5.68}$$

An aqueous solution having a pH of 7 at 298 K is neutral, ie $[H^+] = [OH^-]$; if $pH > 7$, the solution is alkaline, ie $[OH^-] > [H^+]$ and if $pH < 7$, the solution is acid, ie $[H^+] > [OH^-]$. Pure water is a neutral solution; its pH is 7, which is also its pOH. When water is in equilibrium with the atmosphere it absorbs carbon dioxide, and the equilibrium

$$CO_2 + H_2O \rightleftharpoons H^+ + HCO_3^- \tag{5.69}$$

produces a hydrogen ion concentration of $10^{-5} \, \text{mol dm}^{-3}$ at 298 K. Thus the pH of this 'equilibrium water' is 5.0. From (5.63) we can write, for dilute aqueous solutions

$$pH + pOH = 14 \qquad (5.70)$$

Furthermore, the electroneutrality, or charge balance, of aqueous solutions may be formulated generally by

$$a[A^{a+}] + b[B^{b+}] + c[C^{c+}] + \cdots = l[L^{l-}] + m[M^{m-}] + n[N^{n-}] + \cdot (5.71)$$

It is convenient to classify electrolytes as *strong* or *weak*. A strong electrolyte is highly dissociated in dilute solution, for example hydrogen chloride, sodium sulphate or ammonium nitrate. A weak electrolyte is only slightly dissociated in solution, for example hydrogen fluoride, ethanoic acid, ammonia or pyridine (C_5H_5N). Most, but not all, salts are strong electrolytes; mercuric chloride is one of the exceptions. It is noteworthy that the solvent is important in determining the strength of an electrolyte. Thus hydrogen chloride is a strong electrolyte in water but a weak electrolyte in glacial ethanoic acid. This is due to the very different abilities of the two solvents to accept a proton.

$$HCl + H_2O \rightleftharpoons H_3O^+ + Cl^- \qquad (5.72)$$

$$HCl + CH_3CO_2H \rightleftharpoons CH_3CO_2H_2^+ + Cl^- \qquad (5.73)$$

The strength of an acid or base is not the same as its concentration; it refers to the extent of dissociation in solution at the given concentration. We may note that pure nitric acid is a very weak acid, and is not strong enough to protonate the water formed in its self-dissociation, (5.66). The dissociation of a base also produces solvated ions, so that again the solvent determines the strength of the base in that solution. For example, pyridine is a strong base in ethanoic acid, but a weak base in water

$$C_5H_5N + CH_3CO_2H \rightleftharpoons C_5H_5NH^+ + CH_3CO_2^- \qquad (5.74)$$

$$C_5H_5N + H_2O \rightleftharpoons C_5H_5NH^+ + OH^- \qquad (5.75)$$

The terms strong and weak are not synonymous with concentrated and dilute, and should not be so used.

We consider next the relationship between the pH of a strong acid or a strong base and its stoichiometric concentration. If we assume complete dissociation of an acid in water, then if its concentration is c mole dm^{-3}, the concentration of H^+ due to the added acid is also c mole dm^{-3}. The following equations may be set up:

$$[H^+][OH^-] = K_w \qquad (5.76)$$

$$[H^+]_{total} = [X^-] + [OH^-] \text{ (electroneutrality)} \qquad (5.77)$$

$$= c + [OH^-] \qquad (5.78)$$

From (5.76) and (5.78) we obtain

$$[H^+]^2 - c[H^+] - K_w = 0 \qquad (5.79)$$

$$[H^+] = \frac{c}{2} \pm \left\{ \left(\frac{c}{2} \right)^2 + K_w \right\}^{\frac{1}{2}} \qquad (5.80)$$

If $c \gg 10^{-14}$, then K_w in (5.80) is negligible with respect to $(c/2)^2$, so that

$$[H^+] = c \tag{5.81}$$

$$pH = -\log c \tag{5.82}$$

$[H^+]$ is given by (5.81) and pH by (5.82). If $c^2 \ll 10^{-14}$, then $[H^+]$ and pH are given by,

$$[H^+] = K_w^{\frac{1}{2}} \tag{5.83}$$

$$pH = \tfrac{1}{2}pK_w \tag{5.84}$$

Figure 5.5 shows the variation of pH with $[H^+]$ for a strong acid. It is convenient to write Equation (5.79)

$$c = [H^+] - \frac{K_w}{[H^+]} \tag{5.85}$$

in order to obtain this curve. The graph shows clearly that in calculating the pH of a strong acid, the contribution to the total hydrogen ion concentration from the ionization of water may be neglected until c is less than about 10^{-6} mol dm^{-3}.

Examples

 (a) Calculate the pH for 0.005 mol dm^{-3} hydrochloric acid
  c/mol dm^{-3} = 0.005 assuming complete dissociation.
  From (5.67)
  $pH = -\log 0.005 = \bar{3}.6990 = -2.3010$
  $pH = 2.30$

Figure 5.5. pH of a strong acid as a function of concentration.

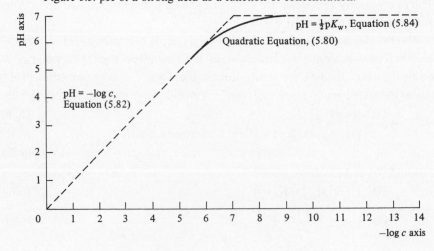

(b) Calculate the pH for 10^{-7} mol dm^{-3} nitric acid.

From (5.80)

$$[H^+] = 0.5 \times 10^{-7} \pm (0.25 \times 10^{-14} + 1.00 \times 10^{-14})^{\frac{1}{2}}$$

$$= 1.62 \times 10^{-7} \text{ mol dm}^{-3} \text{ (the negative root was ignored)}$$

$$pH = 6.79$$

(c) Calculate the pH for 0.01 mol dm^{-3} sodium hydroxide, assuming complete dissociation.

$$[OH^-] = 0.01$$

$$[H^+] = \frac{K_w}{[OH^-]} = 10^{-12}$$

$$pH = 12.00$$

Weak acids, weak bases and their salts

A weak acid is incompletely dissociated; its equilibrium may be represented by

$$HA \rightleftharpoons H^+ + A^- \tag{5.86}$$

where HA is the weak acid. Since the acid is weak, the law of mass action may be applied and the equilibrium constant formulated in terms of concentrations by

$$K_a = \frac{[H^+][A^-]}{[HA]} \tag{5.87}$$

K_a is the dissociation constant of the acid, at a given temperature. The following equations may be written for an aqueous solution of a weak acid:

$$K_a[HA] = [H^+][A^-] \tag{5.88}$$

$$K_w = [H^+][OH^-] \tag{5.89}$$

$$[H^+] = [A^-] + [OH^-] \text{ (electroneutrality)} \tag{5.90}$$

$$[A^-] + [HA] = c \text{ (mass balance)} \tag{5.91}$$

c is the total concentration of HA considered. The general solution is given by

$$[H^+]^3 + K_a[H^+]^2 - (cK_a + K_w)[H^+] - K_a K_w = 0 \tag{5.92}$$

Certain simplifying assumptions can be made for this system. Since the solution is acidic, $[OH^-]$ is very much less than $[A^-]$. Hence from (5.90) and (5.88)

$$[H^+] = [A^-] \tag{5.93}$$

and

$$K_a = \frac{[H^+]^2}{c - [H^+]} \tag{5.94}$$

may be derived. If further we assume that, because the acid is weak, $[H^+]$ is small compared with c, then from (5.94) we derive

$$[H^+] = (K_a c)^{\frac{1}{2}} \tag{5.95}$$

and

$$pH = \frac{1}{2}pK_a - \frac{1}{2}\log c \tag{5.96}$$

For ethanoic acid, $K_a = 1.8 \times 10^{-5}$ and pK_a is 4.75. The variation of pH with c is illustrated by Figure 5.6. In order to calculate these curves, (5.92) and (5.94) are conveniently re-cast in terms of c, in the manner of (5.85). The circumstances requiring the accurate formulation of the dependence of pH upon concentration can be appreciated readily from the curves.

The system weak base–water can be treated in a similar manner. We shall consider ammonia and pyridine (Py) as examples of weak bases.

$$NH_3 + H_2O \rightleftharpoons NH_4^+ + OH^- \tag{5.97}$$

$$Py + H_2O \rightleftharpoons PyH^+ + OH^- \tag{5.98}$$

The general base can be represented by

$$B + H_2O \rightleftharpoons BH^+ + OH^- \tag{5.99}$$

Equations for $[OH^-]$ and $[H^+]$, corresponding to (5.92), (5.94) and (5.95) can be readily derived. Considering $0.1 \, mol \, dm^{-3}$ ammonia solution we may use

$$[OH^-]^2 = 0.1 K_b \tag{5.100}$$

to deduce that the pH of the solution is 11.13. The more accurate form of the equation, that is

$$K_b = \frac{[OH^-]^2}{c - [OH^-]} \tag{5.101}$$

leads to a pH of 11.12 for the same solution of concentration $c = 0.1 \, mol \, dm^{-3}$.

Salt hydrolysis

The salts of a strong base and a weak acid and of a strong acid and a weak base are both hydrolyzed by water and the resulting solutions are alkaline and acidic

Figure 5.6. Variation of pH of ethanoic acid with concentration.

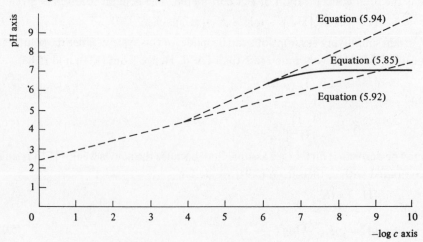

respectively. Evidently, A^- acts as a weak base and BH^+ acts as a weak acid.

$$A^- + H_2O \rightleftharpoons HA + OH^- \tag{5.102}$$

$$BH^+ + H_2O \rightleftharpoons B + H_3O^+ \tag{5.103}$$

The Brønsted–Lowry definition of acids and bases incorporates this more general implication: *an acid is a proton donor and a base is a proton acceptor*. The removal of a proton from any acid produces its conjugate base; eg the A^- species is the conjugate base of the acid HA. Consider the equilibrium

$$NH_4^+ + H_2O \rightleftharpoons NH_3 + H_3O^+ \tag{5.104}$$

$$K = \frac{[NH_3][H_3O^+]}{[NH_4^+][H_2O]} \tag{5.105}$$

The equilibrium constant, given by (5.105), may be related to K_w and K_b by multiplying the right-hand side of (5.105) by $[OH^-]/[OH^-]$. From (5.46) and (5.63) we can deduce

$$K_h = K[H_2O] = \frac{[NH_3][H_2O][H^+][OH^-]}{[NH_4^+][OH^-]} = \frac{K_w}{K_b} \tag{5.106}$$

K_h has been called the hydrolysis constant of the 'acid' NH_4^+. In a similar manner the hydrolysis constant of the 'base' Eth^- (ethanoate),

$$Eth^- + H_2O \rightleftharpoons HEth + OH^- \tag{5.107}$$

is given by

$$K_h = K_w/K_a \tag{5.108}$$

Example

The pK_a of methanoic acid (HMeth) is 3.75. What is the pH of a 0.001 mol dm^{-3} solution of sodium methanoate?

$$Meth^- + H_2O \rightleftharpoons HMeth + OH^- \tag{5.109}$$

We can construct five equations to represent this system:

$$K_a[HMeth] = [H^+][Meth^-] \tag{5.110}$$

$$K_w = [H^+][OH^-] \tag{5.111}$$

$$[Na^+] = 0.001 \tag{5.112}$$

$$[Meth^-] + [HMeth] = 0.001 \text{ (mass balance)} \tag{5.113}$$

$$[H^+] + [Na^+] = [OH^-] + [Meth^-] \text{ (electroneutrality)} \tag{5.114}$$

From (5.112)–(5.114), we may write

$$[H^+] + [HMeth] = [OH^-] \tag{5.115}$$

Certain simplifications, may be introduced: since the solution is basic, $[H^+]$ is very much less than $[Meth^-]$. Hence, from (5.115) we may write

$$[HMeth] = [OH^-] \tag{5.116}$$

Again [HMeth] is very much less than $[Meth^-]$ and thus, from (5.113) we may equate $[Meth^-]$ to 0.001. Finally, from (5.109)

$$K_h = \frac{K_w}{K_a} = K[H_2O] = \frac{[H^+][OH^-][HMeth]}{[H^+][Meth^-]} \tag{5.117}$$

$$K_h = \frac{[OH^-]^2}{0.010} \tag{5.118}$$

$$pK_w - pK_a = 2pOH + \log(0.001)$$
$$pH = 14 - (7.00 - 1.875 + 1.5)$$
$$= 7.37(5)$$

The full calculation of the pH of this system involves solving (5.110)–(5.114); we can derive

$$[H^+]^3 + (c + K_a)[H^+]^2 - K_w[H^+] - K_a K_w = 0 \tag{5.119}$$

If $[H^+]^3 \ll K_a K_w$ we can rewrite (5.119) as a quadratic in $[H^+]$:

$$(c + K_a)[H^+]^2 - K_w[H^+] - K_a K_w = 0 \tag{5.120}$$

which may be solved for $[H^+]$. Hence, pH = 7.40.

5.7.3 *Buffer solutions*

A study of weak electrolytes leads to a consideration of buffer solutions. In many chemical and biological investigations, it is often necessary to maintain a nearly constant pH during the course of a reaction which may produce, or may utilize, hydrogen ions. In these circumstances, the reaction is carried out in a buffer solution. A buffer solution contains a weak base and its salt, or a weak acid and its salt. In terms of the Brønsted–Lowry definition of acid and base, we conclude that a buffer solution is a mixture of a weak acid and its conjugate base or of a weak base and its conjugate acid. Two well-known examples are the ammonia and ethanoic acid systems NH_4^+–NH_3 and HEth–Eth$^-$. Buffer solutions permit the addition of small amounts of strong acid or strong base to them with only very slight change in the pH of the mixture. We consider first the system HEth–Eth$^-$.

Let the solution have the concentration 0.05 mol dm^3 in both ethanoic acid and sodium ethanoate; $K_a = 1.8 \times 10^{-5}$ mol dm^{-3}. Equations similar to (5.110)–(5.114) may be set up for this system. Again, we may make certain simplifications leading to

$$[Eth^-] = [HEth] = 0.05 \tag{5.121}$$

$$[H^+] = K_a = 1.8 \times 10^{-5} \tag{5.122}$$

$$pH = pK_a = 4.74(5) \tag{5.123}$$

Consider next the addition of 1 cm^3 of 1 mol dm^{-3} hydrochloric acid to 100 cm^3 of the buffer solution. Hydrogen ions from the hydrochloric acid will react with the ethanoate ions from the dissociated sodium ethanoate to give ethanoic acid (non-ionized). Since 0.001 mole of hydrogen ion is added the values of $[Eth^-]$ and

[HEth] are decreased and increased respectively by 0.001 mole. In addition the volume of solution has increased from 100 cm^3 to 101 cm^3. The new concentrations of ethanoate and ethanoic acid are given by

$$[\text{Eth}^-] = (0.050 - 0.001) \times 100/101 = 0.0485 \tag{5.124}$$

$$[\text{HEth}] = (0.050 + 0.001) \times 100/101 = 0.0504 \tag{5.125}$$

Using (5.87), the pH of the solution is given by:

$$\text{pH} = pK_a - \log(0.0504/0.0485) = 4.62(8) \tag{5.126}$$

We see that $\Delta \text{pH} = -0.117$ units. If there had been no buffer action, the pH would have been that of a solution of $0.001 \text{ mol dm}^{-3}$ of hydrochloric acid in 101 cm^3, that is of $0.0099 \text{ mol dm}^{-3}$ and the $\text{pH} = 2$. Equation (5.126) may be generalized as

$$\text{pH} = pK_a + \log([\text{salt}]/[\text{acid}]) \tag{5.127}$$

In this form it is often known as Henderson's equation. For the system $NH_4^+ - NH_3$, for example, a similar equation,

$$\text{pOH} = pK_b + \log([\text{salt}]/[\text{base}]) \tag{5.128}$$

may be derived leading to

$$\text{pH} = pK_w - pK_b - \log([\text{salt}]/[\text{base}]) \tag{5.129}$$

The simple equations apply well, within the hydrogen ion concentration ranges $10^{-3} - 10^{-6} \text{ mol dm}^{-3}$ and $10^{-8} - 10^{-11} \text{ mol dm}^{-3}$ respectively. Equation (5.127), for example, suggests that the pH is independent of concentration as long as the ratio [salt]/[acid] is a constant. This cannot be true at extreme dilution, since $\text{pH} \rightarrow 7.00$ as the concentrations tend to zero.

The Van Slyke buffer index

The capacity of a buffer to resist the addition of strong acid or strong base can be measured by the amount of strong acid or strong base required to produce a given change in pH. The Van Slyke *buffer index* β is given by

$$\beta = -\frac{dC_a}{d\text{pH}} = \frac{dC_b}{d\text{pH}} \tag{5.130}$$

The addition of C_a mole of strong acid to 1 dm^3 of buffer solution decreases the pH and increases the acid component of the system by dC_a mole at the expense of the conjugate base. Without proof here, the equation for β developed in terms of a weak acid and its salt is

$$\beta = 2.303 \left\{ \frac{K_w}{[\text{H}^+]} + [\text{H}^+] + \frac{cK_a[\text{H}^+]}{(K_a + [\text{H}^+])^2} \right\} \tag{5.131}$$

The first two terms on the right-hand side of Equation (5.131) are due to the buffering action of water; the third term is due to the acid–conjugate base pair. In the first example of a buffer system, we selected ethanoic acid and sodium ethanoate of the same concentration 0.05 mol dm^{-3}. Thus, c in Equation (5.131), which represents the total ethanoate concentration, is 0.1 mol dm³. Hence $\beta =$ $-dC_a/dpH = -\Delta C_a/\Delta pH$. From (5.131), $\beta = 0.576 \times 10^{-2}$ and $\Delta pH = -0.174$, in good agreement with the previous result. The equation for β developed for a weak base and its salt is given by (5.132).

$$\beta = 2.303 \left\{ \frac{K_w}{[H^+]} + [H^+] - \frac{cK_b^2}{\left(\frac{K_w}{[H^+]} + K_b\right)^2} + \frac{cK_b}{\left(\frac{K_w}{[H^+]} + K_b\right)} \right\} \qquad (5.132)$$

The variation of β with pH is illustrated by Figures 5.7–5.9: ammonia, pyridine, ethanoic acid and phenol are represented at various concentrations c. The maximum useful buffer capacity occurs at the pK_a or the pK_b of the system under consideration; at these values of pH, the concentrations of salt and acid or of salt and base respectively are equal. A ten-fold decrease in the salt concentration produces about a ten-fold decrease in the value of β. Minimum buffer action is found in the pH regions corresponding to the acid and the conjugate base. For example, with 0.05 mol dm⁻³ ethanoic acid and 0.05 mol dm⁻³ sodium ethanoate, the minima in β occur at pH 3.03 for the acid, HEth, and at pH 8.72 for the conjugate base, Eth⁻. A 0.05 mol dm⁻³ solution of ethanoic acid has a buffer index of about 5×10^{-3} and a 0.05 mol dm⁻³ solution of sodium ethanoate has a buffer index of 2×10^{-5}. The pH of a solution of sodium ethanoate is thus very sensitive to the presence of atmospheric carbon dioxide, which would invalidate its accurate measurement. The system HEth–Eth⁻ cannot provide a

Figure 5.7. Buffer index for (a) 0.05 mol dm⁻³ ammonia–0.05 mol dm⁻³ hydrochloric acid, (b) 0.005 mol dm⁻³ ammonia–0.005 mol dm⁻³ hydrochloric acid.

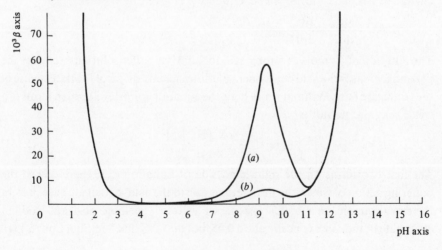

solution of pH outside the approximate range 3–9, unless strong acid or strong base is added. This is indicated by the rapid rise in β outside these limits (Figure 5.8). Simple deductions may be made in respect of the other systems from the curves in Figures 5.7 and 5.9.

5.7.4 *Weak acid–weak base*

If the acid and the base forming a salt are both weak, for example ammonium ethanoate, then both the acid and the conjugate base interact with

Figure 5.8. Buffer index for 0.05 mol dm^{-3} ethanoic acid–0.05 mol dm^{-3} sodium hydroxide.

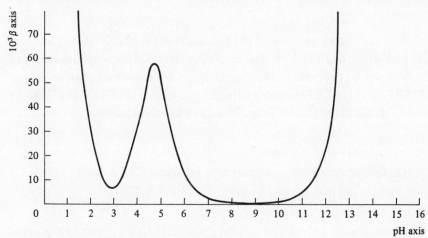

Figure 5.9. Buffer index for (*a*) 0.05 mol dm^{-3} phenol–0.05 mol dm^{-3} sodium hydroxide, (*b*) 0.005 mol dm^{-3} phenol–0.005 mol dm^{-3} sodium hydroxide.

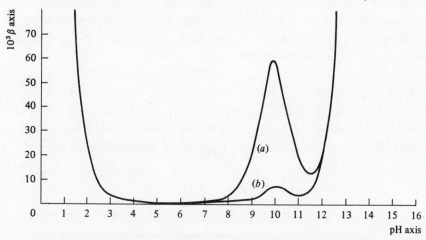

the solvent, as shown by

$$NH_4^+ + H_2O \rightleftharpoons H_3O^+ + NH_3 \tag{5.133}$$

$$Eth^- + H_2O \rightleftharpoons OH^- + HEth \tag{5.134}$$

These equilibria may be combined to give

$$NH_4^+ + Eth^- + 2H_2O \rightleftharpoons H_3O^+ + NH_3 + OH^- + HEth \tag{5.135}$$

or more simply

$$NH_4^+ + Eth^- \rightleftharpoons NH_3 + HEth \tag{5.136}$$

Assuming dilute solutions, the equilibrium constant for (5.136) may be formulated by

$$K_h = [NH_3][HEth]/[NH_4^+][Eth^-] \tag{5.137}$$

From (5.48), (5.63) and (5.87) we may derive

$$K_h = K_w/K_a K_b \tag{5.138}$$

Let the original concentration of the salt, ammonium ethanoate in this example, be c mol dm^{-3}. In general, the degrees of dissociation of the conjugate base and of the acid will be different; in this example they are assumed equal, say α. From (5.137) [NH$_3$] and [HEth] are equal, and equal to αc, and [NH$_4^+$] and [Eth$^-$] are equal and equal to $(1-\alpha)c$. Thus, from (5.138), we may derive

$$K_h = \frac{\alpha^2 c^2}{(1-\alpha)c} = \frac{\alpha^2 c}{(1-\alpha)} \tag{5.139}†$$

Certain further approximations may now be made: if $\alpha \ll 1$ then $(1-\alpha) \to 1$; α is now equal to $K_h^{\frac{1}{2}}$ and we may deduce

$$pH = \tfrac{1}{2}pK_w + \tfrac{1}{2}pK_a - \tfrac{1}{2}pK_b \tag{5.140}$$

For ethanoic acid and for ammonium ethanoate, $K_a \approx K_b$, and so the pH of a solution of ammonium ethanoate is approximately 7.0, although hydrolysis has taken place.

From (5.140), it seems that the pH is independent of the concentration of the solution. This is strictly true only if the solutions are dilute, ie if the activity coefficients are unity and if they are neutral. If $K_a < K_b$, the solution is alkaline, and if $K_a > K_b$, the solution is acid.

5.7.5 Dibasic acids

A dibasic acid can dissociate in two stages and two dissociation constants determine the composition of the system at equilibrium. We will consider the general case of an acid H$_2$A, which dissociates according to

$$H_2A \rightleftharpoons H^+ + HA^- \tag{5.141}$$

$$HA^- \rightleftharpoons H^+ + A^{2-} \tag{5.142}$$

† This type of equation is credited to Ostwald; (5.94) and (5.101) can be expressed in a similar way.

Five equations may be written to represent this system,

$$K_1[H_2A] = [H^+][HA^-] \tag{5.143}$$

$$K_2[HA^-] = [H^+][A^{2-}] \tag{5.144}$$

$$K_w = [H^+][OH^-] \tag{5.145}$$

$$c = [H_2A] + [HA^-] + [A^{2-}] \text{ (mass balance)} \tag{5.146}$$

$$[H^+] = [HA^-] + 2[A^{2-}] + [OH^-] \text{ (electroneutrality)} \tag{5.147}$$

and may be solved by the following stages:

$$c - [H^+] = [H_2A] - [OH^-] - [A^{2-}] \tag{5.148}$$

$$[H_2A] = c - [H^+] + \frac{K_w}{[H^+]} + [A^{2-}] \tag{5.149}$$

$$= c - [H^+] + \frac{K_w}{[H^+]} + \frac{K_2[HA^-]}{[H^+]} \tag{5.150}$$

$$= \frac{[H^+][HA^-]}{K_1} \tag{5.151}$$

$$[HA^-] = \frac{\left(c - [H^+] + \dfrac{K_w}{[H^+]}\right)}{\left(\dfrac{[H^+]}{K_1} - \dfrac{K_2}{[H^+]}\right)} \tag{5.152}$$

$$[A^{2-}] = \frac{K_2}{[H^+]} \frac{\left(c - [H^+] + \dfrac{K_w}{[H^+]}\right)}{\left(\dfrac{[H^+]}{K_1} - \dfrac{K_2}{[H^+]}\right)} \tag{5.153}$$

$$[H^+] = \frac{\left(c - [H^+] + \dfrac{K_w}{[H^+]}\right)}{\left(\dfrac{[H^+]}{K_1} - \dfrac{K_2}{[H^+]}\right)} + \frac{2K_2\left(c - [H^+] + \dfrac{K_w}{[H^+]}\right)}{[H^+]\left(\dfrac{[H^+]}{K_1} - \dfrac{K_2}{[H^+]}\right)} + \frac{K_w}{[H^+]} \tag{5.154}$$

The results of the application of (5.154) to the dibasic acids carbonic and oxalic are shown in Table 5.9, where the pH is listed as a function of the total acid concentration c. If K_2 is much smaller than K_1, and if the pH is less than about 5.5, then (5.154) may be approximated by

$$[H^+]^2 + K_1[H^+] - cK_1 = 0 \tag{5.155}$$

This equation is similar to (5.94) with $K_1 = K_a$; the dibasic acid is effectively considered as a monobasic acid. This simplification can be applied to carbonic acid but not to oxalic acid below a concentration of 10^{-2} mol dm^{-3}, as Table 5.10 shows. The dissociation constants used are listed in Appendix A12.

5.7.6 *Indicators*

An acid–base indicator is a weak acid or a weak base, and is slightly dissociated in solution. The acid and its conjugate base (or the base and its conjugate acid) are differently coloured. This dissociation can be represented by equations such as

$$HIn \rightleftharpoons H^+ + In^- \tag{5.156}$$

$$InOH \rightleftharpoons OH^- + In^+ \tag{5.157}$$

where HIn represents a weak acid indicator, such as methyl red, and InOH represents a weak base indicator, such as phenolphthalein. We shall consider the first of these in more detail.

The equilibrium constant for (5.156) is given by

$$K_{In} = \frac{[H^+][In^-]}{[HIn]} \tag{5.158}$$

We see that $[H^+]$, and thus pH, governs the ratio of $[In^-]/[HIn]$ for the system. In the case of methyl red, HIn is the red form and In^- is the yellow form. If a

Table 5.9. *Evaluation of pH for carbonic acid and oxalic acid at 298 K using (5.154)*

Carbonic acid		Oxalic acid	
$c/\text{mol dm}^{-3}$	pH	$c/\text{mol dm}^{-3}$	pH
1.0	3.18	1.0	0.67
10^{-1}	3.18	10^{-1}	1.28
10^{-2}	3.68	10^{-2}	2.06
10^{-3}	4.19	10^{-3}	2.98
10^{-4}	5.20	10^{-4}	3.88
10^{-5}	6.31	10^{-5}	4.75
10^{-6}	6.83	10^{-6}	5.70
0	7.00	0	7.00

Table 5.10 *Evaluation of Equation (5.155) for carbonic and oxalic acids at 298 K*

Carbonic acid		Oxalic acid	
$c/\text{mol dm}^{-3}$	pH	$c/\text{mol dm}^{-3}$	pH
10^{-1}	3.68	10^{-1}	1.28
10^{-2}	4.18	10^{-2}	2.05
10^{-3}	4.68	10^{-3}	3.01
10^{-4}	5.18	10^{-4}	4.00

fraction α of the total indicator present is dissociated, then $[H^+]$ from Equation (5.158) is given by

$$[H^+] = K_{In} \frac{(1-\alpha)}{\alpha} \tag{5.159}$$

If we assume that when α is less than 0.09, the In^- colour is just not visible to the eye, then the highest $[H^+]$ at which In^- can be detected is given by

$$[H^+] = K_{In} \frac{0.91}{0.09} \approx 10 K_{In} \tag{5.160}$$

The corresponding pH is $pK_{In} - 1$. Further, when 91% of the indicator is dissociated, the colour due to HIn is just not visible to the eye. In this circumstance we can deduce that $pH = pK_{In} + 1$. With these working assumptions we derive the pH range of an acid–base indicator as $pK_{In} \pm 1$. In Table 5.11 a number of acid–base indicators is listed. The pH ranges have been determined experimentally, using buffer solutions. They compare favourably with the values of $pK_{In} \pm 1$. In the case of phthalein indicators such as phenolphthalein the colour change is not simply due to ionization. The indicator exists in two tautomeric forms, the quinonoid forms (II) and (III) predominating in alkaline solution:

(I)   (II)   (III)

Here, $K_{In} = K_1 K_2$. The course of an acid–base titration may be followed by measuring the conductance of the solution (Section 7.4) or the potential of an electrode reversible to hydrogen ions (Section 7.15).

Table 5.11. *Data for acid–base indicators*

	pH range	pK_{In}
Methyl orange	2.9–4.0	3.7
Bromophenol blue	3.0–3.6	4.0
Bromocresol green	3.8–5.4	4.7
Methyl red	4.4–6.0	5.1
Bromothymol blue	6.0–7.6	7.0
Phenol red	6.8–8.4	7.6
Phenolphthalein	8.3–10.0	9.4
Thymolphthalein	9.3–10.5	9.9

5.7.7 *Titrations*

The changes in pH during the titration of an acid with a base can be calculated by means of equations already developed. Attention is drawn to Figures 5.5–5.8 with the relevant theory. We shall consider the titrations of acids and bases of different strengths in solutions of initial concentrations of $1 \, mol \, dm^{-3}$ and $0.01 \, mol \, dm^{-3}$.

Strong acid–strong base

The titration curves of $50 \, cm^3$ of $1 \, mol \, dm^{-3}$ hydrochloric acid with $1 \, mol \, dm^{-3}$ sodium hydroxide and of $50 \, cm^3$ of $0.01 \, mol \, dm^{-3}$ hydrochloric acid with $0.01 \, mol \, dm^{-3}$ sodium hydroxide are shown in Figure 5.10. The pH of the system changes rapidly in the neighbourhood of the equivalence point (in this case neutrality ie pH = 7). For practical purposes, the pH changes rapidly from three to eleven and from five to nine, for the $1 \, mol \, dm^{-3}$ and $0.01 \, mol \, dm^{-3}$ solutions respectively. Suitable indicators for the titrations are shown on the figure. We can see that the use of methyl orange with the $0.01 \, mol \, dm^{-3}$ solutions would lead to an erroneous result; the colour change of the indicator would occur before the equivalence point.

Weak acids and bases

We shall consider the pairs ethanoic acid–sodium hydroxide ($K_a = 1.8 \times 10^{-5}$)

Figure 5.10. pH titration curves for strong acid–strong base (*a*) $1 \, mol \, dm^{-3}$ solutions, (*b*) $0.01 \, mol \, dm^{-3}$ solutions.

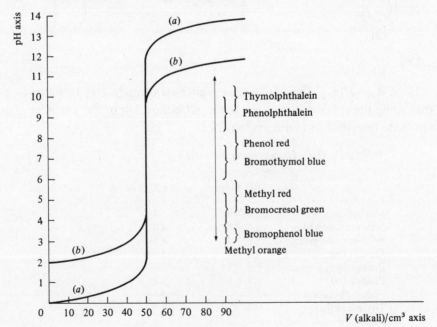

and pyridine–hydrochloric acid ($K_b = 1.7 \times 10^{-9}$). The titration curves are illustrated in Figure 5.11.

On addition of $0.1\,\text{mol}\,\text{dm}^{-3}$ sodium hydroxide to $0.1\,\text{mol}\,\text{dm}^{-3}$ ethanoic acid, the pH increases slowly at first. The slope then decreases; the pH is now passing through the buffer working-range. A rapid change in pH occurs between 7.5 and 9.5 and indicators such as phenolphthalein or thymolphthalein would be suitable for this titration. For concentrations other than $0.1\,\text{mol}\,\text{dm}^{-3}$, the change in pH near the equivalence point must be evaluated; a shortening of this range, for example with $0.01\,\text{mol}\,\text{dm}^{-3}$ solutions, would mean that phenolphthalein was the most suitable indicator. Continued addition of sodium hydroxide, after the equivalence point has been reached, produces only a slow increase in the pH of the solution – a solution of sodium hydroxide is effectively a buffer solution. The buffer index curve for this system (Figure 5.8) should be studied in conjunction with Figure 5.11.

The system pyridine–hydrochloric acid at $0.1\,\text{mol}\,\text{dm}^{-3}$ concentration shows a rapid decrease from pH 9 with addition of the acid. This decrease slows up through the buffer working range (pH 4–6, Figure 5.11). The change at the equivalence point is not very marked and no indicator is quite satisfactory for this titration; the best indicator is benzylaniline–azobenzene.

Figure 5.11. pH titration curves for $0.1\,\text{mol}\,\text{dm}^{-3}$ solutions (*a*) weak acid (ethanoic acid)–strong base, (*b*) weak base (pyridine)–strong acid.

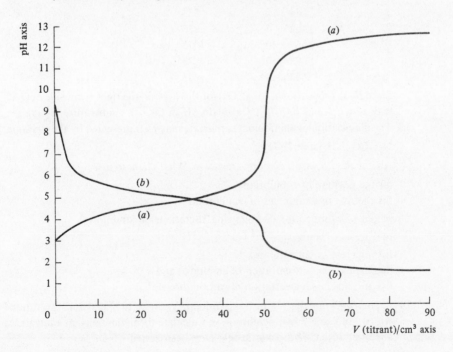

The system weak acid–weak base has not been considered in this section. It is of little importance in titrimetric analysis because the change in pH at the equivalence point is so gradual that no indicator is useful. Such a titration would not be performed with an indicator. However, it may be performed conductimetrically (see p. 249).

Problems 5

5.1 The following data refer to the equilibrium $N_2O_4 \rightleftharpoons 2NO_2$ at 298 K and 308 K, and P is the total pressure. Deduce the expression for K_p, evaluate K_p at both temperatures and calculate ΔH for the reaction in the given temperature interval, using (5.15).

	298 K		308 K	
Initial concentration of N_2O_4 mol dm^{-3}	N_2O_4/mol dm^{-3} decomposed at equilibrium	P/atm	N_2O_4/mol dm^{-3} decomposed at equilibrium	P/atm
6.28×10^{-3}	2.38×10^{-3}	0.212	3.11×10^{-3}	0.238
1.26×10^{-2}	3.62×10^{-3}	0.394	4.87×10^{-3}	0.440
1.98×10^{-2}	4.66×10^{-3}	0.600	6.40×10^{-3}	0.662

5.2 The following data refer to the calcium carbonate equilibrium, (5.26):

Interval	$p(CO_2)$/mmHg	T/K
1	7.73×10^{-2}	773
2	1.84	873
3	22.2	973
4	1.67×10^2	1073
5	7.93×10^2	1173
	2.94×10^3	1273

Since K_p is proportional to $p(CO_2)$ for this system, the right-hand side of (5.15) may be written as $\ln(p_2/p_1)$. Calculate ΔH in the five temperature intervals.

5.3 The dissociation of ammonium carbamate may be represented by the equation:

$$NH_4CO_2NH_2(s) \rightleftharpoons 2NH_3(g) + CO_2(g)$$

ΔH for the forward reaction is negative. What changes in

(a) the position of equilibrium

(b) the rate of attainment of equilibrium

would be expected for the following alterations of conditions:

(i)   decrease in pressure,

(ii) increase in temperature,

(iii) increase in concentration of ammonia, and

(iv) increase in concentration of carbon dioxide?

5.4 25 cm^3 of 0.1 mol dm^{-3} aqueous pyridine ($K_b = 1.7 \times 10^{-9}$ mol dm^{-3}) is titrated with 0.1 mol dm^{-3} hydrochloric acid. Calculate the pH initially, at equivalence and after 30 cm^3 hydrochloric acid have been added.

5.5 Given that the solubility product of barium sulphate is $1.1 \times 10^{-10}\,\text{mol}^2\,\text{dm}^{-6}$, calculate the solubility of barium sulphate in

(a) water

and in aqueous solutions of

(b) $10^{-3}\,\text{mol dm}^{-3}$ sodium sulphate

(c) $10^{-2}\,\text{mol dm}^{-3}$ sodium chloride.

5.6 Solve the quadratic equation $ax^2 + bx + c = 0$, with $a = 1$, $b = 123$ and $c = -200$,

(a) using the equation

$$x = \{-b \pm (b^2 - 4ac)^{\frac{1}{2}}\}/2a$$

(b) by successive approximations, using the equation in the form $x = -(c/b) - (a/b)x^2$. For the first approximation, let $x = -(c/b)$ on the right-hand side. Compare the results, and note the usefulness of (b) when $4ac \ll b^2$.

5.7 Calculate the pH of the following aqueous solutions: (a) $0.01\,\text{mol dm}^{-3}$ sodium hydroxide, (b) $0.005\,\text{mol dm}^{-3}$ sulphuric acid, (c) $1\,\text{cm}^3\,2\,\text{mol dm}^{-3}$ hydrochloric acid in $500\,\text{cm}^3$ water, (d) $10^{-9}\,\text{mol dm}^{-3}$ barium hydroxide.

5.8 The values of K_w for water at 273 K, 298 K and 333 K are 1.2×10^{-15}, 1.0×10^{-14} and $9.6 \times 10^{-14}\,\text{mol}^2\,\text{dm}^{-6}$ respectively. Calculate the pH of water at each temperature, and use Equation (5.15) to calculate an average value for the enthalpy of dissociation of one mole of water into its ions.

5.9 Determine the pH of the following aqueous solutions, using the dissociation constants listed in Appendix A12:

(a) $0.005\,\text{mol dm}^{-3}$ ethanoic acid

(b) $0.02\,\text{mol dm}^{-3}$ pyridine

(c) $0.1\,\text{mol dm}^{-3}$ ammonium nitrate

(d) $1.0\,\text{mol dm}^{-3}$ sodium benzoate.

5.10 Calculate the pH of a solution which contains $0.1\,\text{mol dm}^{-3}$ methanoic acid and $0.2\,\text{mol dm}^{-3}$ potassium methanoate. What is the pH change on addition of $2\,\text{cm}^3$ of $1.0\,\text{mol dm}^{-3}$ sodium hydroxide to $100\,\text{cm}^3$ of the solution?

5.11 Calculate the pH of a $0.05\,\text{mol dm}^{-3}$ solution of phosphoric acid; it may be assumed that the third dissociation of phosphoric acid is negligible.

5.12 By differentiating Equation (5.132) with respect to $[H^+]$, find the condition for maximum and minimum values of β. Show that if c is much greater than $[H^+]$, then $[H^+] = K_a$ is one solution (it corresponds, in fact, to the most useful value of β).

6

Physical properties of systems

The physical properties of any system may be classified generally into three main types: additive, constitutive and colligative.

An *additive* property is, for a given system, one which may be represented by the sum of the corresponding properties of its components. Mass is exactly additive but molar volume and ionic radii are only approximately additive. For example, the molar volume V_m (relative molar mass divided by density) increases regularly by ΔV_m along a homologous series for each methylene group increment (for liquids and solids):

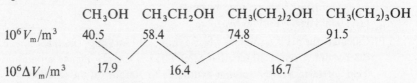

$$
\begin{array}{ccccc}
 & CH_3OH & CH_3CH_2OH & CH_3(CH_2)_2OH & CH_3(CH_2)_3OH \\
10^6 V_m/m^3 & 40.5 & 58.4 & 74.8 & 91.5 \\
10^6 \Delta V_m/m^3 & 17.9 & & 16.4 & 16.7
\end{array}
$$

Again, the interionic distances, r_e, in ionic crystals are additive functions of the appropriate ionic radii (see Chapter 4).

A *constitutive* property depends mainly upon the structural arrangements of atoms or molecules. The boiling point is an example of a constitutive property. The high boiling point of water, as compared to that of hydrogen sulphide for example, is due to the arrangement of the molecules in its hydrogen-bonded structure (see p. 2). Again, both the softness and lubricant property of graphite, and the hardness and abrasive property of diamond are intimately related to the arrangement of carbon atoms in each of these substances.

Colligative properties depend upon the number of molecules, or particles, in the system, rather than upon their nature. The pressure of an ideal gas at constant volume and constant temperature is one example of a colligative property.

In this chapter, we shall be concerned at first with the properties of dilute solutions. A solution is a homogeneous mixture consisting of one condensed phase and two or more components. In a binary solution, it is convenient to refer to the component in excess as the solvent, and to the other component as the solute. We shall see, however, that these terms are completely interchangeable.

6.1 Vapour pressure and Raoult's law

In dealing with mixtures of miscible liquids or with solid–liquid systems, it is convenient to express the composition of the system in terms of the *mole fraction* of each component. The mole fraction $x(A)$ of component A in a mixture containing $n(A)$ mole of species A, $n(B)$ mole of B etc, is given by

$$x(A) = \frac{n(A)}{n(A) + n(B) + \cdots} = \frac{n(A)}{\sum_I n(I)} \qquad (6.1)$$

where the sum is taken over the I components. From (6.1) it follows that

$$\frac{n(A)}{\sum_I n(I)} + \frac{n(B)}{\sum_I n(I)} + \cdots + \frac{n(I)}{\sum_I n(I)} = 1 \qquad (6.2)$$

We shall restrict our discussion mainly to two-component mixtures consisting of $n(A)$ mole of species A and $n(B)$ mole of species B. Hence

$$x(A) = \frac{n(A)}{n(A) + n(B)} \qquad (6.3)$$

$$x(B) = \frac{n(B)}{n(A) + n(B)} \qquad (6.4)$$

$$x(A) + x(B) = 1 \qquad (6.5)$$

The vapour pressure of a solvent is lowered when a solute is dissolved in it. This fact was given a quantitative basis by Raoult, in 1888.

He discovered that for dilute solutions of a solute B in a solvent A at a given temperature, the vapour pressure of A was proportional to the mole fraction of A:

$$p(A) = p_0(A)x(A) \qquad (6.6)$$

where $p_0(A)$ is the vapour pressure of the pure solvent. From (6.6)

$$1 - \{p(A)/p_0(A)\} = 1 - x(A) = x(B)$$

where $x(B)$ is the mole fraction of the solute. Hence

$$\Delta p(A) = p_0(A)x(B) \qquad (6.7)$$

where $\Delta p(A)$ is the lowering of vapour pressure.

Example

The vapour pressure of ethanol at 292 K is 40.00 mmHg, and that of water at the same temperature is 16.48 mmHg. For a solution containing 2 g of water in 50 g of ethanol

$$x(C_2H_5OH) = 0.9072$$

$$x(H_2O) = 0.0928$$

$$p(C_2H_2OH) = p_0(C_2H_5OH)x(C_2H_5OH) = 36.29 \text{ mmHg}$$

$$p(H_2O) = p_0(H_2O)x(H_2O) = 1.529 \text{ mmHg}$$

These are the partial vapour pressures of ethanol and water above the solution, assuming that the mixture behaves ideally, ie it obeys Raoult's law exactly.

In the above problem, the lowering of vapour pressure of ethanol by the addition of water implies a decrease in the *effective* concentration of ethanol in the system, ie a decrease in its activity.

The vapour pressure–composition isotherms for an ideal binary system are shown in Figure 6.1; the vapour pressure of pure B is shown as being greater than that of pure A. As the system is ideal, from (6.7) the variations of both $p(A)$ and $p(B)$ with composition are the straight lines (*a*) and (*b*) terminating in the points $p_0(A)$ and $p_0(B)$ for $x(A)$ and $x(B)$ equal to unity, respectively.

From Dalton's law of partial pressures the total pressure P, at any composition, is given by the sum $[p(A) + p(B)]$. Evidently this is the straight line (*c*), joining the points $p_0(A)$ and $p_0(B)$. A limited number of pairs of liquids are known which exhibit this ideal behaviour; examples are ethyl bromide and ethyl iodide at 303 K, and benzene and ethylene dichloride at 323 K.

6.1.1 *Non-ideal mixtures*

Most pairs of liquids form non-ideal solutions. Sometimes the partial vapour pressure of component A is increased by adding component B; a similar effect is then observed for the partial vapour pressure of component B. This behaviour is described as a positive deviation from Raoult's law and is illustrated in Figure 6.2. The total vapour pressure at all compositions is greater than that corresponding to ideality (except at $x(A) = 1$ and $x(B) = 1$). Positive deviations are exhibited by liquid mixtures in which the molecules repel each other, and so enter

Figure 6.1. Vapour pressure–composition diagram for an ideal binary system.

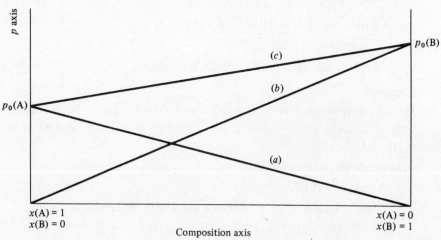

the vapour phase more readily than in the pure liquids. The pairs of liquids pentane–ethanol and heptane–tetrachloromethane are examples of this type of behaviour.

If the molecules of the liquids in a binary system tend to attract each other, their escape into the vapour phase will be retarded. Negative deviations from Raoult's law result, as in Figure 6.3. Examples of such behaviour are shown by ethanone–trichloromethane and pyridine–ethanoic acid mixtures.

We see that the variations of total vapour pressure with composition tell us something about the interaction of the component molecules in a liquid mixture.

Figure 6.2. Vapour pressure–composition diagram for a binary system showing positive deviations from Raoult's law.

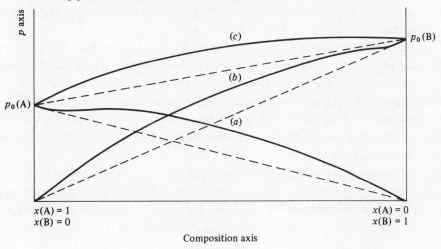

Figure 6.3. Vapour pressure–composition diagram for a binary system showing negative deviations from Raoult's law.

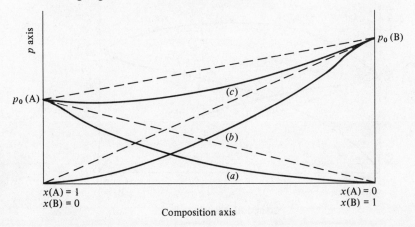

It is evident that two liquids in admixture both exhibit the same type of departure from ideality, at least over part of the range of composition, or none at all.

For a discussion of deviations from Raoult's law in general and of cases where one component shows both positive and negative deviations within different regions of composition, see M. L. McGlashan, *J. Chem. Ed.* (1963), **10**, 516.

The deviations from Raoult's law can be large; the total vapour pressure curve then exhibits a maximum or a minimum. If we increase the deviations shown in Figure 6.2 then a maximum vapour pressure is attained (Figure 6.4). At the composition $x(A)=c$, the mixture has its highest vapour pressure, higher even than that of B, the more volatile of the two components.

Let us consider this situation from another point of view. The vapour pressure at each composition of the mixture will rise with increase in temperature until it attains the same pressure as that of its surroundings, say one atmosphere; the mixture then boils.

Now of all mixtures of A and B, that for which $x(A)=c$ will attain a pressure of one atmosphere at the lowest temperature. The component B attains the ambient pressure at a higher temperature and component A at a higher temperature still. Hence the boiling point–composition curve for mixtures of A and B will exhibit a minimum boiling point, $T(c)$, at the composition $x(A)=c$, Figure 6.5. Pairs of liquids which show this behaviour are ethanol–water and ethyl ethanoate–water, for example.

If we consider enhancement of the negative deviations as shown in Figure 6.6, a minimum vapour pressure (at d) can be obtained, leading to a maximum in the boiling point–composition diagram, Figure 6.7. Examples of this type of behaviour are the systems hydrogen chloride–water and nitric acid–water.

Figure 6.4. Enhanced positive deviations leading to maximum vapour pressure.

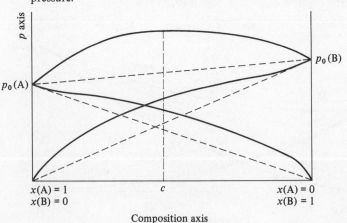

Composition axis

Additional data on maximum and minimum boiling point systems are given in Appendix A16.

We could measure the lowering in vapour pressure in a binary mixture to determine the relative molar mass M_r of one component if that of the other component were known. This method is followed most easily where one component, the solute B, is non-volatile, so that the total vapour pressure is that of the solvent, A. For example, the vapour pressure of an aqueous solution of glucose, $C_6H_{12}O_6$, containing 15 g of glucose in 100 g water, is 17.28 mmHg, at

Figure 6.5. A minimum boiling point due to large positive vapour pressure deviations.

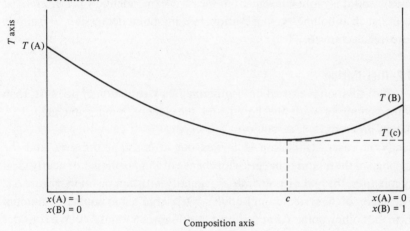

Figure 6.6. Enhanced negative deviations leading to a minimum vapour pressure.

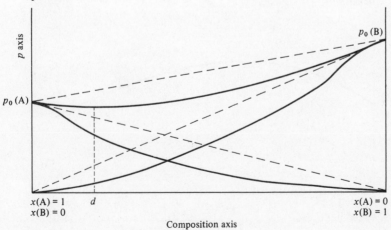

293 K. At the same temperature, the vapour pressure of water is 17.54 mmHg. From (6.7)

$$\frac{p}{p_0(\text{H}_2\text{O})} = x(\text{A})$$

$$\frac{17.28}{17.54} = \frac{(100/18.02)}{(100/18.02) + (15/M_r)}$$

Hence

$$1/M_r = (5.549 - 5.467)/15(0.9852)$$

and

$$M_r = 180.2$$

However, such small changes in vapour pressure are not very easy to measure accurately, and it is more convenient to measure experimentally certain related properties, such as boiling-point elevation, freezing-point depression, or osmotic pressure (see Sections 6.5–6.7).

6.2 Distillation

If distillation is carried out isothermally, by reduction of pressure, then the compositions of both the liquid and the vapour remain unchanged as distillation proceeds.

Usually, however, distillation is carried out at constant pressure, and the boiling point of the mixture changes with change of composition. For nearly ideal binary mixtures, the boiling point varies regularly with composition, Figure 6.8. If the deviation of the system from Raoult's law is large, a maximum or minimum occurs in the boiling point–composition curve, Figures 6.9 and 6.10. We consider three examples of distillation.

Figure 6.7. A maximum boiling point due to large negative vapour pressure deviations.

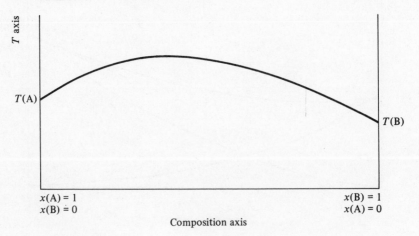

6.2.1 *Regular increase in boiling point*

Figure 6.8 illustrates the temperature–composition diagram for this system. In general the vapour above a mixture has a higher mole fraction of the more volatile component than does the liquid mixture. We can show this as follows.

Suppose liquid of composition $x(A)$ is in equilibrium with vapour of composition $y(A)$; then from Dalton's law of partial pressures

$$y(A) = p(A)/P \tag{6.8}$$

Figure 6.8. Temperature–composition diagram for binary system: regular increase in boiling point.

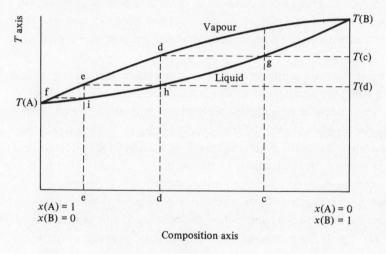

Figure 6.9. A maximum boiling-point system.

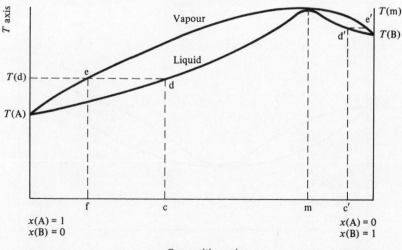

where P is the total vapour pressure.

$$y(A) = x(A)p_0(A)/P \qquad (6.9)$$

and

$$P = p_0(A)x(A) + p_0(B)[1 - x(A)] \qquad (6.10)$$

$$y(A) = \frac{x(A)p_0(A)}{[p_0(A) - p_0(B)]x(A) + p_0(B)} \qquad (6.11)$$

and, if $p_0(A) > p_0(B)$, $y(A) > x(A)$.

Consider the liquid mixture of composition c, Figure 6.8. On heating to temperature T_c the liquid will boil, producing a vapour of composition d which is richer in the lower-boiling component A. This vapour condenses to a liquid also of composition d, which upon distillation yields a vapour of composition e at temperature T_d. By careful distillation we can obtain from such mixtures samples of each of the components: A as the final distillate, and B as the residual liquid in the distillation vessel.

Fractional distillation implies a stepwise progression such as c, g, d, h, e etc leading to the pure component A. The fractionating column is a device which enables this process to be achieved without the separation of each successive distillate. As vapour rises up the fractionating column it meets a downflow of liquid and becomes cooled. From Figure 6.8 we see that if the temperature of the vapour is lowered, it will partially condense giving a liquid richer in B and a vapour richer in A. A series of equilibrium steps follow the same path.

We may note that generally the steeper the boiling point–composition curve, the greater the difference in composition between liquid and vapour at a boiling point. This results from the fact that $dP/dx(A)$ is inversely proportional to

Figure 6.10. A minimum boiling-point system.

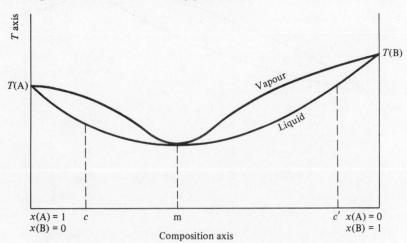

dT/dx(A). Liquid oxygen–liquid nitrogen (liquid air), and trichloromethane–tetrachloromethane, are two systems of this type.

6.2.2 *Maximum and minimum boiling points*

If the mixture shows a maximum or a minimum in the boiling point–composition curve, separation into the two components by distillation at constant pressure is no longer possible. Consider a mixture of A and B which shows a maximum boiling point T(m) at composition m, Figure 6.9. Distillation of a mixture of composition c produces a vapour of composition d at temperature T(d), condensing to a liquid of the same composition. Distillation of this yields a vapour of composition e, and so on. Ultimately the distillate contains pure component A, and the residue in the distillation flask consists of the maximum-boiling mixture m. Note that again the vapour is always richer in the more volatile of the two species A and m.

If we start with a mixture of composition c′, we can obtain samples of B, as distillate, and of m, as residue; a mixture of composition m, in equilibrium with vapour of the same composition, distils unchanged. No mixture of A and B can be separated into these two components.

The same argument applies in the case of a mixture showing a minimum in the boiling point–composition diagram, Figure 6.10. In this case, the mixture m is more volatile than either A or B, and is obtained first, as distillate, from both mixtures c and c′. Again mixture m distils unchanged.

Mixtures such as m, which distil unchanged at a fixed temperature, just as does a pure substance, are called *azeotropes*. That they are not true compounds is shown by the fact that their composition changes if the pressure on the system changes; the boiling point of a pure compound varies with the pressure, but not its composition.

The formation of azeotropes is an important consideration in the separation of the components of a mixture by distillation. For example, the system hydrogen chloride–water forms an azeotrope with a maximum boiling point, depending like its composition on the pressure. This azeotrope, or 'constant boiling point mixture of hydrochloric acid', may be used as a primary standard in acid–base volumetric analysis, from the data in Table 6.1. At 760 mmHg, the boiling points of the components are hydrogen chloride 189.45 K, water 373.15 K. Data for some other azeotropic systems are given in Appendix A16.

6.3 Partially miscible liquids

If a little phenol or ether is added to an excess of water it will dissolve completely. Further addition of the organic compounds leads to the formation of two liquid layers. One layer consists of water saturated with phenol or ether; the other layer consists of the organic compound saturated with water. At a given temperature these two layers are in equilibrium, and are called *conjugate solutions*.

The system phenol–water may be studied at atmospheric pressure. Since there are two components and, in general, three phases (two liquid layers plus vapour), application of the phase rule indicates one degree of freedom:

$$f = c - p + 2 = 1$$

This degree of freedom is usually taken up by variation of the temperature; thus the compositions of the two conjugate solutions are fixed at a given temperature. If we add more water, or more phenol, to the system at equilibrium, only the relative amounts of the two layers changes. At temperature T_1, Figure 6.11, the conjugate solutions have compositions a and b. In this system the mutual solubilities increase with increasing temperature, and the system becomes homogeneous at, or above, temperature T_2; T_2 is called the *consolute temperature*.

Figure 6.12 shows the nicotine–water system; within this range of temperature both components are liquid. The system shows both upper (481 K) and lower (334 K) consolute temperatures, so that outside the region 334–481 K the components are completely miscible.

6.4 Steam distillation

Where the mutual solubilities are very small, for example aniline and water, we consider the liquids to be immiscible. At equilibrium, two liquid phases (virtually the pure components) and a vapour phase coexist. Applying the phase rule

$$p + f = c + 2; \quad f = 1$$

The total vapour pressure is the sum of the vapour pressures of the components, and depends upon the temperature; it is independent of the composition of the mixture. Addition of more aniline or more water merely changes the relative amounts of the two layers.

Table 6.1. *Composition of hydrogen chloride–water constant boiling point mixture*

P/mmHg	B p/K	%HCl (w/w)
400	365.23	21.235
500	370.36	20.916
600	375.36	20.638
700	379.57	20.360
760	381.57	20.222
800	383.16	20.155

The vapour pressure–temperature curves for the aniline–water system are shown in Figure 6.13. The total vapour pressure is 760 mmHg at $T(A+B) = 371.6$ K; $T(A)$ and $T(B)$ are the boiling points of the components, at 760 mmHg. At 371.6 K, the partial vapour pressures are

$$p(H_2O) = 717.6 \text{ mmHg}, \quad p(C_6H_5NH_2) = 42.4 \text{ mmHg}$$

The sum of these partial pressures is 760 mmHg, and the mixture therefore boils at this temperature, 371.6 K. The proportions of aniline and of water in the vapour are in the ratio 42.4:717.6, because both kinds of molecules possess the

Figure 6.11. Portion of the phenol–water diagram.

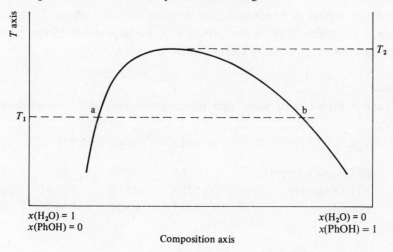

$x(H_2O) = 1$
$x(PhOH) = 0$

$x(H_2O) = 0$
$x(PhOH) = 1$

Composition axis

Figure 6.12. The nicotine–water diagram (under pressure).

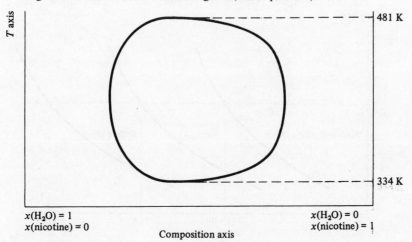

$x(H_2O) = 1$
$x(nicotine) = 0$

$x(H_2O) = 0$
$x(nicotine) = 1$

Composition axis

same kinetic energy E_K at a given temperature and, from (4.29), exert the partial pressures

$$p(H_2O)V = \tfrac{2}{3}E_K n(H_2O)$$
$$p(C_6H_5NH_2)V = \tfrac{2}{3}E_K n(C_6H_5NH_2)$$

Hence

$$\frac{n(H_2O)}{n(C_6H_5NH_2)} = \frac{p(H_2O)}{p(C_6H_5NH_2)}$$

Since the relative molar masses of water and aniline are approximately 18 and 93 respectively, the ratio of the relative weights of these molecules in the vapour is

$$\frac{w(H_2O)}{w(C_6H_5NH_2)} = \frac{717.6 \times 18}{42.4 \times 93} = 3.28$$

Thus if the vapour is condensed, the distillate contains about 23 wt%, or 5.6 mol%, of aniline. This is the principle of the separation of the organic component from a mixture by steam distillation.

Example

The vapour pressures of water and nitrobenzene vary with temperature as follows:

T/K	363	368	371	373
$p(C_6H_5NO_2)$/mmHg	12.9	16.7	19.2	20.9
$p(H_2O)$/mmHg	525.8	633.9	707.3	760.0
P/mmHg	538.7	650.6	726.5	780.9

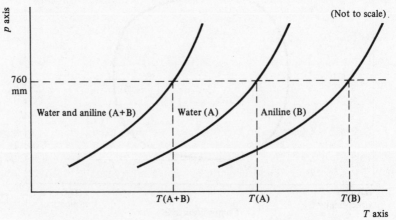

Figure 6.13. Vapour pressure–temperature curves for the aniline–water system.

From Figure 6.14 the boiling point of water–nitrobenzene mixtures at 760 mmHg is 372.4 K. The relative molar masses of the components are 18 and 123; thus

$$w(H_2O)/w(C_6H_5NO_2) = 741 \times 18/20.25 \times 123 = 5.36$$

ie the steam-distillate contains 15.7 wt% of nitrobenzene.

6.4.1 *Solvent extraction*

Consider a solute which is soluble in two 'immiscible' solvents, for example benzoic acid in a water–ether mixture; both the ethereal and aqueous layers contain some solute. At equilibrium at a given temperature there is a fixed ratio between the concentrations of the solute in the two layers, independent of the total amount present, provided that the solute is in the same molecular form in both solvents.

Let w_1 g of solute of relative molar mass M_r be contained in V_1 cm^3 of solution. The initial molarity is thus $1000w_1/M_rV_1$. If this solution is shaken with V_2 cm^3 of an immiscible solvent and w_2 g of solute passes into this second solvent, the molarities of the two layers are

Solvent 1: $1000(w_1 - w_2)/M_rV_1$

Solvent 2: $1000w_2/M_rV_2$

The above proviso ensures that M_r is the same in both solvents.

The ratio of these molarities is $(w_1 - w_2)V_2/w_2V_1$ and is a constant, D, the distribution coefficient for the system at the given temperature; this is the basis of the solvent extraction process.

Consider the extraction by ether of 100 cm^3 (V_2) of aqueous solution containing 5 g (w_1) of solute A. We assume that the ratio of the solute

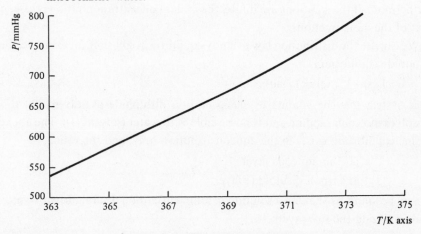

Figure 6.14. Variation of total vapour pressure P with temperature: nitrobenzene–water.

concentrations in ether and water, D, is ten. If the solution is shaken with 15 cm^3 of ether, and w_2 g of A are extracted, then

$$\frac{w_2}{5-w_2} \times \frac{100}{15} = 10$$

from which $w_2 = 3.0$ g. If the extraction is repeated with a second 15 cm^3 portion of ether, a further w_2' of A is extracted where

$$\frac{w_2'}{2-w_2'} \times \frac{100}{15} = 10, \quad w_2' = 1.2 \text{ g}$$

A third similar extraction removes a further 0.48 g of A from the aqueous layer, so that a total of 4.68 g, or 94%, of A has been extracted.

If extraction had been carried out by using the 45 cm^3 of ether in one portion, w_3 g of A would have been extracted, where

$$\frac{w_3}{5-w_3} \times \frac{100}{45} = 10, \quad w_3 = 4.1 \text{ g}$$

or 82% of A. Evidently, it is always advantageous to extract with several small portions of the second solvent, rather than with one portion of the same total volume.

6.4.2 *Distribution law*

The problem just considered is a particular example of the distribution law. For a system of two immiscible liquids, if a third substance is added, soluble in both liquids, then the ratio $c_1/c_2 = D$, the distribution coefficient, is a constant at a given temperature; c_1 and c_2 are the molar concentrations of the solute in the two layers. Consider the data of Table 6.2 for the distribution of iodine between tetrachlorethane (1) and water (2). The value of D is sensibly constant. Strictly, it is the ratio of the activities of the solute in the two solvents which is constant. If the solutions behave ideally the activities may be replaced by molar concentrations; furthermore, if the solutions are dilute, the molar concentration ratio is close to that of the mole fractions.

We can use the distribution law to study equilibria in solution, for example, the tri-iodide equilibrium

$$I_2(s) + I^-(aq) \rightleftharpoons I_3^-(aq)$$

This system may be studied by using carbon disulphide as solvent (2); this dissolves molecular iodine and is immiscible with water (solvent (1)). The above iodine equilibrium exists in the aqueous solution only, and the ratio

$$\frac{[I_2] \text{ in aqueous layer}}{[I_2] \text{ in carbon disulphide layer}} = D$$

D may be found by shaking a solution of iodine in carbon disulphide with water, and analyzing the two layers.

If iodine is distributed between a potassium iodide solution of molarity c and carbon disulphide, the iodine will be present both as free iodine I_2 and as tri-iodide ions I_3^- in the aqueous layer 1, but only as free iodine in the carbon disulphide layer 2. If the total iodine concentration in layer 1 is c_1, and that in layer 2 is c_2, then applying the distribution law, the concentration of free iodine in layer 1 is Dc_2, and the concentration of the tri-iodide is $c_1 - Dc_2$. The concentration of iodide ions I^- in layer 1 is thus $c - (c_1 - Dc_2)$. If K is the equilibrium constant for the iodine equilibrium

$$K = \frac{[I_3^-]}{[I_2][I^-]} = \frac{c_1 - Dc_2}{Dc_2(c - c_1 + Dc_2)}$$

At 298 K the value of K, also called the *stability constant* of the tri-iodide ion, is 955.

6.5 Elevation of the boiling point

For the case of an involatile solute B in dilute solution in a volatile solvent A, the vapour pressure–temperature relationship is shown in Figure 6.15. The pure solvent boils, under atmospheric pressure, at $T(A)$; the solution has a lower vapour pressure and hence a higher boiling point, T_b. The increase $T_b - T(A)$ is the boiling-point elevation ΔT_b for the solution.

For the equilibrium

$$\text{solution} \rightleftharpoons \text{vapour}$$

following (4.77)

$$d \ln p / dT = \Delta H_e / RT^2 \tag{6.12}$$

where ΔH_e is the molar enthalpy of evaporation of the solvent from the solution. Assuming that ΔH_e does not vary with temperature over the temperature range $T_A - T_b$

$$\int_{p_0}^{p} d \ln p = (\Delta H_e / R) \int_{T(A)}^{T_b} T^{-2} dT \tag{6.13}$$

Table 6.2. *Distribution of iodine between tetrachloroethane and water*

c_1/mol dm^{-3}	c_2/mol dm^{-3}	D
0.020	0.000235	85.1
0.040	0.000469	85.2
0.060	0.000703	85.4
0.080	0.000930	86.0
0.100	0.00114	87.5

whence

$$\ln(p/p_0) = -(\Delta H_e/R)[(1/T(A)) - (1/T_b)] \tag{6.14}$$

$$= -\frac{\Delta H_e}{R}\frac{\Delta T_b}{T(A)T_b} \tag{6.15}$$

For a dilute solution, $T(A)T_b \approx T(A)^2$, and ΔH_e may be taken as the enthalpy of evaporation of the pure solvent. From Raoult's law (6.6)

$$p/p_0 = x_1 = 1 - x_2 \tag{6.16}$$
$$\ln(p/p_0) = \ln(1 - x_2) \approx -x_2 \tag{6.17}$$

since x_2, the mole fraction of solute, is small. Hence

$$\Delta T_b = \frac{RT^2(A)x_2}{\Delta H_e} \tag{6.18}$$

For dilute solutions,

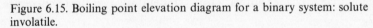

and if w_1, the weight of solvent, is fixed at 1000 g, then $x_2' = m_2 M_{r_1}/1000$, where m_2 is the *molal* concentration (number of moles of solute dissolved in a kilogram of solvent; it is temperature independent) of the solute and

$$\Delta T_b = \frac{RT^2(A)}{\Delta H_e}\frac{m_2 M_{r_1}}{1000} = \frac{RT^2(A)}{1000l_e}m_2 = k_b m_2 \tag{6.20}$$

where l_e is the enthalpy of evaporation per gram of solvent. For a given solvent the term $RT^2(A)/1000l_e$ is a constant, and is called the *ebullioscopic constant* of the solvent, k_b; evidently k_b is the boiling-point elevation for an ideal solution of unit

Figure 6.15. Boiling point elevation diagram for a binary system: solute involatile.

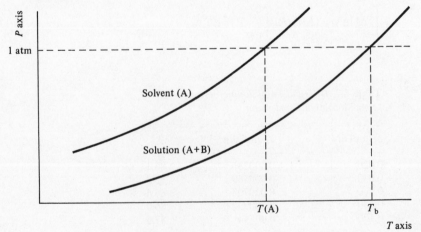

molality in that solvent. Values of k_b for some common solvents are given in Appendix A14.

The boiling points of solvent and solution are measured under the same pressure by a sensitive thermometer, such as the Beckmann thermometer shown in Figure 6.18. For the solvent boiling point, the thermometer is immersed in the vapour; for the solution boiling point, it is lowered into the liquid.

A modern boiling-point elevation apparatus, based upon measurements of the resistance of a temperature-sensitive resistor – a thermistor – has been designed by C. Heitler (Figure 6.16).

A known weight of solvent is boiled in the borosilicate glass bulb by an electric heater; a cold-finger reflux condenser prevents the loss of solvent by evaporation. The boiling liquid is pumped over the tip of the thermistor probe P by a Cottrell pump and the resistance of the thermistor is measured by a Wheatstone bridge. A weighed pellet of the solute is introduced, and the resistance found at the elevated temperature. This will be lower than the previous resistance: the thermistor

Figure 6.16. Heitler semi-micro ebulliometer (courtesy Gallenkamp).

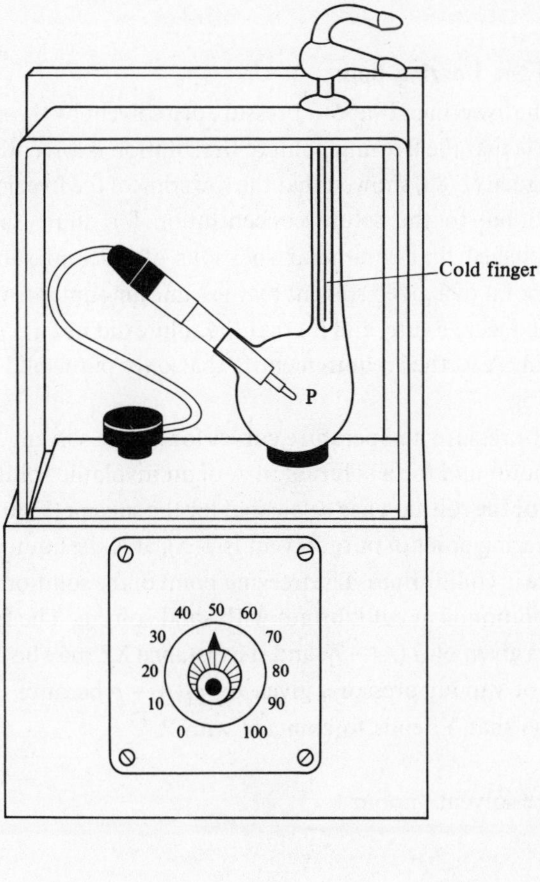

contains a speck of semiconductor material, for which d(resistivity)/dT is positive. From a calibration curve the change in resistance is related to the increase in boiling point.

Further weighed pellets may be added successively; the theory of (6.20), however, depends upon the total concentration of the solution being low.

Example

A solution of 2.001 g urea, $CO(NH_2)_2$, in 0.125 kg of water boiled at 373.139 K; at the same pressure, water boiled at 373.0 K. If ΔH_e is 40.7 kJ mol^{-1}, calculate k_b, and estimate the effect of the approximation involved in (6.20).

$$k_b = RT^2(H_2O)/1000l_e = 8.316 \times (373.0)^2 \times 18.016/(40\,700/1000)$$
$$= 0.512 \text{ K kg mol}^{-1}$$

The approximations involved may be estimated by calculating k_b from (6.20), taking M_{r_2} for urea as 60.20:

$$k_b = \Delta T/m_2, = 0.139/(8 \times 2.001/60) = 0.523 \text{ K kg mol}^{-1}$$

so that the approximations increase k_b by 2.1%.

In deriving (6.20), we have assumed that the solute does not associate or dissociate in solution.

6.6 Depression of the freezing point

One result of the lowering of vapour pressure of a solvent by the addition of a non-volatile solute is that the freezing point of the solution is lower than that of the pure solvent. Blagden (1788) showed that the lowering of the freezing point was generally proportional to the solute concentration for dilute solutions. Raoult (1878–86) established that equimolar solutions of different substances depressed the freezing point of a given solvent by the same amount, provided (a) that the solutions formed were dilute, and (b) that the solute did not associate or dissociate in the solvent. A further requirement is that only pure solid solvent separates out on freezing.

If we plot the vapour pressure–temperature curves for a pure solvent A in the vicinity of its melting point and for a solution in A of an involatile solute B, we observe that the curve for the solution lies *below* that for the solvent (Figure 6.17). On this diagram the freezing point of pure solvent is $T(A)$; at X the liquid solvent and the solid solvent are in equilibrium. The freezing point of the solution is T_f; at this temperature the solution is in equilibrium with solid solvent. The freezing-point depression, ΔT_f, is given by $T(A) - T_f$, and the distance XZ may be taken as equal to the lowering of vapour pressure, given by $p_0(A) - p$ because in *dilute* solution ΔT_f is small, so that Y tends to coincide with Z.

For the equilibrium

solid solvent $\rightleftharpoons$ solvent vapour

in the vicinity of the freezing point:

$$d \ln p/dT = \Delta H_f/RT^2$$

where ΔH_f is the molar enthalpy of freezing of the solvent. Assuming ΔH_f to remain constant over the temperature range $T(A) - T_f$:

$$\int_{p_0(A)}^{p} d \ln p = \frac{\Delta H_f}{R} \int_{T(A)}^{T_f} T^{-2} dT \tag{6.21}$$

$$\ln[p/p_0(A)] = -(\Delta H_f/R)(1/T(A) - 1/T_f) \tag{6.22}$$

$$= \frac{\Delta H_f}{R} \frac{T_f}{T(A)T_f} \tag{6.23}$$

Assuming the solution is dilute, and following through the argument of (6.15)–(6.20), we obtain

$$\Delta T_f = \frac{RT^2(A)}{\Delta H_f} \frac{m_2 M_1}{m_1} = \frac{RT^2(A) m_2}{1000 l_f} \tag{6.24}$$

where l_f is the enthalpy of freezing per gram of solvent, and m_1 has been taken as 1000 g. Thus, if the factor $RT^2(A)/1000 l_f$ is replaced by k_f, the *cryoscopic constant* for the solvent:

$$\Delta T_f = k_f m_2 \tag{6.25}$$

Example

1.821 g of tetrachloromethane in 100 g benzene gave a freezing-point depression of 0.603 K; k_f for benzene is 5.11 K kg mol^{-1}. Thus

$$m_2 = \Delta T_f/k_f = 0.118$$

$$M_{r_2} = 1.821/0.118 = 154.3$$

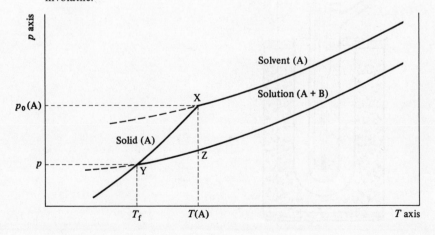

Figure 6.17. Freezing-point depression diagram for a binary system: solute involatile.

The cryoscopic method for determining molecular weights will be described with reference to Figure 6.18. A weighed quantity of solvent whose cryoscopic constant is known is placed in the tube A and a Beckmann thermometer B, preset to the required temperature range, and a stirrer C are fitted. Again, the thermometer may be replaced by a thermistor. The tube is fitted into a wider tube D in order to provide an air space, to aid slow cooling. This assembly is contained in a vessel E in which a freezing mixture is placed. The freezing mixture should be at a temperature about 5° below the freezing point of the solvent. Another stirrer F is provided; both stirrers have thermally insulated handles.

The temperature of the liquid is allowed to fall to about $\frac{1}{2}°$ below its freezing point. Vigorous stirring with the stirrer C induces crystallization and the temperature rises to the freezing point. The solvent is allowed to melt and a weighed quantity of solute in the form of a pellet is added and dissolved; the new freezing point is determined and the molecular weight evaluated. The following points should be noted:

(*a*) The solution should be dilute.

(*b*) The solute should not associate or dissociate in the solvent.

(*c*) Pure solid solvent *only* must separate from solution.

(*d*) Crystallization should commence at about $\frac{1}{2}°$ below the freezing point.

Figure 6.18. A Beckmann freezing-point apparatus.

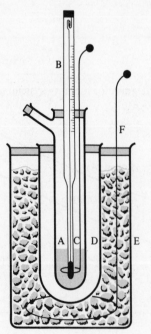

Figure 6.19(*a*) and (*b*) illustrates supercooling with solvent and with solution. When a supercooled solvent crystallizes, the latent heat of fusion is released and the temperature rises rapidly to the freezing point and remains constant until complete melting has taken place. The system solid solvent–liquid solvent consists of two phases and one component; if the pressure is constant, the system is invariant. The temperature thus varies with the composition of the solution, as we have shown. In this case, the initial temperature rise may continue above the freezing point; T_f is fixed by constructing the horizontal line ab on the cooling curve, Figure 6.19(*b*), so as to equalize the areas s and s'.

6.7 Osmosis and osmotic pressure

It was discovered in the middle of the eighteenth century that if alcohol and water were separated by an animal membrane, e.g. a pig's bladder, the water passed through the membrane into the alcohol, but the alcohol did not pass through the membrane into the water. Subsequently, this property was observed with solutions in general. The essential property of the membrane is that it allows the passage of water (or solvent) molecules through it but not that of solute molecules or ions; the term *semipermeable* is used to describe these membranes.

A semipermeable membrane can be formed in the walls of a porous pot by electrolyzing solutions of copper sulphate and potassium ferrocyanide which are separated by the pot (Figure 6.20). Copper ferrocyanide is formed in the pores of the pot and then acts as a good semipermeable membrane. Another good membrane in osmometry is a thin sheet of cellulose.

Osmosis is the passage of solvent from a dilute solution through a semipermeable membrane into a more concentrated solution. It can be demonstrated with the apparatus of Figure 6.21. As water enters the porous pot, the pressure rises; this is indicated by the difference in levels in the pressure gauge, M. Osmotic pressure can be *demonstrated* by this apparatus, but it cannot be *measured* in this way, for the solution changes slightly in concentration during the experiment. In the absence of an external pressure (in the case of Figure 6.21 a hydrostatic

Figure 6.19. Supercooling: (*a*) solvent only, (*b*) solution.

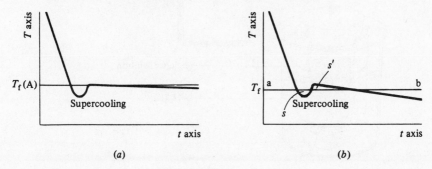

(*a*)   (*b*)

pressure), osmosis could continue until the solutions on each side of the semipermeable membrane have the same osmotic pressure, ie until they are *isotonic*.

. The osmotic pressure may be defined as the external pressure which must be applied to a concentrated solution separated from a dilute solution by a semipermeable membrane, at a particular temperature, to prevent osmosis taking place. Early measurements of osmotic pressure used the Berkeley and Hartley apparatus (Figure 6.22). A known variable external pressure, P, balances the osmotic pressure of the solution–solvent system.

Figure 6.20. Preparation of a semipermeable membrane of copper ferrocyanide $Cu_2Fe(CN)_6$.

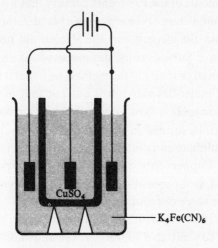

Figure 6.21. Demonstration of osmosis.

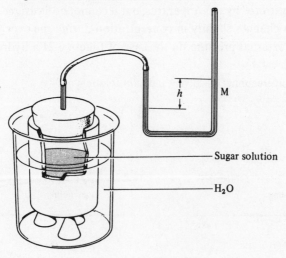

The osmotic pressure is related to the lowering of vapour pressure of the solvent. We give here a descriptive treatment of this relationship. The semipermeable membrane (in the porous pot) separates a dilute solution 1 from the pure solvent A, Figure 6.23. The apparatus is confined at constant temperature and constant pressure under a bell-jar. When equilibrium is established, the osmotic pressure of the ambient solution is given approximately by the height, h, of the column of solution.

The vapour pressure of the solution at Y must be equal to that of the solvent at Y at equilibrium, assuming the solute to be involatile. If the vapour pressure of

Figure 6.22. The Berkeley–Hartley apparatus.

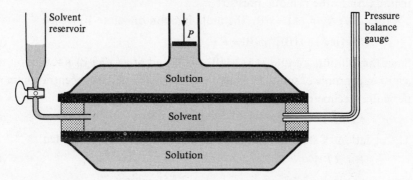

Figure 6.23. Relationship between osmotic pressure and vapour-pressure lowering.

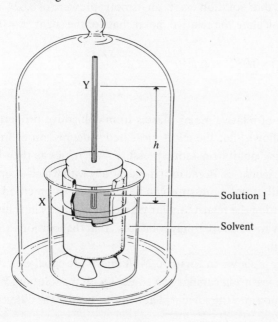

the solvent at X is $p_0(A)$, that at Y is $p' = [p_0(A) - h\rho_0 g]$, where ρ_0 is the density of solvent vapour and g is the acceleration due to gravity. At a pressure $p_0(A)$ and a temperature T, the molar volume of the solvent vapour is $RT/\rho_0(A)$, assuming the vapour behaves as an ideal gas, so that $p_0 = Mp_0(A)/RT$, where M is the molar mass of the solvent in the vapour phase. If the density of the solution is ρ, then the osmotic pressure π is approximately $h\rho g$; this is a good approximation if the solution is dilute. Eliminating hg:

$$\pi = \rho(p_0(A) - p')/\rho_0 = \rho \, \Delta p/\rho_0$$

or

$$\Delta p/p_0(A) = M\pi/\rho RT \tag{6.26}$$

relating Δp to the osmotic pressure π.

From (6.7) $\Delta p/p_0(A) = x(B)$, the mole fraction of solute B. Hence

$$M\pi[n(A) + n(B)]/\rho n(B) = RT \tag{6.27}$$

Since the solution is dilute, $n(A) + n(B) \approx n(A)$ and $Mn(A)/\rho n(B)$ is V, the volume of solution per mole of solute B. Thus $V = 1/c$, where c is the concentration in mole per cubic decimetre. Hence

$$\pi = cRT \tag{6.28}$$

The equation of state for one mole of an ideal gas may be written

$$P = RT/V \tag{6.29}$$

and the similarity between (6.28) and (6.29) is evident.

Example

0.450 g of a sugar in 100 dm^3 solution exerts an osmotic pressure of 0.624 atm at 293 K. We require to calculate the relative molar mass of the sugar; $R = 0.0831$ dm^3 atm mol^{-1} K^{-1}.

$$0.624 = (4.5/M_r) \times 0.0821 \times 293.16$$

whence $M_r = 173.56$.

In the determination of relative molar masses from colligative properties of solutions, it has been shown that the effect measured – depression of freezing point, elevation of boiling point or osmotic pressure – *decreases* as the relative molar mass of the solute increases. Because of the accuracy with which a capillary rise, produced by a small osmotic pressure, can be measured, this method is the most suitable for measuring the relative molar masses of polymers in solution – particularly as polymers are often sparingly soluble, so that the solutions are very dilute (see Problem 6.10).

A modern osmometer is shown in Figure 6.24. A and B are capillary tubes of identical bore. A is the measuring capillary, in contact with the solution; B is the reference capillary, dipping into the solvent. The osmotic head is the difference in the meniscus heights in A and B. This arrangement allows for the effect of surface tension on the capillary rise.

6.8 Anomalous behaviour

A solution containing 0.35 g of potassium chloride in 100 g of water exhibits a freezing-point depression of 0.143 K. We know that the 'relative molar mass' of potassium chloride is 74.53 and, since the cryoscopic constant for water is 1.86 K kg mol^{-1}, we can calculate the expected freezing-point depression from (6.24). This gives a value of 0.087 K. Thus the observed depression is about 1.64 times greater than the expected value. The ratio

$$\frac{\text{observed freezing-point depression}}{\text{calculated freezing-point depression}}$$

is called the *i* factor, a term introduced by van't Hoff in 1886 to explain apparent anomalies, or departures from ideal behaviour.

Figure 6.24. A modern osmometer. (This apparatus is manufactured by Polymer Consultants Limited, High Church Street, New Basford, Nottingham.)

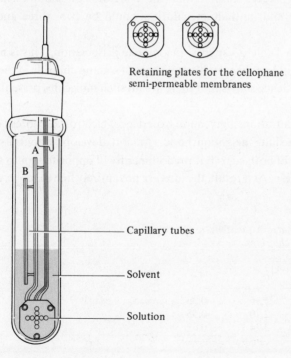

Retaining plates for the cellophane semi-permeable membranes

A

B

Capillary tubes

Solvent

Solution

Measurements of the i factors for a series of chlorides, at a given concentration, lead to the typical results in Table 6.3. These chlorides are electrovalent compounds and dissociate into ions in aqueous solution:

$$MCl_x \rightarrow M^{x+} + xCl^- \qquad (6.30)$$

If the initial concentration of the electrolyte is $c \ mol^{-1} \ kg^{-1}$ and if we *assume* a degree of dissociation 'α', then at equilibrium the concentrations of the species in (6.30) are, in order, $c(1-\alpha)$, $c\alpha$ and $xc\alpha$. The total concentration of all three species is thus $c(1+x\alpha)$. Now the freezing-point depression, and indeed any other colligative property, depends upon the total number of particles in solution. Hence

$$i = \frac{\text{total number of particles in solution after dissociation}}{\text{total number of particles in solution before dissociation}}$$

In the derivation of (6.6), (6.7), (6.20), (6.24) and (6.28) the solute was considered neither to dissociate nor to associate. Thus

$$i = \frac{c(1+x\alpha)}{c} = 1 + x\alpha \qquad (6.31)$$

So from the value of i, an apparent degree of dissociation 'α' may be calculated; this quantity is listed in Table 6.3.

We have seen in Chapter 4 that ionic compounds, such as the chlorides in Table 6.3, are ionized in the solid state, and upon dissolution in water the ions separate. We would expect at first that the i factors for sodium chloride, magnesium chloride and lanthanum chloride would be two, three and four respectively.

The values of 'α' in Table 6.3 suggest incomplete dissociation. This is not the true explanation; we did not begin with molecules of sodium chloride in the solid, and experimental evidence shows that there are no such molecules present in the solution.

At any finite concentration, a given ion experiences electrostatic forces due to the ions in its immediate neighbourhood. An ion develops about itself an 'atmosphere' of ions, of both signs but predominantly of opposite sign to that of the central reference ion. As a result, the ions are not entirely free to move, and so

Table 6.3. i *Factors for some electrovalent chlorides*

	i	'α'
NaCl	1.90	0.90
$MgCl_2$	2.10	0.55
$LaCl_3$	1.24	0.08

do not exert their full effect upon the lowering of solvent vapour pressure according to their mole fraction, as Raoult's law requires.

The extent of this interionic attraction increases with increasing concentration and with increasing ionic charge. The difference between this correct explanation and that of incomplete dissociation, for strong electrolytes, should be noted carefully, as both postulates imply the same type of anomaly in colligative properties.

Let us reconsider the solution of potassium chloride containing 0.35 g in 100 g of water. The molality of this solution is 0.047; the stoichiometric molalities of the potassium and of chloride ions are each 0.047. The properties of this solution show that the effective concentrations, or activities, of the ions are lower, due to interionic attraction. We express this fact by the mean ionic activity coefficient, which is 0.82 for potassium chloride at a stoichiometric concentration of 0.047 molal. Hence the effective concentration is only 0.047×0.82 molal, ie 0.039 molal. From (6.25) we calculate the freezing-point depression at the new effective concentration as 0.0716 K. Hence i would be $0.143/0.0716 = 1.99$ (ie approximately 2).

R. J. Gillespie made extensive use of the i (or v) factor in calculating the extent of ionization of solutes in sulphuric acid, and of the autoprotolysis of the solvent. Sulphuric acid, freezing at 283.8 K, and with a cryoscopic constant of 6.12 K kg mol^{-1}, is a convenient solvent for cryoscopic studies; see R. J. Gillespie & E. A. Robinson, in *Advances in Inorganic Chemistry and Radiochemistry*, ed. H. J. Emeleus & A. G. Sharpe, Vol. I, p. 385–423, Academic Press, London, 1959.

Some solutes become *associated* in solution. The aliphatic carboxylic acids show a strong tendency to form dimers by intermolecular hydrogen bonding:

This tendency is enhanced in non-polar solvents, such as benzene; presumably there is little solvent–solute interaction and the intermolecular hydrogen bonding predominates. *p*-Cresol and nitrobenzene are other examples of substances which associate in benzene or in toluene.

Association leads to i factors which are less than unity:

$$2RCO_2H \rightleftharpoons (RCO_2H)_2$$

Initial concentration	c	0
Equilibrium concentration	$c(1-\beta)$	$c\beta/2$

Hence

$$i = \frac{c(1-\beta)+(c\beta/2)}{c} = 1 - \frac{\beta}{2} \tag{6.32}$$

where β is the degree of association.

Example

A solution of 1.425 g of ethanoic acid in 100 g benzene shows a freezing-point depression of 0.608 K. Since k_f for benzene is 5.11 K, the apparent relative molar mass is 119.8. Thus, as the true value for ethanoic acid is 60.05, ethanoic acid is almost completely dimerized in benzene at this concentration; i from (6.32) is 0.5, and β is unity.

So far we have considered the depression of freezing point of a liquid solvent. The melting point of a solid solvent is lowered in a similar manner, by dissolving a solute in it to form a solid solution which, like a liquid solution, is completely homogeneous.

The cryoscopic constant for some solids is large and so the freezing-point depression may be measured using a thermometer graduated in 0.1 K. This leads to a simple method for determining approximate relative molar masses of organic compounds. Camphor has a cryoscopic constant of about 40 K (see Appendix A14) and may be used for this purpose; its melting point is 450 K however, and camphene (melting point 315.7 K, $k_f = 35$ K) is often preferred.

Example

0.01 g naphthalene mixed with 0.1 g of camphene melted at 313 K. Thus

$$2.7 = \frac{35 \times 0.01 \times 1000}{M_r \times 0.1}$$

whence the relative molar mass (M_r) of naphthalene is 129.6 (true value for naphthalene 128.2).

6.9 Relative molar masses

Two modern methods for determining these quantities will be considered.

6.9.1 *Mass spectrometry*

Traces of the vapour of the substance under examination are ionized by heating or by electron bombardment in a mass spectrometer. The positive ions so formed are accelerated by an electric field V. If the ion is singly charged, it acquires a kinetic energy given by

$$\tfrac{1}{2}mv^2 = Ve \tag{6.33}$$

where m and v are the mass and velocity of the ion. The beam of ions is narrowed by passage through a slit system and is then introduced into a semicircular magnetic field of strength H. The ions are caused to move in a curved path, of radius of curvature r, where

$$e/m = v/Hr \tag{6.34}$$

From (6.33) and (6.34)

$$r^2 = 2mV/eH^2 \qquad (6.35)$$

Thus if H and V are constants, r is proportional to $(m/e)^{\frac{1}{2}}$ for a number of different ions. The deflected beams are intercepted by an electronic detector, and their position is directly related to their mass-to-charge ratio. A very high precision of measurement of m is attainable; for example, the relative molar mass $(N_A \times m)$ of the organic compound diosgenin iodoethanoate, $C_{29}H_{43}O_4I$, has been measured with an accuracy of 1 part in 10^6 as 585.2201.

6.9.2. *X-ray methods (for crystalline solids)*

From the relationship: density = mass/volume we can write

$$D_m = \frac{Z \times M_r \times 1.661 \times 10^{-27}}{V \times 10^{-27}}$$

where D_m is the measured density in $kg\,m^{-3}$, M_r is the relative molar mass, Z is the number of molecules in the unit cell of volume $V\,nm^3$ and $1.661 \times 10^{-27}\,kg$ is the atomic mass unit. The volume of the unit cell is evaluated from the unit cell dimensions a, b and c; for an orthorhombic unit cell, $V = abc$. If M_r is known approximately, a value for Z can be found. Since Z must be an integer, the appropriate number can be used to evaluate M_r more precisely. The error in M_r depends upon errors in measurement of the unit cell dimensions and of the density of the material. This technique is particularly useful for crystals containing solvent of crystallization, since they can be examined directly in the solid state.

6.10 Simple eutectic mixtures and salt hydrates

We conclude this chapter with a short discussion on some simple hydrate systems which are relevant to the subject matter of the chapter. Consider part of the system potassium iodide–water, Figure 6.25. No compounds (ie no crystalline hydrates) are formed in this system. Ice separates from the solution on cooling, along the line AC, whereas along the line BC potassium iodide separates from solution. The line BC represents the solubility of potassium iodide in water at varying temperatures; its steepness indicates that the solubility increases slowly with temperature. The line AC is the freezing-point depression curve for potassium iodide. The point C is a *eutectic* point; at this composition and temperature, a mixture of ice and potassium iodide separate out together and the temperature remains constant at 250 K until all the liquid phase has disappeared.

It is important to note that the eutectic is a *mixture*. The phase rule may be expressed

$$p = c + 1 - f$$

since we are considering a condensed system at constant pressure. At C the system is invariant, and since there are two components there are three phases in equilibrium: solid potassium iodide, ice and solution. As in the case of an azeotrope, p. 203, the composition and temperature of the eutectic can be changed by varying the pressure on the system, indicating that the eutectic is a mixture and not a compound.

A portion of the sodium sulphate–water system is shown in Figure 6.26. Several hydrates are formed in this system. Along the line AC ice is in equilibrium with liquid solution of varying composition, but along CB the solid phase present is sodium sulphate decahydrate. At B a transition point is reached and crystals of

Figure 6.25. The potassium iodide–water phase diagram.

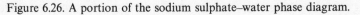

Figure 6.26. A portion of the sodium sulphate–water phase diagram.

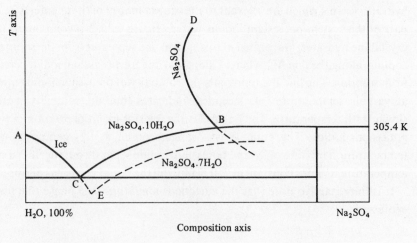

anhydrous sodium sulphate separate out. The point B is called an *incongruent melting point*: the solid and liquid obtained from it by melting do not have the same composition. B is an invariant point and sodium sulphate decahydrate, anhydrous sodium sulphate and liquid are in equilibrium here at a temperature of 305.4 K. The first portion of the curve BD shows a negative temperature coefficient of solubility.

Figure 6.27 shows a portion of the ferric (iron (III))/chloride–water system. Again, several hydrates are formed. The line AB represents ice in equilibrium with solution, forming a eutectic at B with ferric chloride dodecahydrate. At the point C, the liquid and solid phases in equilibrium have the same composition; C is

Figure 6.27. A portion of the ferric chloride–water phase diagram.

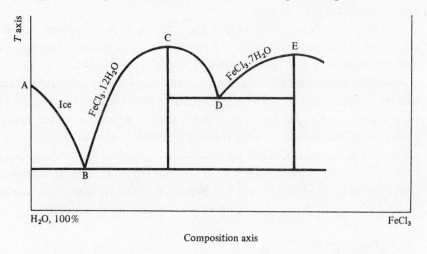

Figure 6.28. Isothermal (298 K) dehydration of the copper sulphate–water system.

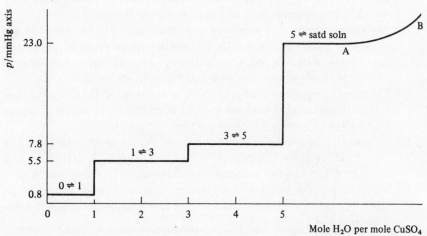

termed a *congruent melting point*. Another congruent melting point is shown at E. The eutectic between ferric chloride dodecahydrate and ferric chloride heptahydrate is represented by D.

The variation of the vapour pressure above a salt hydrate is illustrated in Figure 6.28 with reference to copper sulphate, a much studied example, at 298 K. If a concentrated solution of copper sulphate is evaporated isothermally, the vapour pressure falls along the line BA. At the point A, the solution is saturated with respect to the pentahydrate which then crystallizes. The vapour pressure remains constant until all the saturated solution has crystallized. The phase rule indicates that this is correct:

$$p+f=c+2 \quad \text{or} \quad f=2+2-3=1$$

Since there are three phases, the system is univariant along any horizontal line and is specified by the composition.

At 5 moles water per mole copper sulphate, the vapour pressure decreases sharply on further evaporation to 7.8 mmHg, at which pressure the trihydrate and pentahydrate are in equilibrium.

A study of diagrams such as Figure 6.28 shows how to 'dry' a hydrate to a true stoichiometric composition. In order to obtain a pure sample of copper sulphate pentahydrate, a sample containing water slightly in excess of the required composition, is confined over a mixture of the pentahydrate and trihydrate which are already in equilibrium. The excess moisture is absorbed by the trihydrate, the vapour pressure will equilibrate at 7.8 mmHg and the stoichiometric composition, $CuSO_4 \cdot 5H_2O$, will be attained. We may see from this discussion that a hydrate can rarely, if ever, be used as an analytical standard.

Problems 6

6.1 Explain the terms: (a) mole fraction, (b) colligative property, (c) partial pressure, (d) binary system, (e) critical solution temperature, (f) ideal solution, (g) minimum boiling-point composition, (h) steam distillation.

6.2 The vapour pressures of water and of propanone, at 312.5 K, are 54.30 and 400 mmHg respectively. Calculate the partial vapour pressures of the two components above a solution containing 10 g of water and 20 g of propanone at this temperature, assuming the system to be ideal.

6.3 The vapour pressure of ethanol at 292 K is 40.00 mmHg. If 1.032 g of a non-volatile substance M is dissolved in 98.7 g of ethanol, the lowering of vapour pressure is 0.432 mmHg. Calculate the relative molar mass of M.

6.4 An aqueous solution of a weak monobasic acid containing 0.1 g of acid in 21.7 g of water depresses the freezing point of the solvent by 0.187 K. If the value of k_f for water is 1.86 K, what is the relative molar mass of the monobasic acid?

6.5 Interpret fully the fact that a $0.1 \, \text{mol dm}^{-3}$ aqueous solution of glucose, $C_6H_{12}O_6$, freezes at the same temperature as a solution of calcium chloride containing 0.44 g in 100 g of water.

6.6 A solution of 12.5 g of urea in 170 g of water gave a boiling-point elevation of 0.63 K. Calculate the relative molar mass of urea, taking $k_b = 0.52$ K. Derive any formula used.

6.7 The following data refer to the freezing-point depressions of solutions of tetrachloromethane in benzene; the relative molar mass of tetrachloromethane is 153.8:

$c/\text{mol dm}^{-3}$	0.1184	0.3499	0.8166
T_f/K	0.603	1.761	4.005

In each case deduce an experimental value for k_f and comment upon the results.

6.8 The following measurements of osmotic pressure, at 303 K, were made by Berkeley and Hartley:

Concentration of sucrose solution/mol kg^{-1}	Osmotic pressure π/atm
0.590	15.48
1.081	29.72
1.662	48.81
2.396	74.94
3.331	111.87
4.178	148.46

The following densities of sucrose solutions at this temperature have been made subsequently:

Concentrations of sucrose solution/mol kg^{-1}	Density/g cm$^{-3} = d$
0.000	0.996
0.325	1.034
0.730	1.077
1.252	1.122
1.701	1.160
2.550	1.224
4.660	1.340

Determine the value of R at each concentration using the equation

$$\pi = cRT$$

where c is the concentration in mol dm^{-3}; if the molal concentration is m,

$$c = 1000md/(1000 + m)$$

Extrapolate the results for R to $c = 0$; what are the units of R?

6.9 The mean activity coefficient, $f(\pm)$, for 0.05 molal sodium chloride is 0.821. What would be the freezing-point depression for this aqueous solution if $k_f = 1.86$ K?

6.10 A non-volatile solute of relative molar mass 500 is dissolved in a solvent of cryoscopic constant 6.9 K to form a saturated solution of concentration 0.01 mol kg^{-1}. Calculate

(a) the freezing-point depression

(b) the osmotic pressure

assuming that the molar volume is 22.4 dm^3.

Verify that, for a sparingly-soluble compound of high relative molar mass, it is more practicable to determine M_r from osmotic-pressure measurements than from the depression of the freezing point. It may be assumed that a temperature change can be measured with a probable error of $\cdot \pm 0.002$ K and an osmotic pressure with an error of ± 2 mmHg.

6.11 The freezing-point depressions for *p*-cresol ($M = 108.1$) solutions in benzene ($k_f = 5.11$ K) are 0.420 K and 5.002 K, for solutions of molality 0.0860 and 1.850, respectively. Calculate the degree of dimerization at the two concentrations.

6.12 100 cm^3 of an aqueous solution of phenol containing 15 g dm^{-3} are shaken with

(a) one 50 cm^3 portion of amyl alcohol

(b) two 25 cm^3 portions of amyl alcohol.

If the ratio of the concentration of phenol in water to that in amyl alcohol is 0.0625 calculate the total weights of phenol extracted in the processes (a) and (b) above.

6.13 In the steam distillation of bromobenzene the boiling point was 369 K at 770 mmHg. If the partial vapour pressure of water is 657.6 mmHg at 369 K, calculate the percentage by weight of bromobenzene in the steam-distillate.

6.14 The antibiotic compound gliotoxin, $C_{13}H_{14}O_4N_2S_2 \cdot xH_2O$, crystallizes with a unit cell of volume 1.451 ± 0.002 nm^3; its measured density is 1543 ± 10 kg m^{-3}. Chemical evidence suggests a relative molar mass of about 330. Determine

(a) the number of molecules in the unit cell, and

(b) the value of M_r, with its estimated standard deviation.

7

Electrochemistry

7.1 Electrical conduction

Two examples of electrical conduction are illustrated by Figures 7.1 and 7.2. In Figure 7.1 the current consists of a stream of electrons through the wires and the metal filament. This is electronic conduction. The electrons are part of the electronic structure of the conductor; the battery B supplies the electromotive force to drive them round the circuit, from a region of high electron potential (the negative terminal of the battery) to one of low potential (the positive terminal). Note that the flow of electrons is from negative to positive, the opposite to the flow of conventional 'electric current'.

In Figure 7.2, the current is carried through the sulphuric acid solution by positive hydrogen ions (cations) and by negative sulphate ions (anions). The cations are discharged by the negative electrode; they gain electrons, and so are reduced to hydrogen atoms, which combine to give hydrogen gas

$$H_3O^+ + e^- \rightarrow \tfrac{1}{2}H_2 + H_2O \tag{7.1}$$

We indicate here that the species reduced is a solvated proton, H_3O^+. The sulphate anions are not discharged; they travel towards the positive electrode, but the electron involved in the reduction (7.1) comes from the oxidation of a water molecule

$$\tfrac{3}{2}H_2O \rightarrow H_3O^+ + \tfrac{1}{4}O_2 + e^- \tag{7.2}$$

We envisage this as the favourable process in comparison with

$$\tfrac{1}{2}SO_4^{2-} \rightarrow \tfrac{1}{2}SO_4 + e^- \tag{7.3}$$

since the sulphate radical (SO_4) is not a stable species.

Solutions which contain ions are able to conduct a current and are called electrolyte solutions; the substance producing these ions in solution is termed an electrolyte. Salts in the liquid (molten) state are also electrolytes, and are good conductors.

7.2 Faraday's laws of electrolysis

The decomposition of an electrolyte by an electric current is called electrolysis, and its quantitative aspect is governed by Faraday's laws of electrolysis. They may be combined to give

$$m(i) = \frac{M_r(i)It}{F|z(i)|} \qquad (7.4)$$

where $m(i)$ is the mass of an ionic species i which is discharged by the passage of a current of I ampere for t second; $M_r(i)$ is the relative molar mass of the ion and $|z(i)|$ its numerical charge. The product It is the charge, in coulomb, passed through the solution; if $It = F$,

$$m(i) = M_r(i)/|z(i)| \qquad (7.5)$$

In other words, F coulomb is the charge required to discharge $1/|z(i)|$ mole of any ion; this quantity is 96 486.5 coulomb, and is termed the Faraday; F therefore has the units $C \, mol^{-1}$. The total charge on a mole of ions is $N_A|z(i)|e$, where e is the electronic charge, and N_A is the Avogadro constant. Hence

$$z(i)F = N_A|z(i)|e$$

or

$$e = F/N_A = 1.6022 \times 10^{-19} C$$

7.3 Electrolyte solutions

The basis of modern electrolyte theory is largely the work of Svante Arrhenius; he assumed that an electrolyte – acid, base or salt – undergoes

Figure 7.1. Electronic conduction.

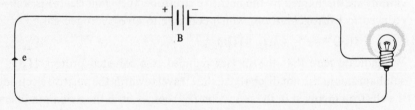

Figure 7.2. Electronic conduction, with electrolytic conduction through a solution of sulphuric acid.

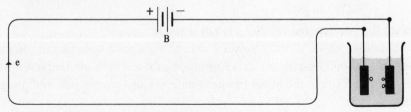

spontaneous dissociation in solution to a greater or lesser degree, forming positive and negative ions. Under the influence of an applied electric field, the cations and anions acquire a component of velocity towards the cathode and anode respectively. This component is small: for example, potassium ions in $0.1 \, \text{mol dm}^{-3}$ potassium chloride at 298 K, under a potential gradient of $1 \, \text{V m}^{-1}$, travel with velocity of $1.034 \times 10^{-7} \, \text{m s}^{-1}$. The random thermal motion of the ions is much greater.

The extent of dissociation of an electrolyte is expressed by α, the fraction of 'molecules' which are dissociated into ions by interaction with the solvent. Suppose a fraction α of an electrolyte $M_a X_b$ to be dissociated:

$$M_a X_b \rightleftharpoons aM^{z+} + bX^{z-} \tag{7.6}$$

If the solution contains x formula weights of the electrolyte per dm^3, the concentrations of the species present are:

$$M_a X_b = x(1-\alpha) \, \text{mol dm}^{-3}$$
$$M^{z+} = \alpha a x \quad \text{mol dm}^{-3}$$
$$X^{z-} = \alpha b x \quad \text{mol dm}^{-3}$$

The dissociation constant, from the law of mass action, is

$$K = \frac{(\alpha a x)^a (\alpha b x)^b}{x(1-\alpha)} = \frac{a^a b^b \alpha^{(a+b)} x^{(a+b-1)}}{(1-\alpha)} \tag{7.7}$$

As we saw in Chapter 6, the ratio

$$\frac{\text{number of moles of all species in solution}}{\text{number of moles if no dissociation were to occur}}$$

is called the van't Hoff i factor. In the present case

$$i = \alpha a x + \alpha b x + x(1-\alpha) = (a+b-1)\alpha + 1 \tag{7.8}$$

or

$$\alpha = \frac{i-1}{(a+b-1)} = \frac{i-1}{v-1} \tag{7.9}$$

where $v = a + b$. For example, assuming lanthanum sulphate, $La_2(SO_4)_3$, to be fully dissociated

$$La_2(SO_4)_3 \rightleftharpoons 2La^{3+} + 3SO_4^{2-}$$

at concentration $x \, \text{mol dm}^{-3}$, the i factor is $4\alpha + 1 = 5$. Assuming the activity coefficients to be unity, a colligative property such as freezing-point depression would be five times as great as expected for a solution of molar concentration (refer also to Section 6.8).

7.4 Electrolytic conductance

The resistivity ρ of a solid conductor is obtained from its resistance R and its physical dimensions, length l and cross-sectional area A:

$$R = \rho l / A \quad \text{or} \quad \rho = A R / l \tag{7.10}$$

R is measured in Ω(ohm), l in m, and A in m^2; thus ρ has the units Ω m.

In considering electrolytes we prefer to use the reciprocal of ρ, the *conductivity* κ, with the units Ω^{-1} m^{-1}

$$\kappa = 1/\rho = l/AR \qquad (7.11)$$

However, for liquids and solutions, l/A cannot in general be obtained from the physical dimensions of a conductivity cell (see Figure 7.3), since the conductance path is no longer normal to the electrodes. We determine experimentally the quantity l/A, called the cell constant of the conductivity cell, by calibration with a solution of known conductivity κ_s (see p. 234). If the cell containing this standard solution has resistance R_s:

$$R_s = \frac{1}{\kappa_s}\frac{l}{A} \qquad (7.12)$$

hence

$$l/A = R_s\kappa_s \qquad (7.13)$$

If the cell containing a solution of conductivity κ_x has resistance R_x, then

$$\kappa_x = \frac{1}{R_x}\frac{l}{A} = \frac{R_s\kappa_s}{R_x} \qquad (7.14)$$

and κ_x is determined.

In order to study the effect of concentration on conductance for various electrolyte solutions we need to define the conductance of one mole of electrolyte at various concentrations. Consider a solution of MCl of concentration c mol dm^{-3}. Then one mole of electrolyte is present in $1/c$ dm^3, or, to be consistent with the units of κ, in $1/1000c$ m^3. If this volume of solution were enclosed by two parallel, plane electrodes separated by the unit distance 1 m, the area of each electrode would be $1/1000c$ m^2. Each cubic metre of solution would have a conductivity κ, so that the conductance of one mole of MCl, at a

Figure 7.3. Wheatstone bridge circuit for the measurement of the resistance R of an electrolyte solution.

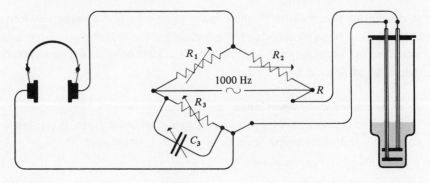

concentration of c mol dm^{-3}, would be $\kappa/1000c\ \Omega^{-1}$ mol^{-1} m^2 (or κ/c with c in mol m^{-3}). This is called the *molar conductance* at concentration c, Λ_c.

Because the great majority of electrochemical literature was, and still is, written in terms of the cgs system of units, we give in Appendix A13 a comparison of the relative values of some electrochemical quantities in the two systems.

If the electrolyte considered is MCl$_z$, dissociating into M^{z+} and zCl$^-$ ions, then for comparison with MCl etc, we compare the conductance of $1/z$ moles of MCl$_z$, ie

$$\Lambda_{1/z} = \kappa/1000cz\ \Omega^{-1}\ \text{mol}^{-1}\ \text{m}^2 \tag{7.15}$$

7.4.1 *Measurement of conductivity*

In the Wheatstone bridge type of measurement, shown in Figure 7.3, the quantity which is directly measured is the resistance of the conductivity cell filled with the solution. The electrodes of the cell are connected to the 'unknown' arm of the bridge, and balance is obtained by adjusting both R_3 and C_3 for a sharply-defined minimum in the response of the detector. In order to avoid electrolyzing the solution and polarizing the electrodes the bridge is operated with an alternating (AC) emf supply; this is why C_3 is necessary to balance the capacitance of the cell. If the frequency of the AC supply is about 1 kHz, a pair of headphones, or a cathode ray oscilloscope, may be employed as the detector.

The Wheatstone bridge is often replaced nowadays by a transformer ratio-arm bridge, Figure 7.4; in this case the conductance of the cell is measured directly. G_s and C_s are two components from a range of standard conductors and capacitors

Figure 7.4. Transformer bridge circuit – resistive and reactive components of bridge impedances.

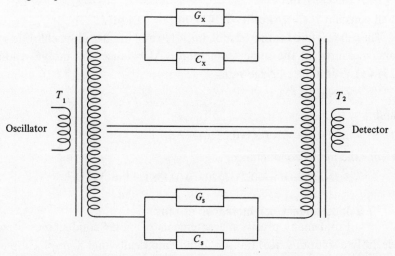

which may be switched into the circuit; A_x is the composite admittance $(G_x + j\omega C_x)$ where $j = \sqrt{-1}$, of the cell in the 'unknown' arm of the bridge. Balance is obtained by selecting appropriate values of G_s and C_s, and by adjustment of the fine control – the ratio of the numbers of turns tapped from the transformers T_1 and T_2 – until, at its maximum gain, the detector registers zero output across the secondary winding of T_2. The conductance G_x of the cell is then

$$G_x = G_s \times \text{turns-ratio } T_1/T_2$$

In the Wayne–Kerr series of conductance bridges the frequency of the oscillatory supply is 1592 Hz, facilitating AC calculations, since $\omega = 2\omega f = 10^4$ Hz.

Since G_x has been measured directly, from

$$\kappa_x = G_x l/A \tag{7.16}$$

l/A is the cell constant, obtained by calibration with a solution of known conductivity, as already mentioned. Such calibrating solutions are generally aqueous solutions of potassium chloride; conductivities at three concentrations (to suit different conductivity cells) and at two temperatures are given in Table 7.1. A good account of the calibration of these solutions is given in R. A. Robinson & R. H. Stokes, *Electrolyte Solutions*, Academic Press, New York (1959). The table shows the necessity for strict control of temperature in making conductance measurements; the conductivity cell must be immersed in a thermostat bath, preferably of light oil, maintained at a temperature constant to within at least ± 0.1 K.

Example

A conductivity cell was calibrated with 0.1 mol dm^{-3} potassium chloride at 298 K, for which $\kappa = 0.1411\,\Omega^{-1}\,\text{m}^{-1}$. At balance, the conductance was $141.1\,\Omega^{-1}$ at a bridge ratio of 10:1.

$$G(\text{cell}) = 14.11\,\Omega^{-1}$$

Cell constant $l/A = \kappa/G(\text{cell}) = 0.1411/14.11 = 0.01\,\text{m}^{-1}$.

The cell was filled with 0.02 mol dm^{-3} (20 mol m^{-3}) sodium chloride solution, and measured at the same temperature. At balance, the bridge reading was $231.6\,\Omega^{-1}$ at a 10:1 bridge ratio,

$$G(\text{NaCl}) = 23.16\,\Omega^{-1}$$

and

$$\kappa(\text{NaCl}) = G \times l/A = 0.2316\,\Omega^{-1}\,\text{m}^{-1}$$

Hence the molar conductance,

$$\Lambda_c(\text{NaCl}) = \kappa/c = 0.2316/20 = 0.01158\,\Omega^{-1}\,\text{mol}^{-1}\,\text{m}^2$$

7.5 Independent conductances of ions

From many precise measurements of molar conductance of aqueous electrolyte solutions, Kohlrausch found empirically that a graphical plot of Λ_c

against $c^{\frac{1}{2}}$ was linear, at low concentrations, for strong electrolytes. The molar conductances of weak electrolytes such as ethanoic acid and ammonia showed marked curvature of similar plots (Figure 7.5).

Where the plots are linear they may be extrapolated to limiting values at $c = 0$; these are the molar conductances at zero concentration (infinite dilution), designated Λ_0. In this limit there is no interaction between the ions, and they move independently with characteristic maximum velocities. It might be thought

Table 7.1. *Conductivity of potassium chloride solutions*

	$\kappa/\Omega^{-1} \, m^{-1}$	
$c/mol \, dm^{-3}$	291 K	298 K
0.010	0.1223	0.1411
0.100	1.119	1.289
1.00	9.820	11.17

Figure 7.5. Molar conductivity of electrolyte solutions at 298 K.

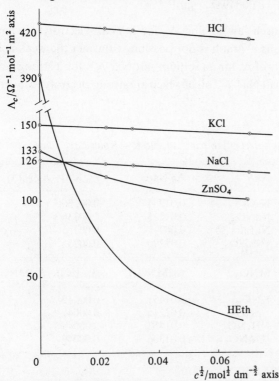

that Λ_0 would simply be a measure of the conductance of the solvent, but this is not so; we may regard Λ_0 as the mathematical limit of Λ_c as $c \rightarrow 0$, ie

$$\Lambda_0 = \underset{c \to 0}{\text{Limit}} \; \kappa/c \qquad\qquad (7.17)$$

As $c \rightarrow 0$, κ also tends to zero, and the ratio κ/c has a finite limit, characteristic of the strong electrolyte.

The concept of molar conductance at zero concentration leads to Kohlrausch's postulate of the independent conductances of ions (Table 7.2). The difference between Λ_0 for K^+A^- and Na^+A^- is seen to be constant, and independent of the anion species A^-; similarly, the difference in Λ_0 of M^+Cl^- and $M^+NO_3^-$ is independent of the cation species M^+. These relationships are examples of Kohlrausch's law of independent conductance of the ions in an electrolyte; the law is exact only for Λ_0, ie at $c=0$, where there can be no interaction between the ions. We can summarize:

$$\Lambda_0(MA) = \lambda_0(M^+) + \lambda_0(A^-) \qquad\qquad (7.18)$$

where λ_0 is the molar conductance of the ion at zero concentration. Evidently, from Table 7.2

$$\lambda_0(K^+) - \lambda_0(Na^+) = 0.00236 \, \Omega^{-1} \, mol^{-1} \, m^2$$
$$\lambda_0(Cl^-) - \lambda_0(NO_3^-) = 0.00049 \, \Omega^{-1} \, mol^{-1} \, m^2$$

We can apply the Kohlrausch law to find Λ_0 for a weak electrolyte, where extrapolation of the Λ_c versus $c^{\frac{1}{2}}$ graph is not possible. Consider the weak acid HA; we make measurements of Λ_c for its sodium salt Na^+A^-, for a strong acid H^+Y^-, and for the sodium salt Na^+Y^-, all of which are strong electrolytes, over a

Table 7.2. $\Lambda_0/\Omega^{-1} \, mol^{-1} \, m^2$ *for electrolyte pairs*[a] *at 298 K – Kohlrausch's law*

KA	$\Lambda_0(KA)$	NaA	$\Lambda_0(NaA)$	$\Lambda_0(KA) - \Lambda_0(NaA)$
KCl	0.01500	NaCl	0.01264	0.00236
KNO$_3$	0.01451	NaNO$_3$	0.01215	0.00236
KEth^3	0.01453	NaEth^3	0.00917	0.00236
$\frac{1}{2}$K$_2$SO$_4$	0.01535	$\frac{1}{2}$Na$_2$SO$_4$	0.01299	0.00236
MCl	$\Lambda_0(MCl)$	MNO$_3$	$\Lambda_0(MNO_3)$	$\Lambda_0(MCl) - \Lambda_0(MNO_3)$
KCl	0.01500	KNO$_3$	0.01451	0.00049
NaCl	0.01264	NaNO$_3$	0.01215	0.00049
NH$_4$Cl	0.01501	NH$_4$NO$_3$	0.01452	0.00049
$\frac{1}{2}$CaCl$_2$	0.01350	$\frac{1}{2}$Ca(NO$_3$)$_2$	0.01306	0.00049

[a] Eth$^-$ represents the anion $CH_3CO_2^-$.

range of concentration. By extrapolation of each plot we obtain the corresponding Λ_0. Then, from the Kohlrausch law

$$\Lambda_0(Na^+A^-) = \lambda_0(Na^+) + \lambda_0(A^-)$$
$$\Lambda_0(H^+Y^-) = \lambda_0(H^+) + \lambda_0(Y^-) \tag{7.19}$$
$$\Lambda_0(Na^+Y^-) = \lambda_0(Na^+) + \lambda_0(Y^-)$$

Hence

$$\Lambda_0(Na^+A^-) + \Lambda_0(H^+Y^-) - \Lambda_0(Na^+Y^-) = \lambda_0(H^+) + \lambda_0(A^-)$$
$$= \Lambda_0(HA) \tag{7.20}$$

In this way, Λ_0 for ethanoic acid, at 298 K, is found to be $0.0396\,\Omega^{-1}\,mol^{-1}\,m^2$ (see Problem 7.5). Even at the low concentration of $10^{-3}\,mol\,dm^{-3}$, Λ_c for ethanoic acid is only $0.00515\,\Omega^{-1}\,mol^{-1}\,m^2$; we compare the molar conductances of a strong and a weak acid in Table 7.3.

7.6 Dissociation constant

Arrhenius suggested that the ratio of Λ_c, the molar conductance of an electrolyte in solution of molarity c, to Λ_0, its molar conductance at $c=0$, was a measure of α, the fraction of the electrolyte molecules which dissociate into ions at this concentration. If we replace α in the Ostwald equation (p. 186) by this ratio, the expression for the dissociation constant K_a becomes

$$K_a = \frac{c(\Lambda_c/\Lambda_0)^2}{(1 - \Lambda_c/\Lambda_0)} = \frac{c\Lambda_c^2}{\Lambda_0(\Lambda_0 - \Lambda_c)} \tag{7.20}$$

The dissociation constant K_a for ethanoic acid has been calculated in this way, as recorded in Table 7.3. It will be seen that K_a is reasonably constant over this range of concentration, so that the assumption $\Lambda_c/\Lambda_0 = \alpha_c$ is valid for ethanoic acid and similar weak electrolytes; incomplete dissociation at finite concentrations explains why the Λ_c versus $c^{\frac{1}{2}}$ plots are curved. From (7.20) we can derive

$$\Lambda_c = \frac{K_a\Lambda_0 + \{K_a^2\Lambda_0^2 + 4cK_a\Lambda_0^2\}^{\frac{1}{2}}}{2c}$$

so that Λ_c is evidently not a linear function of $c^{\frac{1}{2}}$.

A similar calculation for hydrochloric acid, Table 7.4, shows that the assumption is not valid for strong electrolytes, and another explanation is required for their much smaller change of Λ_c and its linearity in $c^{\frac{1}{2}}$. We shall return to the consideration of strong electrolytes in Section 7.9.

7.7 Mobilities of ions

The molar conductance of an ion in solution of molarity c is related to its mobility $u(i)$ in that solution – ie its velocity towards an electrode due to an applied electric field of gradient $1\,V\,m^{-1}$. The relationship is

$$\lambda(i) = z(i)Fu(i) \tag{7.21}$$

where F is the Faraday. To derive this expression, consider the imaginary cell of Figure 7.6 in which c moles of an electrolyte MA are enclosed between planar electrodes of area 1 m^2, separated by a distance of 1 m. The concentration of MA is thus c mol m^{-3}. If a current of I A flows when the potential difference of the electrodes is 1 V, then from Ohm's law

$$I = V/R = 1/R = \kappa \qquad (7.22)$$

since the conductance $1/R$ is, from the geometry of the cell, the conductivity κ. The conductance is made up of the flow of positive and negative ions. We imagine three planes parallel to the electrodes; B is a central reference plane, and the planes A and C are on either side of it, separated by distances numerically equal to $u(+)$ and $u(-)$ respectively, where $u(+)$ and $u(-)$ the mobilities of the ions, are also their speeds, in m s^{-1}, under the 1 V m^{-1} field gradient. Hence in one

Table 7.3. *Ostwald's law applied to ethanoic (H Eth) acids, at 298 K*

c/mol dm^{-3}	Λ_c(HEth)/Ω^{-1} mol^{-1} m^2	Λ_c/Λ_0	K_a
0.10	0.000523	0.0132	1.765×10^{-5}
0.01	0.00168	0.0427	1.905×10^{-5}
0.001	0.00515	0.130	1.942×10^{-5}
0.00	0.0396	1.00	

Table 7.4. *Ostwald's law applied to hydrochloric acid*

c/mol dm^{-3}	Λ_c/Ω^{-1} mol^{-1} m^2	Λ_c/Λ_0	Apparent K
0.10	0.03913	0.9185	1.035
0.01	0.0412	0.967	0.283
0.10	0.0421	0.988	0.081

Figure 7.6. Schematic diagram to illustrate the calculation of ion mobility.

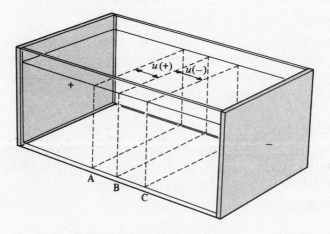

second all the positive ions in the volume $u(+) \, m^3$ reach, or cross, the plane B; similarly, all the negative ions in a volume $u(-) \, m^3$ reach, or cross, B from the opposite direction. If we consider here that MA is a symmetrical electrolyte, for which $|z(+)| = |z(-)| = z$, i.e. $MA \rightleftharpoons M^{z+} + A^{z-}$ (the argument may be extended to the general case), then the charge in coulomb crossing the plane B per second is Q, where

$$Q = zF\{n(+) + n(-)\} \tag{7.23}$$

where $n(+)$ and $n(-)$ are the numbers of moles of cations and anions in volumes $u(+)$ and $u(-)$ respectively. Then

$$dQ/dt = zF\alpha c\{u(+) + u(-)\} \tag{7.24}$$

where α is the fraction of free ions of the electrolyte MA at molarity c. Since dQ/dt is the current I, from (7.22)

$$\kappa = zF\alpha c\{u(+) + u(-)\} \tag{7.25}$$

and by definition

$$\Lambda_c = \kappa/c$$

So that

$$\Lambda_c = zF\alpha\{u(+) + u(-)\} \tag{7.26}$$

At $c = 0$, $\alpha = 1$, and (7.26) becomes

$$\Lambda_0 = zF\{u(+) + u(-)\} \tag{7.27}$$

Equation (7.27) may be subdivided into the ionic conductances $\lambda_0(+) = zFu_0(+)$, etc. In general

$$\lambda(i) = z(i)Fu(i) \qquad \lambda_0(i) = z(i)Fu_0(i) \tag{7.21}$$

7.8 Transport numbers

In using the law of independent conductances of ions to obtain Λ_0 for a weak electrolyte, we did not require the individual conductances $\lambda_0(i)$. These could be obtained from measurement of their mobilities via (7.21); in practice, however, $\lambda(i)$ is obtained indirectly by introducing the concept of *transport number* (in American literature, transference number).

The transport number $t(i)$ is defined as the fraction of the charge passed through an electrolyte which is carried by a particular ionic species i. Thus, if $q(i)$ is the charge carried by ion i through a solution of sodium chloride:

$$\left. \begin{array}{l} t(Na^+) = q(Na^+)/\{q(Na^+) + q(Cl^-)\} \\ \text{and} \qquad t(Cl^-) = q(Cl^-)/\{q(Na^+) + q(Cl^-)\} \end{array} \right\} \tag{7.28}$$

so that $t(Na^+) + t(Cl^-) = 1$. In a mixture of sodium chloride and hydrochloric acid solutions:

$$t(Na^+) + t(H^+) + t(Cl^-) = 1$$

The total charge passed through an electrolyte solution is not in general equally divided between the anions and cations. For an electrolyte $A^{z+}B^{z-}$ for which $|z(+)| = |z(-)| = z$, from (7.21)

$$\lambda(A) = zFu(A) \qquad \lambda(B) = zFu(B)$$

hence
and

$$\left. \begin{array}{l} t(A) = zFu(A)/\{zFu(A) + zFu(B)\} = u(A)/\{u(A) + u(B)\} \\ t(B) = zFu(B)/\{zFu(A) + zFu(B)\} = u(B)/\{u(A) + u(B)\} \end{array} \right\} \quad (7.29)$$

Thus the fraction of the charge which is carried by an ion, ie its transport number, depends upon its relative mobility. A solution in which $u(A^+)$ and $u(B^-)$ are numerically different still remains electrically neutral when a charge is passed through it, but the concentration does not remain uniform throughout the solution. Specific changes occur in the number of ions of each type in the vicinity of the electrodes. Hittorf (in 1853) interpreted these changes in terms of the transport numbers of the ions.

Consider the passage of 1 Faraday of charge through aqueous copper sulphate between two copper electrodes, in the transport cell of Figure 7.7 the original composition of the solution is known. If the transport numbers of the cations and anions are $t(+)$ and $t(-)$ respectively, the charge transported by the copper ions is $t(+)F$, and that transported by sulphate ions is $t(-)F$. Thus for the passage of a total charge of $1F$, $\frac{1}{2}t(+)$ mole of copper ions leave the anode compartment and

Figure 7.7. Hittorf's apparatus for the determination of transport numbers. The clips are open during the experiment, and are closed when the contents of the outer compartments are removed for analysis, so as to prevent siphoning from the centre compartment.

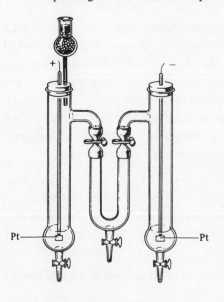

enter the centre compartment; a similar quantity of copper ions move from the centre compartment into the cathode compartment. At the cathode $\frac{1}{2}$ mole of copper ions are discharged and deposited on the cathode

$$\tfrac{1}{2}Cu^{2+} + e^- \rightarrow \tfrac{1}{2}Cu \tag{7.30}$$

Thus the net loss of copper ions from the cathode compartment is $\frac{1}{2} - \frac{1}{2}t(+)$ mole, $= \frac{1}{2}\{1 - t(+)\} = \frac{1}{2}t(-)$ mole, since $t(+) + t(-) = 1$. The cathode compartment also loses $\frac{1}{2}t(-)$ mole of sulphate ions, so that $\frac{1}{2}$ mole of copper sulphate is removed from this compartment. Considering now the anode compartment, $\frac{1}{2}t(+)$ mole of copper ions leave, and $\frac{1}{2}t(-)$ mole of sulphate enter, having 'passed through' the centre compartment. The sulphate anion is not discharged: the electrode reaction at this electrode is

$$\tfrac{1}{2}Cu \rightarrow \tfrac{1}{2}Cu^{2+} + e^- \tag{7.31}$$

The anode compartment, therefore, gains $\frac{1}{2} - \frac{1}{2}t(+) = \frac{1}{2}t(-)$ mole of copper sulphate.

In practice only a small fraction of a Faraday is passed, measured by a coulometer connected in series with the transport cell; this is to ensure that the changes at the electrodes are confined within their respective compartments. The centre compartment should be checked to make sure it has remained unchanged in composition. Note that the whole of each compartment must be analyzed after the electrolysis; it is not the change of concentration, but of the number of moles within each compartment, which is required (see example below).

If in the electrolysis of aqueous copper sulphate the electrodes were of platinum instead of copper, the reaction at the cathode would have been the same as (7.30) and the change within this compartment would again be a loss of $\frac{1}{2}t(-)$ mole of copper sulphate per Faraday passed. At the anode, however, the oxidation process (7.31) would be replaced by

$$\tfrac{3}{2}H_2O \rightarrow H_3O^+ + \tfrac{1}{4}O_2 + e^- \tag{7.32}$$

If this compartment is analyzed for copper ions, a loss of $\frac{1}{2}$ mole of copper ions per Faraday will be recorded; if we analyze for sulphate ions, a gain of $\frac{1}{2}t(-)$ mol F^{-1} will be observed. The difference, $\frac{1}{2}t(-) - \{-\frac{1}{2}t(+)\} = \frac{1}{2}$ mole, represents the hydroxonium ions, H_3O^+, from the anode oxidation. This illustrates the fact that we must know the electrode reactions before se can interpret the changes in electrolyte content in terms of transport numbers.

Example

A solution containing 10.850 g of silver nitrate in 1000 g water was electrolyzed between silver electrodes. A series coulometer recorded the charge passed through the solution as 142.0 C (ie 1.47×10^{-3} F). The anode and centre compartments were analyzed after the electrolysis:

Mass of anode compartment $= 54.900$ g

Mass of silver nitrate present $= 0.722$ g

Mass of centre compartment $= 46.200$ g

Mass of silver nitrate present $= 0.496$ g

Mass of water in the centre compartment $= 45.704$ g

Mass of silver nitrate originally present in this amount of water $=$
$45.704 \times 10.850/1000 = 0.495$ g

Thus the centre compartment has remained unchanged.

Mass of water in the anode compartment $= 54.178$ g

Mass of silver nitrate originally present $= 54.178 \times 10.85/1000 = 0.5878$ g

Mass of silver nitrate after electrolysis $= 0.722$ g

So the anode compartment has gained 0.134 g silver nitrate, or $0.134/169.9 =$
7.88×10^{-4} mole for the passage of 1.47×10^{-3} F, ie a gain of 0.536 mol F^{-1} of
silver nitrate. With silver electrodes the anode reaction is

$$Ag \rightarrow Ag^+ + e^-$$

and the anode compartment will gain $t(-)$ mol F^{-1}. Therefore $t(-)$, ie $t(NO_3^-)$,
$= 0.536$, and $t(Ag^+) = 0.464$.

Some results of transport number measurements are shown in Table 7.5. Since
the transport number $t(i)$ or $t(j)$ is a ratio $t(i, \text{ or } j)/(t(i) + t(j))$, its variation with c is
much less than that of $\lambda(i)$ or of $\Lambda(ij)$. From the transport numbers, the individual
conductances of the ions are found from

$$\lambda(i) = t(i)\Lambda(ij) = t(i)(\lambda(i) + \lambda(j))$$

where $t(i)$ is the transport number of ion i in the electrolyte containing that ion, of
molar conductance $\Lambda(ij)$. Some values of ion conductances are given in Table 7.6.
The high conductances of the hydrogen and hydroxyl ions in aqueous solution,
compared with those of other ions, are the result of the relationship of these ions
as the self-dissociation products of the solvent

$$2H_2O \rightleftharpoons H_3O^+ + OH^- \tag{7.33}$$

For these ions in the strongly hydrogen-bonded solvent water, a chain process of
conduction is envisaged as in Figure 7.8; the movement of hydroxyl ions may be
considered as the movement of protons in the opposite direction. This 'chain
mechanism' of conduction was first suggested by Grotthus: although proton
transfer will occur only when an adjacent water molecule is suitably orientated,
the process is faster than diffusion through the randomly-orientated solvent
molecules. An extreme example of chain conduction is that of the bisulphate ion
in sulphuric acid. This ion is a product of the self-dissociation (autoprotolysis) of
the solvent:

$$2H_2SO_4 \rightleftharpoons H_3SO_4^+ + HSO_4^- \tag{7.34}$$

In this viscous medium the chain process of conduction is almost the only mechanism; thus, the transport number of the bisulphate ion in sulphuric acid solutions is in the range 0.96–0.99. (Such 'insoluble' salts as lead sulphate and barium sulphate dissolve readily in 100% sulphuric acid as the corresponding bisulphates:

$$Ba^{2+}SO_4^{2-} + H_2SO_4 \rightarrow Ba^{2+} + 2(HSO_4)^-$$

and the transport number $t(Ba^{2+})$ is about 0.009; see R. J. Gillespie & S. Wasif, *J. Chem. Soc.* (1953), 209.)

Table 7.5. *Cation transport numbers* $(t(+))$ *at 298 K*

	$c/\text{mol dm}^{-3}$			
Electrolyte	0.01	0.05	0.10	0.50
HCl	0.825	0.829	0.831	0.835
KCl	0.490	0.490	0.490	0.489
NaCl	0.392	0.388	0.385	0.381
AgNO$_3$	0.465	0.466	0.468	0.470

Table 7.6. *Molar conductances of ions at 298 K*

	$\lambda_0(+)/\Omega^{-1}\,\text{mol}^{-1}\,\text{m}^2$		$\lambda_0(-)/\Omega^{-1}\,\text{mol}^{-1}\,\text{m}^2$
H_3O^+	0.035	OH^-	0.0192
K^+	0.00745	I^-	0.00768
NH_4^+	0.00745	Cl^-	0.00755
Ag^+	0.00635	NO_3^-	0.00706
Na^+	0.00509	$CH_3CO_2^-$	0.00408
Li^+	0.00387	$\frac{1}{2}SO_4^{2-}$	0.00790
$\frac{1}{2}Ca^{2+}$	0.0060	$\frac{1}{4}Fe(CN)_6^{4-}$	0.0111

Figure 7.8. Diagrammatic representation of a chain mechanism for proton transfer.

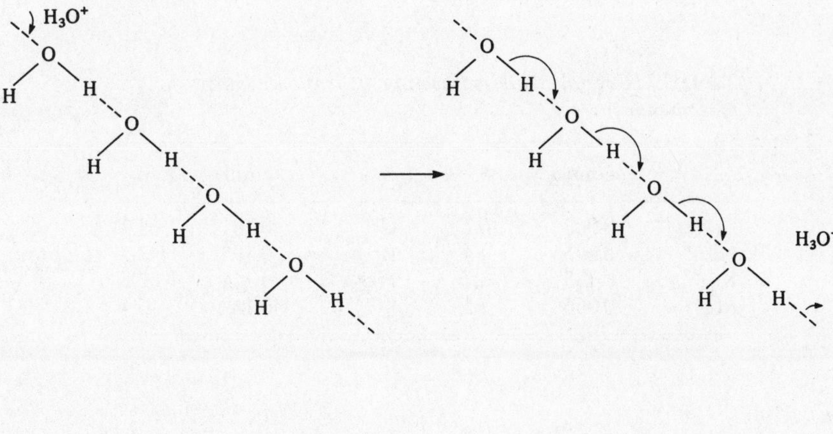

For a sphere of radius a moving through a medium of viscosity η under a force f, G. G. Stokes derived an expression for the uniform velocity v

$$v = f/6\pi\eta a \tag{7.35}$$

Thus, under a given force f, smaller particles move more rapidly than larger particles. For the ions of the alkali metals we would expect their velocities, and therefore their conductances, in aqueous solution, to increase in the order $K^+ < Na^+ < Li^+$, which is the reverse of the order shown in Table 7.6. We conclude that the species moving through the solution is the solvated (in this case, hydrated) cation, and that the degree of hydration is greater for the smaller ions, so that the lithium ion is effectively the largest cation. Comparing the conductances and other physical properties of the species $Li^+(H_2O)_m$, $Na^+(H_2O)_n$ and $K^+(H_2O)_p$, where m, n and p are the numbers of water molecules of solvation, their primary hydration shells, we deduce that $m > n > p$, Table 7.7.

7.9 Conductances of strong electrolytes

For a strong electrolyte such as potassium chloride, dissociation into ions is considered to be complete at all concentrations in aqueous solution, as it is in the solid. The reason for the decrease in Λ_c with c, linear in $c^{\frac{1}{2}}$, is the interionic attraction between cations and anions, which reduces the velocities and therefore the conductances $\lambda(i)$ of the ions. This is shown in Table 7.8; over the range 0–0.10 mol dm^{-3}, the velocities of the ions in ethanoic acid remain practically constant, whereas those of sodium chloride decrease by about 15%.

Debye and Hückel, in 1923, showed that this reduction in ionic velocities leads to a reduction in conductance proportional to $c^{\frac{1}{2}}$ for very dilute solutions. This was the result found experimentally by Kohlrausch. If we write it in the form

$$\Lambda_c = \Lambda_0 - Sc^{\frac{1}{2}} \tag{7.36}$$

then S is the slope of the Λ_c versus $c^{\frac{1}{2}}$ plot. The theory of Debye and Hückel interpreted S as the sum of two terms

$$S = A\Lambda_0 + B \tag{7.37}$$

Table 7.7. *Crystal radii, and average hydration numbers, $\bar{n}$, of some ions*

	r/nm	$\bar{n}$		r/nm	$\bar{n}$
Li^+	0.060	6.2	Cl^-	0.181	3.7
Na^+	0.095	4.4	Br^-	0.193	3.1
K^+	0.133	3.7	NO_3^-	0.124	2.5
Mg^{2+}	0.066	7.5	SO_4^{2-}	0.148	5.4

$A\Lambda_0$ is the relaxation term, related to how quickly an ion returns to its equilibrium state after being perturbed, and B is the electrophoretic term, related to the effect of the solvent upon the moving ion. A is a function of the relative permittivity ε_r of the solvent, and of the temperature; B is a function of ε_r, T and also the viscosity η of the solution – or, for the dilute solutions to which the theory is applicable, to the solvent viscosity. In general:

$$A = 0.8204 \times 10^6/(\varepsilon_r T)^{\frac{3}{2}} \tag{7.38}$$

$$B = 8.2501 \times 10^{-3}/(\varepsilon_r T)^{\frac{1}{2}}\eta \tag{7.39}$$

For dilute aqueous solutions at 298 K, $\varepsilon_r = 78.54$ and $\eta = 8.9 \times 10^{-4}$ kg m^{-1} s^{-1}. Substituting these values in (7.38) and (7.39):

$$A = 0.2289 \qquad B = 6.024 \times 10^{-4}$$

The molar conductance at c mol dm^{-3} is given by

$$\Lambda_c = \Lambda_0 - (A\Lambda_0 + B)c^{\frac{1}{2}} \tag{7.40}$$

which is usually known as the Debye–Hückel limiting law of conductance, since the various approximations made in its derivation are less significant at extreme dilution.

Table 7.9 illustrates the fit of the conductance data for sodium chloride to this equation; agreement is similar for most uni–univalent electrolytes in aqueous solution, in this range of concentration.

7.10 Ion association

In solvents of relative permittivity less than about 50, uni–univalent electrolytes exhibit lower conductances than would be expected from (7.40), due to ion association. Even in water, with ε_r nearly 80, bi–bivalent electrolytes such as $Mg^{2+}SO_4^{2-}$ show this effect. The force of attraction between ions M^{z+} and A^{z-} in a medium of relative permittivity ε_r, and separated by a distance r, is

$$f = |z(+)z(-)|e^2/4\pi\varepsilon_0\varepsilon_r r^2 \tag{7.41}$$

where e is the electronic charge and ε_0 is a constant (the permittivity of a vacuum).

Table 7.8. *Variation of ionic velocity with concentration*

	c/mol dm^{-3}	Λ_c/Ω^{-1} mol^{-1} m^2	α	$10^6 u(+)$/m s^{-1}	$10^6 u(-)$/m s^{-1}
CH_3CO_2H	0	0.0398	1.00	0.363	0.425
	0.01	0.00182	0.0472	0.362	0.423
	0.10	0.000417	0.0117	0.359	0.421
NaCl	0	0.01265	1.00	0.517	0.795
	0.01	0.01185	1.00	0.481	0.747
	0.10	0.01067	1.00	0.428	0.680

Thus the attraction between the ions increases with the product of their charges $z(+)z(-)$, and with decreases in ε_r. An equilibrium is set up

$$M^{z+}(\text{solv}) + A^{z-}(\text{solv}) \rightleftharpoons (M^{z+}A^{z-})^{|z(+)-z(-)|}(\text{solv}) \qquad (7.42)$$

where (solv) indicates solvation. With symmetrical electrolytes, $|z(+)| = |z(-)|$, the ion-pair on the right-hand side of (7.42) is uncharged, and is probably unsolvated; thus it does not contribute to the conductance of the solution. In Figure 7.9 the molar conductances Λ_c are plotted against $c^{\frac{1}{2}}$ for the electrolytes potassium

Table 7.9. *Molar conductance of aqueous sodium chloride at 298 K*

$c/\text{mol dm}^{-3}$	$\Lambda_c/\Omega^{-1}\,\text{mol}^{-1}\,\text{m}^2$ (from (7.40))	$\Lambda_c/\Omega^{-1}\,\text{mol}^{-1}\,\text{m}^2$ (exp.)	$\delta\%$
0	—	0.01265^a	—
5×10^{-4}	0.01245	0.01245	-0.02
1×10^{-3}	0.01237	0.01237	-0.07
5×10^{-3}	0.01202	0.01207	-0.37
1×10^{-2}	0.01176	0.01185	-0.76

^{a} From extrapolation of experimental data to $c=0$

Figure 7.9. Variation of molar conductance with the square root of concentration.

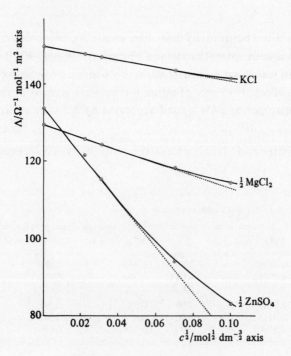

chloride, magnesium chloride and zinc sulphate in aqueous solution at 298 K. The negative slope increases as the force of attraction between cation and anion increases with the product $z(+)z(-)$. In the case of zinc sulphate this is sufficient for the equilibrium

$$Zn^{2+}(aq) + SO_4^{2-}(aq) \rightleftharpoons (Zn^{2+}SO_4^{2-})$$

to be established. The ion association constant K_A, an equilibrium constant, has the value 4.9×10^{-3} mol^{-1} dm^3. We limit our present discussion to the case of pairwise association of ions, but we note that more extensive association may occur with highly charged ions in media of very low relative permittivity. (An interesting example is shown by the behaviour of electrolytes in liquid hydrogen chloride. See T. C. Waddington, *J. Chem. Soc.* (1963), 69.)

We may treat ion-pair formation in the same way as the incomplete dissociation of weak electrolytes. If α is the fraction of the electrolyte in the form of free ions, $(1 - \alpha)$ is the fraction in the form of ion-pairs, and an ion association constant K_A may be defined

$$K_A = (1 - \alpha)/c\alpha^2 f^2(\pm) \tag{7.43}$$

assuming the ion-pair to be uncharged. From conductance measurements α and K_A can be determined, and thus the concentrations of the species present in solution. The concentration of the ion-pairs may be very important in a system where this species shows more or less pronounced chemical or biological activity as compared with the free solvated ions.

To take ion-pair formation into account we must modify (7.40) by the introduction of α

$$\Lambda_c = \alpha\Lambda_0 - \alpha(A\Lambda_0 + B)(\alpha c)^{\frac{1}{2}} \tag{7.44}$$

This equation may be solved for α by a process of iteration; if we put $\alpha = 1$ in the square-root term, a first approximation to α is

$$\alpha_1 = \Lambda_c/\{\Lambda_0 - (A\Lambda_0 + B)c^{\frac{1}{2}}\} \tag{7.45}$$

Putting α_1 in (7.44) we obtain a closer approximation to α

$$\alpha_2 = \Lambda_c/\{\Lambda_0 - (A\Lambda_0 + B)(\alpha_1 c)^{\frac{1}{2}}\} \tag{7.46}$$

and this process can be continued $n + 1$ times until α_n and α_{n+1} agree to within the precision of the experimental data. K_A is then obtained from (7.43). Some values of K_A are listed in Table 7.10. For further data, and for other methods of determining K_A, see C. W. Davies, *Ion Association*, Butterworth, London, 1962.

7.11 Applications of conductance measurements

7.11.1 *Determination of solubility products*

If the solubility in water of a sparingly soluble electrolyte such as silver bromate, $AgBrO_3$, is s mol dm^{-3} the molar conductance of the saturated solution at this temperature is Λ_s:

$$\Lambda_s = \kappa_s/1000s \tag{7.47}$$

where κ_s is the conductivity of the saturated solution, in the usual units of $\Omega^{-1}\,m^2$. (We recall that in $\Lambda_c = \kappa/c$, c must be expressed in $mol\,m^{-3}$.) Since the electrolyte is sparingly soluble the saturated solution is very dilute, and Λ_s will approximate to Λ_0, which is known from the conductances of its ions:

$$\Lambda_s \approx \Lambda_0 = [\lambda_0(Ag^+) + \lambda_0(BrO_3^-)] \tag{7.48}$$

Thus if we measure κ_s and refer to the literature for the ion conductances, s may be found from (7.47), and hence the solubility product S.

Example

The conductivity of a saturated aqueous solution of silver chloride at 298 K is $1.7574 \times 10^{-4}\,\Omega^{-1}\,m^{-1}$; the solvent water has a conductivity of $2.01 \times 10^{-5}\,\Omega^{-1}\,m^{-1}$. The solute ion conductances are $\lambda_0(Ag^+) = 0.00543\,\Omega^{-1}\,mol^{-1}\,m^2$, $\lambda_0(Cl^-) = 0.00655\,\Omega^{-1}\,mol^{-1}\,m^2$. Thus $\Lambda_0 = 0.01198\,\Omega^{-1}\,mol^{-1}\,m^2 = \Lambda_s/s$.

$$s = \kappa_s/\Lambda_0 = 1.5564 \times 10^{-4}/0.01198 = 0.01297\,mol\,m^{-3}$$

or

$$1.297 \times 10^{-5}\,mol\,dm^{-3}$$

Solubility product $= s^2 = 1.68 \times 10^{-10}\,mol^2\,dm^{-6}$.

7.11.2 *Determination of charge on a complex ion*

For singly-charged ions in water at 298 K, the average molar conductance λ_0, excluding the hydroxonium and hydroxide ions, is $0.0065\,\Omega^{-1}\,mol^{-1}\,m^2$ (Table 7.6). For divalent ions λ_0 is about $0.012\,\Omega^{-1}\,mol^{-1}\,m^2$, and correspondingly greater for ions of higher charge. Thus a comparison of Λ_c for a complex electrolyte with the values for combinations of known species of ions at

Table 7.10. *Association constants for ion-pair formation at 298 K*

Electrolyte	Solvent	ε_r	$\log K_A$
$Ca(OH)_2$	H_2O	78.5	1.3
$CaSO_4$	H_2O	78.5	2.28
$Ca_2Fe(CN)_6$	H_2O	78.5	3.77
$MgSO_4$	H_2O	78.5	2.23
$CuMal^a$	H_2O	78.5	5.70
$(CH_3)_4NClO_4$	CH_3OH	32.6	1.75
CsI	C_2H_5OH	24.3	2.15
$Cu(ClO_4)_2$	CH_3CN	36.2	2.28

[a] Mal is the malonate (propane dicarboxylate) ion, $CH_2(CO_2)_2^{2-}$

the same concentration, usually taken as 2^{-10} mol dm^{-3} or 1 mole in 1024 dm^3, tells us the charges on the ions of the complex species. Some examples are shown in Table 7.11.

7.11.3 *Acid–base titrations monitored by conductance*

If the conductivity of an acid solution is monitored during the course of titration by a base, an indication of the equivalence point of the titration is given from the conductivity versus titre plot; the shape of this curve depends upon the relative strengths (not concentrations) of the acid and the base.

If a strong base such as sodium hydroxide is added to a strong acid the highly conducting hydroxonium ions ($\lambda_0 = 0.0350\,\Omega^{-1}$ mol^{-1} m^2) are replaced by sodium ions ($\lambda_0 = 0.0035\,\Omega^{-1}$ mol^{-1} m^2) and the conductivity of the solution decreases until the equivalence point is reached; only in this case is the solution neutral at equivalence. Further addition of the base introduces excess sodium and hydroxide ions ($\lambda_0(\text{OH}^-) = 0.0190\,\Omega^{-1}$ mol^{-1} m^2) and the conductivity increases. If the volume change is kept small, e.g. by titrating 0.01 mol dm^{-3} acid with 1.0 mol dm^{-3} base, the changes in conductivity are practically linear, as shown in Figure 7.10(a).

If a weak acid is titrated by a strong base, there is an initial decrease in conductivity; the anion formed represses dissociation of the acid by the common-ion effect. With further addition of the base the conductivity increases due to

$$\text{HA} + \text{Na}^+\text{OH}^- \rightleftharpoons \text{H}_2\text{O} + \text{A}^- + \text{Na}^+$$

Beyond the equivalence point the conductivity rises more rapidly, due to excess hydroxide ions (Figure 7.10(b)). If the acid is very weak so that its initial conductivity is inappreciable, the conductivity of the solution increases throughout the titration.

The titration of a weak acid by a weak base can be monitored by change of pH (ie by following the emf of a hydrogen electrode, p. 254) but the equivalence point cannot be defined since the pH changes gradually even near equivalence; for the

Table 7.11. *Ionization of complex electrolytes*

		$\Lambda_{1024}/\Omega^{-1}$ mol^{-1} m^2	Mode of ionization
Electrolyte type	AB	0.0120–0.0140	A^+B^-
	$\text{AB}_2, \text{A}_2\text{B}$	0.0240–0.0280	$\text{A}^{2+}2\text{B}^-, 2\text{A}^+\text{B}^{2-}$
	$\text{AB}_3, \text{A}_3\text{B}$	0.0350–0.0430	$\text{A}^{3+}3\text{B}^-, 3\text{A}^+\text{B}^{3-}$
	AB_4	0.0450–0.0520	$\text{A}^{4+}4\text{B}^-$
Examples of	$\text{Pt(NH}_3)_6\text{Cl}_4$	0.0500	$\text{Pt(NH}_3)_6^{4+}4\text{Cl}^-$
complex ions	$\text{K}_3\text{Co(NO}_2)_6$	0.0418	$3\text{K}^+, \text{Co(NO}_2)_6^{3-}$
	$\text{Co(NH}_3)_3(\text{NO}_2)_3$	0.00015	non-ionic

same reason there is no sharp colour change of an indicator. It is possible to carry out this titration by observing the change in conductivity at the equivalence point (Figure 7.11).

The titration of a mixture of a strong and a weak acid by a strong base shows two points of inflection of the conductivity versus titre plot, corresponding to the

Figure 7.10. (*a*) Titration of 100 cm^3 of 0.01 mol dm^{-3} hydrochloric acid with $1.038 \text{ mol dm}^{-3}$ sodium hydroxide. (*b*) Titration of 100 cm^3 of 0.01 mol dm^{-3} salicylic acid with $1.000 \text{ mol dm}^{-3}$ sodium hydroxide.

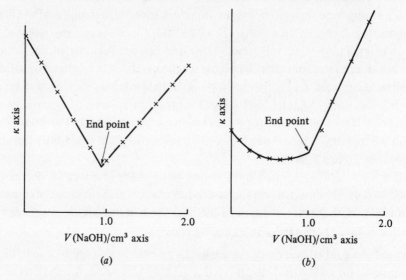

(*a*)   (*b*)

Figure 7.11. Titration of 100 cm^3 of 0.01 ethanoic acid (HEth) with $1.000 \text{ mol dm}^{-3}$ ammonia solution ($NH_3 \cdot H_2O$).

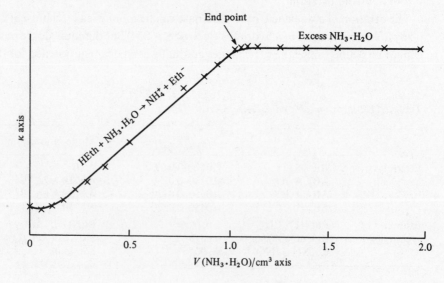

two equivalence points. The weak acid is effectively prevented from dissociating until the strong acid has been used up (Figure 7.12). Precipitation reactions in which there is a conductivity change at the end point may be followed similarly.

7.12 Electromotive force (emf): the emfs of voltaic cells

Metals have a tendency to dissolve in water, producing the metal cations, and leaving their valence electrons on the metal. Thus the metal acquires a negative potential, preventing the release of further cations, and an equilibrium is established

$$M \rightleftharpoons M^{z+} + ze^- \tag{7.49}$$

This tendency to dissolve is diminished, and may even be reversed, if there are already cations of the metal in solution. In thermodynamic terms, ΔG for the

Figure 7.12. Titration of a mixture of $50 \, cm^3$ of $0.020 \, mol \, dm^{-3}$ hydrochloric acid and $50 \, cm^3$ of 0.020 ethanoic acid with $1.000 \, mol \, dm^{-3}$ sodium hydroxide. The first (sharper) end point corresponds to the neutralization of the strong acid.

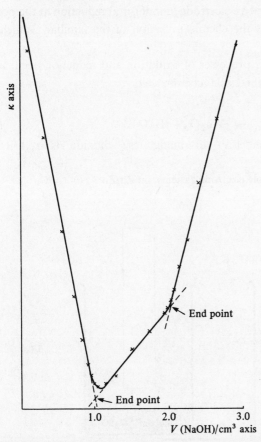

dissolution of a metal in water is negative, but becomes less negative, and may become positive as the concentration (or more correctly, the activity) of its cation increases. For zinc, with a high tendency to oxidize by (7.49), the reversal of sign of ΔG would occur at about 25 mol dm^{-3} zinc ions; for copper the concentration would be 10^{-12} mol dm^{-3}. Thus at all practical concentrations, zinc tends to oxidize

$$\tfrac{1}{2}Zn \rightarrow \tfrac{1}{2}Zn^{2+}(aq) + e^- \tag{7.50}$$

and copper ions tend to be reduced:

$$\tfrac{1}{2}Cu^{2+} + e^- \rightarrow \tfrac{1}{2}Cu \tag{7.51}$$

as illustrated in Figure 7.13.

If the two systems Zn/Zn^{2+} and Cu/Cu^{2+} are joined electrically as in Figure 7.14 where the two solutions are in electrical contact but are prevented from mixing freely, electron flow will occur; electrons are transferred from zinc to copper through the external circuit. The current flow is maintained by the transfer of anions from the copper ion solution to the zinc ion solution and of cations in the reverse direction, through the tap T. This is a voltaic cell, in which oxidation occurs at the negative electrode (anode), and reduction at the positive electrode (cathode). This is the chemist's version of the familiar Daniell cell, Figure 7.15.

In all voltaic cells similar processes of oxidation and reduction occur at the electrodes; for example in the Leclanché dry cell:

$$\tfrac{1}{2}Zn \rightarrow \tfrac{1}{2}Zn^{2+} + e^- \tag{7.52}$$

$$MnO_2 + H_3O^+ + e^- \rightarrow \tfrac{1}{2}Mn_2O_3 + \tfrac{3}{2}H_2O \tag{7.53}$$

At the cathode manganese(IV) in manganese dioxide is reduced to

Figure 7.13. Reversible electrode systems: (*a*) Zn/Zn^{2+}, (*b*) Cu/Cu^{2+}.

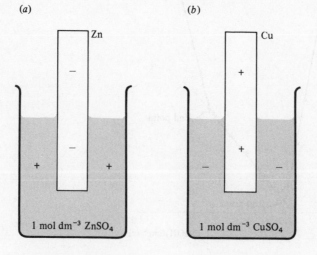

manganese(III) in dimanganese trioxide at an inert (carbon) electrode. The hydrogen ions are supplied by a paste of ammonium chloride:

$$NH_4^+ + 2H_2O \rightleftharpoons NH_3H_2O + H_3O^+ \qquad (7.54)$$

The sum of the potential differences between each electrode and its solution represents the emf of a voltaic cell. Here, emf has its usual meaning of the potential difference between two points in the absence of an $I \times R$ voltage drop, ie when the current I is zero. A conventional voltmeter draws an appreciable current from the circuit, so that emf is measured by a potentiometer (Figure 7.16) or an electrometer.

The individual potential difference between an electrode and its solution

Figure 7.14. An emf cell.

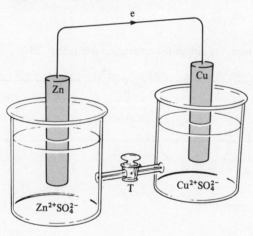

Figure 7.15. A Daniell cell.

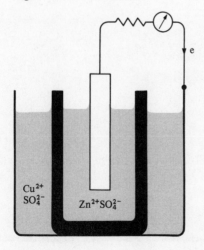

cannot be measured, because there is no way of making contact with the solution without introducing another electrode potential.

7.12.1 *Measurement of electrode potentials*
In order to compare individual electrode potentials, a reference electrode is introduced. This is the *standard hydrogen electrode* (SHE), shown in Figure 7.17. It consists of a platinum electrode, coated with finely-divided platinum (platinum black) in an aqueous solution containing hydrogen ions at unit activity; hydrogen gas is bubbled through the solution over the electrode at a partial pressure of one atmosphere. An equilibrium is set up:

$$\tfrac{1}{2}H_2 + H_2O \rightleftharpoons H_3O^+ + e^- \tag{7.55}$$

The platinum black acts as a catalyst for the rapid establishment of this equilibrium. At 298 K the potential of the SHE is defined as zero volts.

Figure 7.16. A potentiometer circuit for measuring emf (see p. 257).

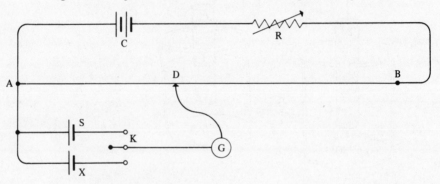

Figure 7.17. A hydrogen electrode.

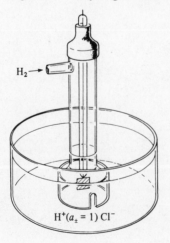

The potential of any electrode system can be determined by measuring the emf of a voltaic cell in which that electrode is combined with the SHE (Figure 7.18(*a*) and (*b*)). In (*a*) the SHE is the cathode, attracting electrons from the zinc anode:

$$\tfrac{1}{2}Zn + H_3O^+ \rightarrow \tfrac{1}{2}Zn^{2+} + \tfrac{1}{2}H_2 + H_2O \tag{7.56}$$

In (*b*) the SHE is the anode, supplying electrons to the copper cathode:

$$\tfrac{1}{2}H_2 + \tfrac{1}{2}Cu^{2+} + H_2O \rightarrow H_3O^+ + \tfrac{1}{2}Cu \tag{7.57}$$

At 298 K, and with unit activity of zinc ions, the emf of the cell (*a*) is 0.763 V; we adopt the convention: emf of any cell = potential of right-hand electrode − potential of left-hand electrode. Writing E for cell emf and π for electrode potential, $E = \pi(SHE) - \pi(Zn^{2+}/Zn) = 0.763$ V. Hence, $\pi(Zn^{2+}/Zn) = -0.763$ V, which is the *standard reduction potential* for the zinc electrode, and is written $\pi^{\ominus}$. At 298 K in a solution of copper ions of unit activity, the emf of cell (*b*) is 0.345 V. Thus $E = \pi(Cu^{2+}/Cu) - \pi(SHE) = 0.345$ V. Hence, $\pi(Cu^{2+}/Cu) = 0.345$ V, which under these standard conditions is the standard reduction potential, $\pi^{\ominus}$, of the copper electrode. Note that when the spontaneous electrode process is an oxidation, as in

$$\tfrac{1}{2}Zn \rightarrow \tfrac{1}{2}Zn^{2+} + e^- \quad (a(Zn^{2+}) = 1.0)$$

the electrode potential is negative; when the spontaneous electrode process is a reduction, as in

$$\tfrac{1}{2}Cu^{2+} + e^- \rightarrow \tfrac{1}{2}Cu$$

the electrode potential is positive. This explains the designation of these potentials as reduction potentials. A list of standard reduction potentials is given in Appendix A15.

Figure 7.18. Comparison of electrode potentials with an SHE: (*a*) Zn/Zn^{2+}, (*b*) Cu/Cu^{2+}; P is a potentiometer.

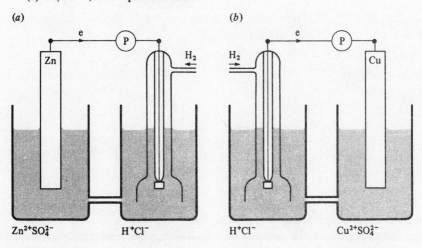

The hydrogen electrode is not a very convenient practical standard, and a subsidiary reference electrode is generally used. This is the calomel electrode (Figure 7.19), 'calomel' being the old name for mercury(I) chloride, Hg_2Cl_2. In this electrode a paste of mercury and calomel is in contact with a solution of potassium chloride; the electrode potential has been standardized against the SHE for three concentrations of potassium chloride, as given in Table 7.12. The position of the equilibrium at the calomel electrode depends upon the activity of the chloride ions in solution:

$$\tfrac{1}{2}Hg_2Cl_2 + e^- \rightarrow Hg + Cl^- \tag{7.58}$$

The electrode potential is given by

$$\pi = \pi^{\ominus} - (RT/F)\ln a(Cl^-) \tag{7.59}$$

where $\pi^{\ominus}$ (ie for $a(Cl^-) = 1$, $T = 298$ K) is 0.267 V. The activity of the chloride ions in saturated potassium chloride is 2.66, and the variation of π with temperature is given by

$$\pi = 0.242 - 7 \times 10^{-4}(T - 298)$$

Another practical reference electrode is the silver–silver chloride electrode,

Table 7.12. *Potentials of the calomel electrode, at 298 K*

$c/(KCl)/mol\ dm^{-3}$	π/V
0.1	0.334
1.0	0.281
4.5 (satd)	0.242

Figure 7.19. A saturated calomel reference electrode.

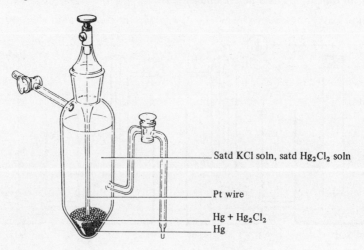

Satd KCl soln, satd Hg_2Cl_2 soln

Pt wire

$Hg + Hg_2Cl_2$
Hg

consisting of a silver wire coated with silver chloride by anodic deposition from a chloride solution, in a solution of chloride ions. The electrode reaction

$$Ag + Cl^- \rightarrow AgCl + e^- \tag{7.60}$$

is reversible, and the position of equilibrium again depends upon $a(Cl^-)$. For this electrode $\pi^\ominus = 0.222$ V.

7.12.2 *Measurement of emf*

The measurement of emf must be made under reversible conditions if the result is to be interpreted thermodynamically (Section 7.12.3). Thus no unidirectional chemical reaction, and hence current flow, must occur during the measurement. A potentiometer is generally used, as shown in Figure 7.16. The storage battery C is connected via the variable resistor R to a length of uniform resistance wire AB, wound on a drum. The sliding contact D connects a point on this wire through the galvanometer G to the standard cell S or to the cell to be measured, X, according to the setting of the switch K. Both cells are connected so as to oppose the emf of cell C. With K in the 'standardize' position D is fixed at a position on AB so that the voltage across AB corresponds to E_S, the emf of the standard cell. R is now adjusted until no current flows through G; then $V(AD) = E_S$, and the bridge wire is calibrated. K is now changed to the 'measure' position; E_X replaces E_S and D is now adjusted for zero current. E_X is given by $E_X/E_S = AD/AB$, and is read directly in volts. The resistance R adjusts for small changes in the emf of cell C.

The standard cell S is the Weston cell, which may be represented in the conventional way (in this convention, we must always write the cell with the anode (the negative electrode of a voltaic cell) on the left. The emf is then $\pi(RH$ electrode) $- \pi(LH$ electrode), as already stated):

$$12.5\% \; Cd/Hg | CdSO_4(satd), Hg_2SO_4(satd) | Hg \tag{7.61}$$

The left-hand electrode is a 12.5% cadmium amalgam covered with a layer of crystalline cadmium sulphate, $3CdSO_4 \cdot 8H_2O$; the right-hand electrode is of mercury covered by a paste of mercury(I) sulphate. The cell contains an aqueous solution saturated with respect to both salts. The cell reaction is reversible:

$$Cd(in \; Cd/Hg) + Hg_2^{2+} \rightleftharpoons Cd^{2+} + 2Hg \tag{7.62}$$

The emf of the Weston cell is 1.0183 V at 293 K, and its variation with temperature is given by

$$E_T = 1.0183 - 4.06 \times 10^{-5}(T - 293) - 9.5 \times 10^{-7}(T - 293)^2 \; V \tag{7.63}$$

The reason for the choice of components in this cell is to enable it to be reproduced exactly; the electrodes are both liquids, so that there can be no 'strain potentials' as are sometimes set up in metals, and the composition of the solution is fixed at a given temperature, since both solutes are in contact with the saturated solution.

7.12.3 *Thermodynamics of voltaic cells*

The electrical energy generated by a voltaic cell working in a thermo-dynamically-reversible manner is equal to the corresponding free energy decrease, $-\Delta G$. Consider the Daniell cell, which we may represent conventionally

$$Zn|Zn^{2+}SO_4^{2-} :: Cu^{2+}|Cu$$

where the symbol :: implies that the two solutions are in electrical contact (e.g. through a sintered glass disc) but are not freely miscible. For the passage of one Faraday through the cell, the reaction is

$$\tfrac{1}{2}Zn + \tfrac{1}{2}Cu^{2+} \rightarrow \tfrac{1}{2}Zn^{2+} + \tfrac{1}{2}Cu \tag{7.64}$$

The electrical work, w_e, done is nFE, where n is the number of Faradays, and E the emf of the cell, ie

$$w_e = nFE = -\Delta G \tag{7.65}$$

and for (7.64), where $n = 1$,

$$w_e = FE = -\Delta G$$

Under standard conditions, ie $T = 298$ K and $a(\pm) = 1$ for each solution, E is the standard emf, $E^\ominus$, of the cell, and

$$FE^\ominus = -\Delta G^\ominus \tag{7.66}$$

Note that a spontaneous change is denoted by a positive emf E and a negative ΔG. For the Daniell cell,

$$E^\ominus = \pi^\ominus(Cu^{2+}/Cu) - \pi^\ominus(Zn^{2+}/Zn)$$

$$= 0.345 - (-0.761) = 1.106 \text{ V}$$

Hence

$$\Delta G^\ominus = -FE^\ominus = -96\,486 \times 1.106 = -106.7 \text{ kJ}$$

Since ΔG is an extensive quantity it depends upon the amount of chemical change involved, ie on the number of Faradays passed. Thus, if instead of (7.64) we considered the change

$$Zn + Cu^{2+} \rightarrow Zn^{2+} + Cu \tag{7.67}$$

the electrical work w_e would be $2FE$, and

$$\Delta G^\ominus = -2FE^\ominus = -213.4 \text{ kJ}$$

Again, for the cell

$$Cu|Cu^{2+}SO_4^{2-} :: Ag_2^+SO_4^{2-}|Ag$$

for the passage of one Faraday

$$\tfrac{1}{2}Cu + Ag^+ \rightarrow \tfrac{1}{2}Cu^{2+} + Ag \tag{7.68}$$

with $E^\ominus = 0.454$ V. Thus $\Delta G^\ominus = -FE^\ominus = -43.81$ kJ. However, for the reaction

$$Cu + 2Ag^+ \rightarrow Cu^{2+} + 2Ag$$

implying the passage of two Faradays

$$\Delta G^\ominus = -2FE^\ominus = -87.62 \text{ kJ}$$

The enthalpy change for a cell reaction can be measured by carrying out the reaction in a calorimeter. For the reaction (7.63) $\Delta H^\ominus = -104.9$ kJ, very close to $\Delta G^\ominus$. This led originally to the incorrect view that the electrical work w_e could be equated to ΔH. However, measurements involving other cells showed that the close correspondence of ΔG and ΔH for the Daniell cell was fortuitous. For example, the reaction

$$\tfrac{1}{2}Pb + Ag^+ \rightarrow \tfrac{1}{2}Pb^{2+} + Ag \tag{7.69}$$

may be studied calorimetrically, or by setting up the cell†

$$Pb|PbCl_2, HCl, AgCl|Ag \tag{7.70}$$

with $E^\ominus = 0.490$ V. For this reaction, $\Delta H^\ominus$, as determined calorimetrically is -52.6 kJ mol^{-1} (ie mol^{-1} of electrons), but $-FE^\ominus = \Delta G^\ominus = -47.25$ kJ. The difference is $T\,\Delta S^\ominus$, from (3.97)

$$\Delta G^\ominus = \Delta H^\ominus - T\,\Delta S^\ominus \tag{7.71}$$

Hence the standard entropy change $\Delta S^\ominus$ for (7.69) is

$$-52.6 - (-47.25) \times 1000/298.15 = -17.94 \text{ J K}^{-1}\text{ mol}^{-1}$$

For the Daniell cell reaction the corresponding $\Delta S^\ominus$ is 4.36 J K^{-1} mol^{-1}.

The emf of a voltaic cell varies with temperature, as shown by the results for the Daniell cell and for the cell (7.70) in Figure 7.20. The variation of ΔG with temperature of a reversible reaction is expressed by the Gibbs–Helmholtz equation

$$(\partial\,\Delta G/\partial T)_P = -\Delta S \tag{7.72}$$

Substituting for ΔG in terms of the cell emf

$$(\partial(-nFE)/\partial T)_P = -\Delta S$$

Figure 7.20. Variation of E with T for the Daniell cell and for cell (7.70).

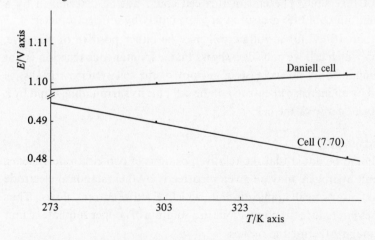

† Cell (7.70) contains only one solution (HCl) saturated with respect to both PbCl$_2$ and AgCl.

or, since n and F are independent of temperature

$$nF(\partial E/\partial T)_P = \Delta S \tag{7.73}$$

Thus the measurement of E at several temperatures and the calculation of the temperature coefficient dE/dT gives us the entropy change of the cell reaction within the range of temperature. Combining (7.71) and (7.73)

$$-nFE = \Delta H - nFT(\partial E/\partial T)_P$$

or

$$E = -\Delta H/nF + T(\partial E/\partial T)_P \tag{7.74}$$

Thus both ΔG and ΔH may be obtained from emf measurements on reversible cells. Note that it is not permissible to extrapolate the plot of Figure 7.20 to $T=0$ to obtain ΔH; in general the enthalpy change of a reaction varies with temperature, although over a limited range the variation of ΔH is negligible compared with that of $T \Delta S$, as the linearity of the graph indicates.

The thermodynamic treatment of this section is applicable to reversible voltaic cells, in which both electrode reactions are thermodynamically reversible. A practical indication that a cell is reversible is given by the potentiometer; if a balance point may be obtained when approached from either side of the position for $I(\text{galvo})=0$, then the cell is reversible. Consider the cell

$$\text{Zn}|\text{Zn}^{2+}::\text{H}^+\text{Cl}^-(\text{aq})|\text{Pt}$$

This is not reversible; the electrode reaction

$$\tfrac{1}{2}\text{Zn} \rightleftharpoons \tfrac{1}{2}\text{Zn}^{2+} + e^-$$

is reversible, but the reaction at the platinum electrode, in the absence of a supply of hydrogen gas, can proceed only in the direction

$$\text{H}_3\text{O}^+ + e^- \rightarrow \text{H}_2\text{O} + \tfrac{1}{2}\text{H}_2$$

and so is not reversible. The emf of this cell could not be determined by a potentiometer, but could be measured approximately by an electrometer.

The fact that dE/dT for a voltaic cell may be either positive or negative, depending on which cell we consider, shows that a spontaneous reaction is not always accompanied by an increase in entropy of the cell system; it is always accompanied by an increase in entropy of the cell plus its surroundings, and by a decrease in free energy of the cell.

7.12.4 *Overvoltage*

It should be noted that the relative positions of two electrode systems, eg a metal and hydrogen, may be given incorrectly by their standard electrode potentials for a practical application in which electric current flows. Thus hydrogen does not reduce the copper ions in a solution of copper sulphate of unit activity, although $\Delta G^{\ominus}$ for the process

$$\tfrac{1}{2}\text{H}_2 + \tfrac{1}{2}\text{Cu}^{2+} + \text{H}_2\text{O} \rightarrow \text{H}_3\text{O}^+ + \tfrac{1}{2}\text{Cu} \tag{7.75}$$

is -33.28 kJ; this is because of the large overvoltage for the oxidation of hydrogen at a copper surface.

We shall not discuss overvoltage in detail here, but we may note that the voltage of a voltaic cell, ie the difference in potential between its electrodes when a current is flowing, is less than its emf, and decreases as I, or more practically σ the current density in ampere per square metre of electrode surface, increases. This is illustrated in Figure 7.21. The difference between the emf ($\sigma = 0$) and the voltage at σ A m^{-2} is called the *overvoltage* of the cell; it may be subdivided into the overpotentials contributed by each electrode.

There are in general three components of overvoltage, which are (1) the internal resistance of the cell, (2) changes of concentration at the electrodes, brought about by the electrode reactions, and (3) the activation energies of slow reaction steps at the electrodes. There is a high overvoltage for the reduction of hydrogen ions on the surface of most metals except for platinized platinum. This is very convenient in the analytical technique of polarography, where metal ions such as zinc and even sodium are reduced at a mercury cathode in preference to hydroxonium ions.

Theoretically a voltage of about 1.2 V should cause the decomposition of an aqueous solution of an acid or base. In practice the 'decomposition voltage' V_d required is much larger – about 1.7 V – depending on the decomposition and the nature of the electrodes. Some overvoltages for hydrogen evolution from dilute sulphuric acid are given in Table 7.13.

7.12.5 *Further thermodynamics of voltaic cells*

For a cell in which the reaction

$$mM + nN \rightarrow pP + qQ$$

Figure 7.21. Current density as a function of voltage, showing a decomposition voltage V_d.

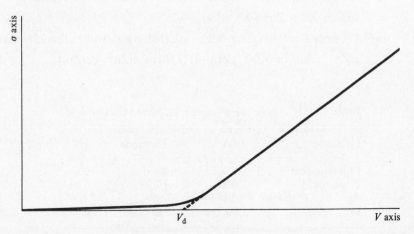

the van't Hoff isotherm is without proof here

$$\Delta G = \Delta G^{\ominus} + RT \ln\left\{\frac{a^p(P)a^q(Q)}{a^m(M)a^n(N)}\right\} \tag{7.76}$$

where $\Delta G^{\ominus}$ is the standard free energy change for the forward reaction, and $a(P)$ etc are the activities of products and reactants.

At equilibrium $\Delta G = 0$, and the activities in (7.76) are now all at their equilibrium values, denoted by 'e' in (7.77)

$$0 = \Delta G^{\ominus} + RT \ln\left\{\frac{a^p(P)a^q(Q)}{a^m(M)a^n(N)}\right\}_e \tag{7.77}$$

The logarithmic term is now $\ln K$, where K is the thermodynamic equilibrium constant, and

$$\Delta G^{\ominus} = -RT \ln K$$

as given earlier (3.103). Substituting in terms of emf from (7.66)

$$E^{\ominus} = RT \ln K \tag{7.78}$$

and

$$E = E^{\ominus} - \frac{RT}{nF} \ln\left\{\frac{a^p(P)a^q(Q)}{a^m(M)a^n(N)}\right\} \tag{7.79}$$

We can subdivide the emf E into the individual electrode potentials π; for example, for the cell

$$Zn\,|\,Zn^{2+}SO_4^{2-} :: Ag^+SO_4^{2-}\,|\,Ag$$

writing the cell reaction

$$Ag^+ + \tfrac{1}{2}Zn \rightarrow \tfrac{1}{2}Zn^{2+} + Ag$$

$$E = (0.799 + 0.763) - \frac{RT}{F} \ln\left\{\frac{a^{\frac{1}{2}}(Zn^{2+})a(Ag)}{a^{\frac{1}{2}}(Zn)a(Ag^+)}\right\} \tag{7.80}$$

The electrode reactions are

cathode: $Ag^+ + e^- \rightarrow Ag$, $\pi^{\ominus} = 0.799$ V

anode: $Zn \rightarrow Zn^2 + e^-$, $\pi^{\ominus} = -0.763$ V

From (7.77), since $E = \pi(\text{RH electrode}) - \pi(\text{LH electrode}) = \pi(\text{cathode}) - \pi(\text{anode})$

$$\pi(Zn^{2+}/Zn) = \pi^{\ominus}(Zn^{2+}/Zn) + (RT/F) \ln\{a(Zn^{2+})/a(Zn)\} \tag{7.81}$$

Table 7.13. *Hydrogen overvoltages η in dilute sulphuric acid*

Electrode	η/V	Electrode	η/V
Pt (platinized)	0.005	Sn	0.53
Pt (smooth)	0.09	Pb	0.64
Ag	0.15	Zn	0.70
Cu	0.23	Hg	0.78

and

$$\pi(Ag^+/Ag) = \pi^{\ominus}(Ag^+/Ag) + (RT/F) \ln a(Ag^+) \qquad (7.82)$$

In deriving (7.81) and (7.82) from (7.80), the activities of the elements zinc and silver have been equated to unity. These equations show how the respective electrode potentials vary with the activity of the cations. For example, at 298 K, if $a(Zn^{2+}) = 0.10$ mol dm^{-3} and $a(Ag^+) = 2.0$ mol dm^{-3}:

$$\pi(Zn^{2+}/Zn) = -0.763 + 0.0252 \times (-2.303) = -0.821 \text{ V}$$

$$\pi(Ag^+/Ag) = 0.799 + 0.0252 \times (0.6931) = 0.81 \text{ V}$$

$$E(\text{cell}) = 0.816 - (-0.821) = 1.637 \text{ V}$$

It is convenient to calculate the term RT/F as a function of temperature, with the following values:

T/K	$(2.303RT/F)/V$
288.15	0.0243
293.15	0.0248
298.15	0.0252

Thus E has increased from $E^{\ominus} = 1.56$ V to 1.637 V due to the changes in the electrode potentials. These changes are in the directions expected from our considerations on p. 252. A decrease in the activity of zinc ions facilitates the dissolution of more zinc; increase in the activity of silver ions facilitates the removal of these ions from solution:

$$Ag^+ + e^- \rightarrow Ag$$

7.12.6 *Liquid junction potentials; salt bridge*

The cells we have been considering incorporate a liquid junction between two solutions of differing composition; in general, ions crossing this boundary will have different velocities, and a liquid junction potential will be set up, adding algebraically to the cell emf. We can estimate this potential only approximately; therefore we reduce its value as far as possible by joining the two half-cells by a salt bridge. The salt bridge consists of a concentrated aqueous electrolyte solution, often in the form of a gel; a little agar–agar is stirred into the heated solution which then gelatinizes on cooling. For a reason which will be apparent from (7.86), the electrolyte is selected so that its ions have nearly equal transport numbers; concentrated solutions of potassium chloride, potassium nitrate, or ammonium nitrate satisfy this condition. If the hot solution is poured into a U-tube, after cooling the inverted tube is convenient for connecting the two half-cells. The salt bridge is indicated on the cell diagram by a double vertical line, $\|$, or by specifying its composition:

$$Zn|Zn^{2+}SO_4^{2-}|KNO_3 \text{ satd}|AgNO_3|Ag$$

7.13 Concentration cells

The cells we have considered so far may be called chemical cells, in which a chemical reaction occurs; ΔG of this reaction determines the cell emf. Now if an electrolyte is transferred from a solution in which the mean ionic activity is a_1 to a solution in which the final activity is a_2, the free energy change per mole transferred is

$$\Delta G = RT \ln(a_2/a_1) \tag{7.83}$$

Note that if $a_2 < a_1$, ΔG is negative, corresponding to the spontaneous change on mixing the two solutions. If the transfer is carried out in a voltaic cell, we can relate the ΔG of transfer to the emf of the cell:

$$\Delta G = -nFE$$

Consider the cell

$$\text{Pt}, \text{H}_2 | \text{H}^+\text{Cl}^-(a_1) :: \text{H}^+\text{Cl}^-(a_2) | \text{H}_2, \text{Pt}$$

With the cathode as the right-hand electrode, the reaction here is

$$\text{H}_3\text{O}^+(a_1) + \text{e}^- \rightarrow \text{H}_2\text{O} + \tfrac{1}{2}\text{H}_2$$

and at the left-hand electrode

$$\tfrac{1}{2}\text{H}_2 + \text{H}_2\text{O} \rightarrow \text{H}_3\text{O}^+(a_2) + \text{e}^-$$

Thus 1 mole of hydroxonium ions passes out of solution into the gas phase at the cathode; but this loss is partially compensated by the transport of $t(+)$ mole of hydroxonium ions towards the cathode, across the common junction of the two solutions. At the anode, 1 mole of hydroxonium ions is formed, but $t(+)$ mole is transported away from the anode, across the liquid junction. At the same time, $t(-)$ mole of chloride ions is transported from the cathode to the anode (note that hydroxonium ions travel from anode to cathode, and chloride ions in the reverse direction, *inside the cell*).

The net result, therefore, is the transfer of $1 - t(+) = t(-)$ mole of both hydroxonium and chloride ions, ie $t(-)$ mole hydrochloric acid from the solution of activity a_1 to that of a_2. The corresponding ΔG is

$$\Delta G = t(\text{Cl}^-)RT \ln\{a(\text{HCl})_1/a(\text{HCl})_2\} \tag{7.84}$$

setting up an emf

$$E = t(\text{Cl}^-)(RT/F) \ln\{a(\text{HCl})_2/a(\text{HCl})_1\} \tag{7.85}$$

In practical terms $a(\text{HCl}) = a(\text{H}_3\text{O}^+)a(\text{Cl}^-) = a^2(\pm)$. So finally

$$E = 2t(\text{Cl}^-)RT \ln\{a(\pm)_2/a(\pm)_1\} \tag{7.86}$$

Note that it is the transport number of the ion to which the electrodes are *not* reversible, ie $t(\text{Cl}^-)$ which appears in (7.86).

Which solution has the higher activity, $(\text{HCl})_1$ or $(\text{HCl})_2$? From the electrode reactions, the anode solution gains hydroxonium ions, and transport carries in an equivalent amount of chloride ions; the cathode solution similarly loses hydrogen chloride. These are spontaneous changes, so that we must have $a(\pm)_2 > a(\pm)_1$.

This type of voltaic cell is termed a *concentration cell with transport*; its source of emf is the difference in mean ion activity of the electrolyte in the two half-cells.

We could set up the same cell with electrodes reversible to chloride ions:

$$Ag|AgCl, HCl(a(\pm)_1) :: HCl, AgCl(a(\pm)_2)|Ag \tag{7.87}$$

We leave it as an exercise to show that in this form

$$E = 2t(H_3O^+)(RT/F)\ln\{a(\pm)_1/a(\pm)_2\} \tag{7.88}$$

and $a(\pm)_1 > a(\pm)_2$ (for E to be positive, the ln term must be greater than unity).

Suppose that we know $a(\pm)_1$, and wish to find $a(\pm)_2$. We set up the cell (7.87) and measure E. But in order to find $a(\pm)_2$ from (7.88) we require $t(H_3O^+)$ at the appropriate concentration; using the cell in its original form with (7.86), we require $t(Cl^-)$. Often the transport data are not available, particularly for more complex ions. Now we have seen that there are electrodes reversible to both hydroxonium and chloride ions, so that we could set up the cell in the form·

$$Ag|AgCl, HCl, H_2(a(\pm)_1)|Pt–Pt|H_2, HCl, AgCl(a(\pm)_2)|Ag \tag{7.89}$$

There is now no common liquid junction and therefore no transport of ions between the half-cells. The passage of one Faraday through the cell results in the transference of one mole of hydrochloric acid from $a(\pm)_2$ to $a(\pm)_1$, so that

$$\Delta G = RT\ln\{a(HCl)_1/a(HCl)_2\} \tag{7.90}$$

$$E = 2(RT/F)\ln\{a(\pm)_2/a(\pm)_1\} \tag{7.91}$$

The transport numbers in the previous equations have been eliminated; this is an example of a *concentration cell without transport*. It is the most theoretically satisfactory form of concentration cell, and has been used wherever possible for the accurate determination of activity coefficients. However, it is often not possible to find electrodes reversible to both ions in solution.

A very convenient way of joining the two half-cells which eliminates the transport number is to use the salt bridge (p. 263). Since this contains an electrolyte for which $t(+) = t(-) = \frac{1}{2}$, (7.86) or (7.88) becomes

$$E = (RT/F)\ln\{a(\pm)_{conc}/a(\pm)_{dil}\} \tag{7.92}$$

This is an example of a concentration cell with a salt bridge connection. This method of joining the half-cells is common in potentiometric titrations.

7.14 Redox equilibrium

Another type of reversible electrode consists of an inert metal, usually platinum, or carbon, in a solution containing an ion in two oxidation states, e.g. $Pt|Fe^{3+}/Fe^{2+}$ or $C|MnO_4^-/Mn^{2+}$ in acid solution. The standard potentials are:

$$\pi^{\ominus}(Fe^{3+}/Fe^{2+}) = 0.771\,V \qquad \pi^{\ominus}(MnO_4^-/Mn^{2+}) = 1.51\,V$$

In the Mn(VII)/Mn(II) electrode, in acid solution, the equilibrium

$$\tfrac{1}{5}MnO_4^- + \tfrac{8}{5}H_3O^+ + e^- \rightleftharpoons \tfrac{1}{5}Mn^{2+} + \tfrac{12}{5}H_2O \tag{7.93}$$

is established. Combination of these two electrode systems in the cell

$$Pt|Fe^{3+}, Fe^{2+}|KCl\ satd|MnO_4^-, Mn^{2+}, H_3O^+|Pt$$

provides information for the reaction

$$\tfrac{1}{5}MnO_4^- + \tfrac{8}{5}H_3O^+ + Fe^{2+} \rightarrow \tfrac{1}{5}Mn^{2+} + Fe^{3+} + \tfrac{12}{5}H_2O \qquad (7.94)$$

The standard emf $E^\ominus$ of the cell is $1.51 - 0.771 = 0.739$ V. Hence $\Delta G^\ominus = -72.8$ kJ for (7.94) and $\Delta G^\ominus = -RT \ln K$, so that $K = 5.6 \times 10^{12}$, and the reaction is complete.

A knowledge of half-cell reactions is useful in constructing equations for more complex redox processes. For example, consider the oxidation of manganese(II) to manganese(VII) by bismuth(V) (bismuthate) in acid solution. The half-cell reactions are

$$\tfrac{1}{5}Mn^{2+} + \tfrac{12}{5}H_2O \rightarrow \tfrac{1}{5}MnO_4^- + \tfrac{8}{5}H_3O^+ + e^-$$

$$\tfrac{1}{2}BiO_3^- + 3H_3O^+ + e^- \rightarrow \tfrac{1}{2}Bi^{3+} + \tfrac{9}{2}H_2O$$

Adding these equations:

$$\tfrac{1}{2}BiO_3^- + \tfrac{1}{5}Mn^{2+} + \tfrac{7}{5}H_3O^+ \rightarrow \tfrac{1}{2}Bi^{3+} + \tfrac{1}{5}MnO_4^- + \tfrac{21}{10}H_2O \qquad (7.95)$$

The standard potential for the Bi(V)/Bi(III) electrode is 1.71 V, so that for (7.95)

$$\Delta G^\ominus = -FE^\ominus = -F(1.71 - 1.51) = -19.7\ kJ$$

and $K = 2.8 \times 10^3$, so that reaction is nearly complete.

7.15 Measurement of pH

Since we cannot measure the activity of a single ionic species in solution, we cannot use the equation

$$\pi = \pi^\ominus + (RT/F) \ln a(H_3O^+)$$

to determine the pH from the definition

$$pH = -\log a(H_3O^+)$$

Instead, a practical definition of pH is based upon emf measurements. We set up a voltaic cell with a hydrogen electrode (p. 254) and a reference electrode, eg calomel or silver–silver chloride. If the emf of this cell containing a buffer solution of defined pH s is E_s, then from $E = \pi_{RHS} - \pi_{LHS}$

$$E_s = \pi_{ref} - \{\pi^\ominus(H) + (RT/F) \ln a(H_3O^+)_s\} \qquad (7.96)$$

where π_{ref} is the potential of the reference electrode, and $\pi^\ominus(H)$ the standard potential of the hydrogen electrode. For the same cell containing a solution of pH x:

$$E_x = \pi_{ref} - \{\pi^\ominus(H) + (RT/F) \ln a(H_3O^+)_x\}$$

Thus

$$E_s - E_x = 2.303\ RT\ \ln\{a(H_3O^+)_x - a(H_3O^+)_s\}$$
$$= 0.0591(pH_s - pH_x) \qquad (7.97)$$

at $T = 298$ K.

$$pH_x = pH_s + (E_s - E_x)/0.0591$$

at 298 K, or

$$pH_x = pH_s + (E_s - E_x)/1.982 \times 10^{-4}T$$

(7.98)

The British standard of reference for pH is a 0.05 mol dm^{-3} solution of potassium hydrogen phthalate, which is defined as having pH $= 4.00$ at 288 K with

$$pH = 4.00 + 0.05\{(T - 288)/100\}^2 \tag{7.99}$$

at temperature T.

The pH of a solution HX is measured directly by a cell of the type

$$Ag \,|\, AgCl, HCl \,\|\, HX \,|\, KCl \text{ satd}, Hg_2Cl_2 \,|\, Hg$$

The left-hand electrode is the *glass electrode*; the symbol $\|$ represents the glass membrane separating hydrochloric acid inside the bulb from the HX external solution.

The glass electrode (Figure 7.22) consists of a thin-walled glass bulb containing a silver–silver chloride electrode in dilute hydrochloric acid. This acts as a concentration half-cell; the bulb is permeable to hydroxonium ions, so that the potential set up depends on the difference in hydroxonium ion activity of the hydrochloric acid in the bulb and the solution in which it is placed.

The potential, and therefore the emf of the complete cell, is established in a circuit which includes a glass membrane of several megohm impedance. Even though the conventional potentiometer draws zero current from the cell at balance, this high impedance makes the instrument too insensitive for a clear

Figure 7.22. Glass electrode.

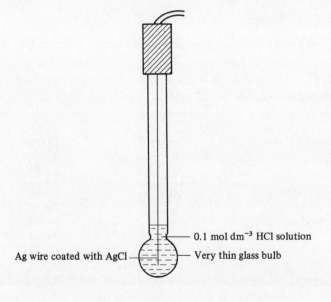

Ag wire coated with AgCl ———

0.1 mol dm^{-3} HCl solution

Very thin glass bulb

balance to be obtained. The glass electrode (and similar electrodes, see next section) must therefore be used in conjunction with an electrometer of very high input impedance; a pH meter is such an electrometer, calibrated in pH units against standard buffer solutions at the appropriate temperature.

7.15.1 *Specific ion electrodes*

The glass electrode is one of a series of electrodes specific to a particular cation or anion, in which a membrane separates the solution under test from a medium with a fixed concentration of that ion. These specific ion (or ion-selective) electrodes do not, of course, measure the single-ion activity; they measure the mean ionic activity of the species at the total ionic strength of the solution. Data for some typical examples are shown in Table 7.14. An important consideration is the extent of interference from other ions in solution. In Table 7.14 the concentration range for species Y, the species being measured, is the range over which the plot of emf versus pY is linear $(pY = -\log[Y]/\text{mol dm}^{-3})$; with suitable calibration [Y] may be measured outside these limits: [X] is the concentration of an ion X which would produce a 10% error in the measurement of [Y], at $a(Y) = 10^{-3}$ mol dm^{-3}.

Similar electrodes are available for the determination of ammonia, carbon dioxide and sulphur dioxide.

For further details of specific ion electrodes and of the determination of selectivity constants, see J. Koryta, *Ion-selective Electrodes*, Cambridge University Press, Cambridge, 1975.

7.16 Potentiometric titrations

7.16.1 *Acid-base titrations*

If we set up the cell

$$Pt|H_2, HX|KCl \text{ satd}|\text{calomel reference half-cell} \tag{7.100}$$

Table 7.14. *Data for some specific ion electrodes*

Species Y	Range of [Y]/mol dm^{-3}	Species X	[X]/mol dm^{-3}
Na$^+$	$3-(1 \times 10^{-6})$	K$^+$	10^{-1}
		Li$^+$	5×10^{-5}
Ca^{2+}	$1-(1 \times 10^{-5})$	Mg^{2+}, Sr^{2+}	8×10^{-3}
Pb^{2+}	$1-(1 \times 10^{-7})$	Ag$^+$	1×10^{-7}
F$^-$	satd$-(1 \times 10^{-6})$	OH$^-$	10^{-1} [F$^-$]
Br$^-$	$1-(1 \times 10^{-6})$	I$^-$	2×10^{-4} [Br$^-$]
NO$_3^-$	$1-(5 \times 10^{-6})$	ClO$_4^-$	1×10^{-7}
		CN$^-$, NO$_2^-$	3×10^{-3}

we can titrate the acid solution HX with a standard solution of a base and follow the course of the titration by the emf of the cell, where

$$E = \pi_{ref} - \{\pi^{\ominus} + RT \ln a(H_3O^+)/a^{\frac{1}{2}}(H_2)\}$$
$$= \text{constant} + 2.303(RT/F)\text{pH} \qquad (7.101)$$

The hydrogen electrode could be the SHE, except that for this application the partial pressure of hydrogen need not be one atmosphere, so long as it remains constant during the titration. Thus E changes during the titration in exactly the same way, apart from a change of scale, as the pH; the shape of the E versus titre curve will depend upon the relative strengths of HX and of the base. A sudden change in E denotes the equivalence point.

Suppose HX consists of 25 cm^3 of hydrochloric acid of concentration 0.1 mol dm^{-3}, and we titrate with sodium hydroxide of concentration 0.5 mol dm^{-3} (in potentiometric titrations, as in conductimetric titrations, the titrant is usually more concentrated than the solution being titrated, to minimize the effect of dilution). Initially the pH is 1.0, and the emf, from (7.101) with $T =$ 298 K, is (constant $+0.059$) V. After the addition of 4.95 cm^3 of base, the pH is

$$\{(25 \times 0.1) - (4.95 \times 0.5)\}/29.95 = 3.08,$$

and the emf is (constant $+0.182$) V. After the addition of 5.05 cm^3 of base, $[OH^-] = \{(5.05 \times 0.5) - (25 \times 0.1)\}/30.05 = 8.3 \times 10^{-4}$ mol dm^{-3}, pH $= 14 - (-\log[OH^-]) = 11.6$, and $E = $ (constant $+0.686$) V. Thus the emf changes by 0.12 V in the first 4.95 cm^3 of titration, and by 0.5 V in the next 0.1 cm^3, giving a clear indication of equivalence, Figure 7.23. In this titration the pH is 7.0 at the end point, so that $E = $ (constant $+0.143$) V at this point.

The SHE is inconvenient for general laboratory use and as the hydrogen electrode in titration cells. A number of more convenient electrodes have been introduced, of which two are in use nowadays: the quinhydrone electrode (particularly used in non-aqueous solutions), and the glass electrode (p. 267).

The quinhydrone system involves the equilibrium

$$+ 2 H_3O^+ + 2e^- \rightleftharpoons \qquad\qquad + 2H_2O \qquad (7.102)$$

Quinone, Q   Hydroquinone, HQ

The position of this equilibrium is determined by the activity of the hydroxonium ions in the aqueous solution.

In practice a few milligrams of the sparingly-soluble solid quinhydrone, a 1:1 molecular complex of Q and HQ, is added to the solution and the equilibrium

(7.102) is established at a platinum electrode; the electrode potential is

$$\pi(Q/HQ) = \pi^{\ominus}(Q/HQ) + (RT/2F)\ln\{a(Q)a^2(H_3O^+)/a(HQ)\} \qquad (7.103)$$

with $\pi^{\ominus} = 0.699$ V. Since $a(Q)/a(HQ) = 1$,

$$\pi = 0.699 + (RT/2F)\ln\{a^2(H_3O^+)\} \qquad (7.104)$$

Use of the quinhydrone electrode is restricted to acid or neutral solutions; quinones decompose rapidly in alkaline solution.

7.16.2 *Halide titrations*

The concentration of halide ions in aqueous solution is usually determined by titration with standard silver nitrate solution. We can carry out this titration potentiometrically in the cell

Reference half-cell $|$ KNO$_3$ satd $|$ MX, AgX $|$ Ag $\qquad (7.105)$

where MX is the halide solution and X$^-$ the halide ion. Note that the salt bridge contains potassium nitrate rather than potassium chloride, to obviate diffusion of chloride ions into the halide solution. The emf of this cell is given by

$$E = \{\pi^{\ominus}(Ag/AgX) - (RT/F)\ln a(X^-)\} - \pi_{\text{ref}}$$

$$= \text{constant} + \{(2.303\,RT)/F\}pX \qquad (7.106)$$

where pX is $-\log[X^-]$, by analogy with pH. The titration curve is thus of the

Figure 7.23. Variation of emf during titration of a strong acid with a strong base.

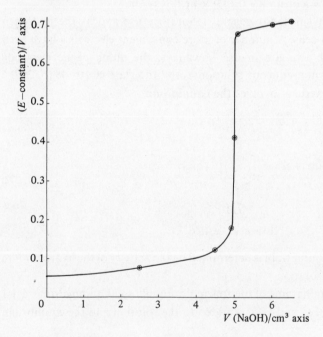

V (NaOH)/cm^3 axis

same shape as the acid–base titration curve, and shows a similar large increment ΔE at the equivalence point. The results for a typical chloride ion versus silver ion titration are recorded in Table 7.15. Before the equivalence point is reached, $a(Cl^-)$ is simply the activity of the unreacted chloride ion in the total volume of solution. At the equivalence point, the solution is saturated with silver chloride, and $a(Cl^-) = S^{\frac{1}{2}}(AgCl) = 10^{-5}$ mol dm^{-3}, where S is the solubility product of silver chloride. Beyond the equivalence point $a(Cl^-)$ is given by

$$a(Cl^-) = S(AgCl)/a(Ag^+)$$

where $a(Ag^+)$ is the activity of the excess silver ions in solution. The titration curve is plotted in Figure 7.24. Other precipitation titrations can be followed in a similar way; for example, sulphate ions may be titrated by a standard solution of

Table 7.15. *Titration of 25 cm^3 0.01 mol dm^{-3} chloride ions by 0.05 mol dm^{-3} silver ions*

Titre, V/cm^3	$a(Cl^-)$/mol dm^{-3}	E/V
0	10^{-2}	0.340
2.0	$\{(25-10)/27\} \times 10^{-2} = 5.5 \times 10^{-3}$	0.355
4.0	$\{(25-20)/29\} \times 10^{-2} = 1.7 \times 10^{-3}$	0.385
4.9	$\{(25-24.5)/29.9\} \times 10^{-2} = 1.7 \times 10^{-4}$	0.444
5.0	10^{-5}	0.517
5.1	$10^{-10}/\{(0.1 \times 0.05)/30.1\} = 4 \times 10^{-7}$	0.599
6.0	$10^{-10}/\{(1 \times 0.05)/31\} = 5 \times 10^{-8}$	0.653

Figure 7.24. Variation of emf during titration of chloride ions by silver ions.

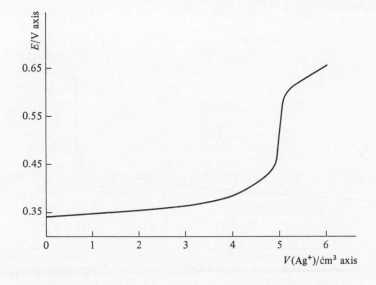

barium chloride, following the change of $[SO_4^{2-}]$ by the potential of a lead–lead sulphate electrode.

7.16.3 *Redox titrations*

The emf of the cell

$$\text{calomel electrode}\,|\,Fe^{3+}, Fe^{2+}, H_3O^+\,|\,Pt \tag{7.107}$$

is given by

$$E = \pi^{\ominus}(Fe^{3+}/Fe^{2+}) + (RT/F)\ln\{a(Fe^{3+})/a(Fe^{2+})\} - \pi_{\text{ref}} \tag{7.108}$$

In an acidified solution of iron(II) sulphate the ratio $a(Fe^{3+})/a(Fe^{2+})$ will be very small. If we titrate this solution with standard potassium permanganate, iron(II) is converted quantitatively into iron(III), and the ratio increases to its maximum value at the equivalence point; addition of excess permanganate causes no appreciable change. Thus from (7.108) the emf will increase to a maximum. Beyond the equivalence point, where both manganese(II) and permanganate ions are present in solution, it is the ratio $a(MnO_4^-)/a(Mn^{2+})$ which determines the redox potential in the cell (7.107); the emf is now given by

$$E = \pi^{\ominus}(MnO_4^-/Mn^{2+}) + \frac{RT}{5F}\ln\left\{\frac{a(MnO_4^-)a^8(H_3O^+)}{a(Mn^{2+})a^{12}(H_2O)}\right\} - \pi_{\text{ref}} \tag{7.109}$$

The half-cell reaction has been written here (cf (7.93))

$$MnO_4^- + 8H_3O^+ + 5e^- \rightarrow Mn^{2+} + 12H_2O$$

The term $a^8(H_3O^+)/a^{12}(H_2O)$ is constant; as excess permanganate is added the emf continues to increase. The standard potentials are

$$\pi^{\ominus}(Fe^{3+}/Fe^{2+}) = 0.771\,V; \qquad \pi^{\ominus}(MnO_4^-/Mn^{2+}) = 1.51\,V$$

Thus the emf increases slowly during the first part of the titration: changes abruptly from 0.771 V to 1.51 V approximately: and continues to increase slowly beyond equivalence (Figure 7.25).

Figure 7.25. Potentiometric redox titration curves.

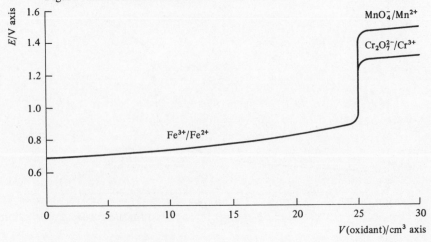

If the iron(II) is oxidized by chromium(VI) as the dichromate ion $Cr_2O_7^{2-}$, the emf change is from 0.771 V to 1.33 V, showing that the permanganate ion is a stronger oxidant than the dichromate ion in acid solutions.

7.16.4 *Redox indicators*

Oxidation of colourless solutions by permanganate ions does not require an indicator; the reagent is self-indicating. Titration by dichromate ions, however, requires an indicator if the end point is to be determined visually. The 'redox' indicator, like an acid–base indicator, should change colour suddenly at the equivalence point. For the dichromate ion oxidation, a solution of the base diphenylamine in sulphuric acid is suitable, changing from colourless to blue. The oxidation of the indicator is

where $H_3SO_4^+$ represents the proton solvated in sulphuric acid.

Figure 7.25 shows that a suitable indicator should respond in the range 0.95–1.25 V. For diphenylamine in sulphuric acid, $\pi^\ominus = 0.76$ V, and phosphoric acid is added to the solution to reduce the activity of iron(III) ions by complex formation. Thus the ratio $a(Fe^{3+})/a(Fe^{2+})$ in (7.108) is decreased, and the first part of the titration curve now occurs at lower emf. The electrode potential of diphenylamine then coincides with the vertical part of the curve, and a sudden change in the position of equilibrium (7.110) gives a corresponding change in colour.

7.17 Measurement of activity coefficients

One of the most convenient methods for determining experimentally the mean activity coefficient $f(\pm)$ of an electrolyte in solution is to set up a cell in which the emf depends upon $f(\pm)$. As an example, consider the cell

$$Pt\,|\,H_2,\,HCl(a(\pm)),\,AgCl\,|\,Ag$$

The emf is given by

$$E = \pi^\ominus(Ag/AgCl) - (RT/F)\ln a(Cl^-) - \pi^\ominus(H^+/H_2) + (RT/F)\ln a(H_3O^+)$$
$$= E^\ominus - (RT/F)\ln c^2(HCl) - (RT/F)\ln f^2(\pm) \qquad (7.111)$$

We measure the emf of the cell for a range of activities of hydrochloric acid. Equation (7.111) may be written

$$E + (2RT/F)\ln c^2(HCl) = E^\ominus - (2RT/F)\ln f^2(\pm) \qquad (7.112)$$

where the left-hand side is an experimental quantity. From the Debye–Hückel equation we expect $\ln f(\pm)$ to be proportional to $c^{\frac{1}{2}}$; if we plot the left-hand side of (7.112) against $c^{\frac{1}{2}}$ and extrapolate to $c = 0$, $\ln f(\pm)$ becomes zero, and the function extrapolates to $E^{\ominus}$. Note that we are not assuming values for the $\pi^{\ominus}$ potentials but are determining the quantity $E^{\ominus}$, $= \pi^{\ominus}(\text{Ag}/\text{AgCl}) - \pi^{\ominus}(\text{H}^+/\text{H}_2)$, experimentally. Once $E^{\ominus}$ is found we can calculate $f(\pm)$ at each value of c.

Some data of Harned and Ehlers (1933) are given in Table 7.16, together with the mean ion activity coefficients derived. The extrapolation procedure is illustrated in Figure 7.26. Values of $f(\pm)$ calculated from the Debye–Hückel limiting law of activity coefficients (Equation (5.52) and from the Davies equation (5.53) are included for comparison.

Table 7.16. *Emf of cell* $Pt|H_2$, $HCl(c)$, $AgCl|Ag$, *and derived activity coefficients;* $T = 298\ K$

$10^2c/\text{mol dm}^{-3}$	E/mV	$f(\pm)$ (exp)	$f(\pm)$ (Debye–Hückel)	$f(\pm)$ (Davies)
0.3215	520.5	0.939	0.936	0.938
0.5619	492.6	0.929	0.916	0.925
0.9138	468.6	0.910	0.894	0.906
1.3407	449.7	0.894	0.873	0.891

Figure 7.26. Extrapolation procedure for the determination of $E^{\ominus}$.

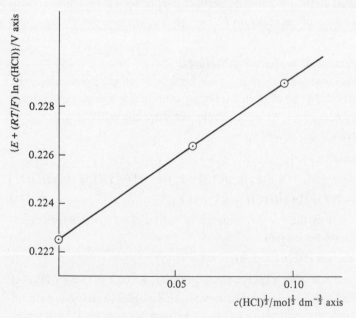

Problems 7

7.1 A water voltameter and a silver voltameter are connected in series. Calculate the weight of silver which is deposited during the time that $80 \, cm^3$ of hydrogen is liberated at 293 K and 755 mmHg (1 dm^3 of hydrogen weighs 0.089 g at STP, ie 760 mmHg and 273.15 K).

7.2 Describe the electrode reactions which take place in the following electrolyses:
(a) copper sulphate in aqueous solution between platinum electrodes
(b) aqueous copper sulphate between copper electrodes
(c) dilute aqueous sodium chloride between carbon electrodes
(d) concentrated aqueous sodium chloride between carbon electrodes.

7.3 At 298 K the resistance of a conductivity cell containing $0.1 \, mol \, dm^{-3}$ potassium chloride solution is 307.62 Ω. The same cell containing $0.1 \, mol \, dm^{-3}$ silver nitrate has a resistance of 362.65 Ω. Refer to the text to find the conductivity of $0.1 \, mol \, dm^{-3}$ potassium chloride and calculate
(a) the cell constant
(b) the molar conductance of the silver nitrate solution.

7.4 The dissociation constant of propanoic acid is 1.34×10^{-5} at 298 K and the limiting conductances of its ions are:
$$\lambda_0(H^+) = 0.0350 \, \Omega^{-1} \, mol^{-1} \, m^2 \qquad \lambda_0(CH_3CH_2CO_2^-) = 0.00358 \, \Omega^{-1} \, mol^{-1} \, m^2$$
Calculate the conductivity ($\Omega^{-1} \, m^{-1}$) of a $0.1 \, mol \, dm^{-3}$ solution of propanoic acid.

7.5 From the following data at 291 K determine the molar conductances at zero concentration, Λ_0, for
(a) sodium chloride
(b) sodium ethanoate (NaEth)
(c) hydrochloric acid
(d) ethanoic acid.

		$\Lambda / \Omega^{-1} \, mol^{-1} \, m^2$		
$c/mmol \, dm^{-3}$	$c^{\frac{1}{2}} mmol^{\frac{1}{2}} dm^{-\frac{3}{2}}$	NaCl	NaEth	HCl
2	1.414	0.01056	0.00745	0.03736
10	3.162	0.01020	0.00712	0.03681
36	6.000	0.00976	0.00671	0.03604
80	8.944	0.00935	0.00634	0.03530
100	10.00	0.00920	0.00621	0.03506

7.6 The conductivities of saturated barium sulphate solution and of the conductivity water (solvent) are $3.59 \times 10^{-4} \, \Omega^{-1} \, m^{-1}$ and $6.180 \times 10^{-5} \, \Omega^{-1} \, m^{-1}$ respectively, at 298 K. From the limiting molar conductances of the ions:
$$\lambda_0(\tfrac{1}{2}Ba^{2+}) = 0.0065 \, \Omega^{-1} \, mol^{-1} \, m^2 \qquad \lambda_0(\tfrac{1}{2}SO_4^{2-}) = 0.0079 \, \Omega^{-1} \, mol^{-1} \, m^2$$
calculate the solubility product of barium sulphate at 298 K. Evaluate $f(\pm)$ for saturated barium sulphate solution from the Debye–Hückel limiting equation (5.52) and comment upon its effect on the calculated solubility product.

7.7 From the molar conductances of ions in Appendix A13 calculate the transport number of the chloride ion in infinitely dilute solutions of the following electrolytes at 298 K:

(a) hydrochloric acid
(b) sodium chloride
(c) potassium chloride
(d) calcium chloride.

Explain the variation in $t_0(Cl^-)$ throughout this series.

7.8 Using the standard electrode potentials $\pi^\ominus$:

$Zn^{2+}/Zn = -0.763$ V

$H^+/\frac{1}{2}H_2 = 0$ V

$Ag^+/Ag = 0.799$ V

set up three standard electrochemical cells, observing the sign conventions. For each cell write down

(a) the electrode reactions
(b) the overall cell reaction
(c) the standard emf $E^\ominus$ of the cell
(d) the standard free energy change for the spontaneous reaction
(e) the direction of electron flow in an external circuit containing the cell.

7.9 Calculate the standard emf and the temperature coefficient of emf for the cell

$Pt|H_2(1\ atm), HCl(a = 1), AgCl(satd)|Ag$

The standard free energies and enthalpies of formation are:

	$\Delta G_f^\ominus/kJ\ mol^{-1}$	$\Delta H_f^\ominus/kJ\ mol^{-1}$
AgCl	− 127.19	− 109.62
HCl(aq)	− 167.36	− 131.38

7.10 The following reduction potentials are reported for reactions of iron in dilute acid at 300 K:

$Fe^{2+} + 2e^- \rightarrow Fe,\quad E^\ominus = -0.440$ V

$Fe^{3+} + e^- \rightarrow Fe^{2+},\quad E^\ominus = 0.771$ V

(a) calculate $\Delta G^\ominus$ and the equilibrium constant for the reaction $Fe + 2Fe^{3+} \rightarrow 3Fe^{2+}$ in acid solution at 300 K

(b) for the reaction $O_2 + 4H^+ + 4e^- \rightarrow 2H_2O$, $E^\ominus = 1.229$ V

Is the iron(II) ion inherently unstable with respect to oxidation by oxygen in acid solution? What is $E^\ominus$ for this reaction?

7.11 The emf of the cell

$Hg|Hg_2Cl_2(satd), KCl(satd), HA(a), quinhydrone(satd)|Pt$

for a certain value of concentration a is 219 mV at 298 K. Calculate the pH of the solution of HA at this concentration, given that the potential of the saturated calomel electrode is 0.244 V and the standard potential of the quinhydrone electrode is 0.700 V, both at 298 K.

7.12 The emf of the cell

$Ag|HCl(a), AgCl(satd), Hg_2Cl_2(satd)|Hg$

is 0.0455 V at 298 K. What is the cell reaction? If the temperature coefficient of emf is 3.4×10^{-4} V K^{-1} calculate the molar enthalpy of reaction at 298 K.

8

Kinetics of chemical reactions

8.1 Kinetics

The usual stoichiometric equation gives no indication of how, or how fast, the represented reaction occurs. For example, the equations

$$H_2 + Cl_2 \rightarrow 2HCl$$
$$H_2 + Br_2 \rightarrow 2HBr$$
$$H_2 + I_2 \rightarrow 2HI$$

are formally similar; yet the details of the reactions are quite different, as is shown by their very different rates of reaction under comparable conditions. An account of the steps by which a reaction occurs is called the *mechanism* of the reaction; the purpose of studying the kinetics of a reaction is to obtain as much information as possible concerning its mechanism. This study enables us to carry out reactions under favourable conditions – so that we obtain rapidly a high yield of the required products, with a minimum of interference from side reactions.

There are a number of factors which determine the rate of a chemical reaction. In this introduction to chemical kinetics we shall consider three: the concentrations of the reactants, the temperature, and the influence of a catalyst.

All reactions, with about three known exceptions, show considerable increases in rate with rise of temperature; the effects of changes in reactant concentration are less uniform. In the original deduction of the law of mass action, it was assumed that the rate of the reaction

$$aA + bB \rightarrow \text{products}$$

was proportional to $[A]^a[B]^b$ so that

$$\text{rate} = k[A]^a[B]^b \tag{8.1}$$

where k is the rate constant for the reaction at the given temperature. This happens to be approximately true for the reaction

$$H_2(g) + I_2(g) \rightarrow 2HI(g) \tag{8.2}$$

The rate of reaction, which we can express in the equivalent forms

$$\text{rate} = d/dt[HI] = -d/dt[H_2] = -d/dt[I_2]$$

is given by

$$d/dt[HI] = k[H_2][I_2] \qquad (8.3)$$

This is the rate equation for (8.2).

It has been shown that the reaction between elementary iodine and hydrogen in the temperature range 633–738 K is not the bimolecular process (8.2) but the termolecular reaction

$$H_2 + 2I \rightarrow 2HI$$

which is kinetically equivalent. The iodine atoms are supplied by the equilibrium

$$I_2 \rightleftharpoons 2I$$

see J. H. Sullivan, *J. Chem. Phys.* (*1967*), **46**, 73. Although (8.2) appears not to occur in the above temperature range, we shall use it as an example in this introduction to reaction kinetics. In general, however, the rate equation cannot be inferred from the stoichiometric equation; the rate equations for the formation of hydrogen chloride and hydrogen bromide are not given by analogous expressions, but by much more complex rate equations.

Again, we might expect, from the equation

$$2N_2O_5 \rightarrow 4NO_2 + O_2$$

that the rate of decomposition of dinitrogen pentoxide would be given by

$$-\frac{d}{dt}[N_2O_5] = k[N_2O_5]^2 \qquad (8.4)$$

the negative sign indicating that $[N_2O_5]$ *decreases* as the reaction proceeds. In fact, the correct form of the rate equation is

$$-\frac{d}{dt}[N_2O_5] = k[N_2O_5] \qquad (8.5)$$

The rate equation must be determined *by experiment*, under well-defined conditions. In the first case, we must show that increase of concentration of *either* hydrogen *or* iodine causes a proportionate increase in the rate of reaction; in the second case, that the rate increases as the *first power*, not as the *square*, of the concentration of dinitrogen pentoxide.

To give a further example, the nitration of an aromatic compound requires both the aromatic molecule and nitric acid as reactants; for example

$$C_6H_5CH_3 + HNO_3 \rightarrow C_6H_4CH_3NO_2 + H_2O$$

Yet, under certain conditions, the concentration of the organic molecule may be increased four-fold without any appreciable change in the rate of its nitration.

Clearly the rate equation

$$\frac{d}{dt}[C_6H_4CH_3NO_2] = k[HNO_3][C_6H_5CH_3]$$

does *not* apply under these conditions.

8.1.1 *Order of reaction*

The dependence of the rate upon the concentration of a particular reactant is expressed by the *order* of the reaction with respect to that component.

This is the power to which the concentration of that reactant is present in the rate equation, as found by experiment. Thus, in the formation of hydrogen iodide, the reaction is *first-order* with respect to both hydrogen and iodine; the decomposition of dinitrogen pentoxide (8.5) is *first-order* with respect to this reactant; and the nitration of toluene is, within a certain range, *zero-order* with respect to (ie independent of) the concentration of toluene.

In the thermal decomposition of hydrogen iodide, the rate equation is

$$-\frac{d}{dt}[HI] = k[HI]^2 \qquad (8.6)$$

and the reaction is *second-order* with respect to this reagent.

A few reactions of nitric oxide with halogens are of the form

$$2NO + X_2 = 2NOX \qquad X_2 = Cl_2 \text{ or } Br_2$$

$$\frac{d}{dt}[NOX] = k[NO]^2[X_2] \qquad (8.7)$$

and the reaction is *second-order* with respect to nitric oxide, and *first-order* with respect to the halogen X_2.

In simple reactions, no order higher than two with respect to a single component is met with, and the sum of the orders with respect to all components never exceeds three. (The more complex equations which one meets in elementary textbooks, for example

$$4Zn + 10HNO_3 = 4Zn(NO_3)_2 + 5H_2O + N_2O$$

actually take place by a series of *simple* steps, one of which is the slowest and determines the overall rate of reaction; it is called the *rate-determining* step.)

It is convenient to group reactions according to their *molecularities*. The molecularity is the number of molecules of all species which are involved in the rate-determining step of a reaction. Thus, the decomposition of dinitrogen pentoxide is a *unimolecular* process (8.5); that of hydrogen iodide is *bimolecular*, (8.6). The formation of a nitrosyl compound is one of the very few examples of a *termolecular* reaction, as in (8.7). In a number of nitration reactions, the rate-determining step is:

$$2HNO_3 \rightarrow NO_2^+ + NO_3^- + H_2O \qquad (8.8)$$

and this is followed by a rapid reaction:

$$NO_2^+ + C_6H_5CH_3 \rightarrow C_6H_4CH_3NO_2 + H^+ \tag{8.9}$$

The reaction is therefore bimolecular, from (8.8).

The rate equation for the decomposition of dinitrogen pentoxide is given by (8.5). The total order is one, so that this is a *first-order* rate equation.

The rate of a reaction is expressed as the change in concentration per unit time, e.g. $mol\,dm^{-3}\,s^{-1}$. The concentration term on the right-hand side of (8.5) is in $mol\,dm^{-3}$, so that the units of k, for a *first-order* reaction, are those of $(time)^{-1}$, e.g. s^{-1}. The rate of reaction *decreases* as the reaction proceeds; this is seen in Figure 8.1, where the concentration of dinitrogen pentoxide is plotted as a function of time. The rate is given by the tangent to the curve, and decreases along the series of increasing times $t_1, t_2, t_3 \ldots$ Suppose the initial concentration of dinitrogen pentoxide is $a\ mol\,dm^{-3}$. If we found the concentration of dinitrogen pentoxide present at times $t = t_1, t = t_2$, etc, we could determine the rate constant k. If x_1 mole of dinitrogen pentoxide has been decomposed at time t_1, x_2 mole at time t_2, etc, then $dx_1/dt = k(a - x_1)$, the rate at time t_1, $dx_2/dt = k(a - x_2)$, the rate at time t_2, etc.

Now these forms of the rate equation are particular examples of the general expression

$$dx/dt = k(a - x) \tag{8.10}$$

Equation (8.10) can be integrated:

$$\int dx/(a-x) = k \int dt$$

or

$$-\ln(a - x) = kt + \text{an arbitrary constant } (\alpha)$$

Figure 8.1. Variation of total pressure with time during thermal decomposition of dinitrogen pentoxide at 329 K. (In a gaseous mixture, concentrations are measured by partial pressures.)

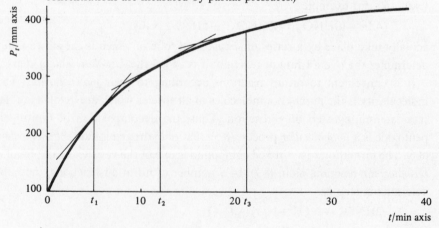

At time $t=0$, $x=0$; therefore

$$-\ln a = \alpha$$

and

$$-\ln(a-x) = kt - \ln a$$

or

$$\ln\{a/(a-x)\} = kt \tag{8.11}$$

Equation (8.11) is the *integrated form* of the rate equation (8.10) for a first-order reaction. Since $\ln a$ is a constant, we can obtain the rate constant k by plotting $\ln(a-x)$ against time t; the slope of this line is $-k$. Notice that, once again, the units of k in (8.11) are reciprocal time since the left-hand side of this equation is dimensionless, and the reaction is only completed (ie $x=a$) when $t=\infty$. However, the reaction is about 99% complete after seven half-lives (see (8.13)).

Half life

Suppose that x, the amount of dinitrogen pentoxide decomposed after time t, is some definite fraction β of a. Then

$$\ln\{a/(a(1-\beta))\} = kt,$$

$$t_\beta = \frac{\ln\{1/(1-\beta)\}}{k} \tag{8.12}$$

where t_β is the time required for the fraction β to be decomposed.

In particular, if $\beta = \frac{1}{2}$,

$$t_{\frac{1}{2}} = \frac{\ln 2}{k} = \frac{0.693}{k} \tag{8.13}$$

$t_{\frac{1}{2}}$ is called the *half-life* or *half-time* of the reaction; it is independent of a, so that, for example, it requires the same period of time, $t_{\frac{1}{2}}$, for 1 kg of dinitrogen pentoxide to become 0.5 kg as for 1 g to become 0.5 g.

In general, the half-life of a reaction is inversely proportional to the initial concentration a of reactant to the power $(n-1)$, where n is the order of the reaction; for a reaction $A^n \rightarrow$ products

$$t_{\frac{1}{2}} \propto \frac{1}{a^{n-1}} \tag{8.14}$$

For a first-order process, $n=1$, and $t_{\frac{1}{2}}$ is independent of a, as we have seen, (8.13).

The decay of a radioactive isotope provides an example of a truly first-order process. From the rate equation

$$k = \frac{1}{t}\ln\left(\frac{a}{a-x}\right)$$

$$= \frac{1}{t}\ln\left(\frac{a}{a_t}\right)$$

Therefore

$$a_t = a \exp(-kt)$$

where k is the *decay constant* for the isotope. At time $t_\frac{1}{2}$, $a_t = a/2$, so that

$$0.5 = \exp(-kt_\frac{1}{2})$$

Therefore

$$-kt_\frac{1}{2} = \ln 0.5$$

or

$$t_\frac{1}{2} = \frac{0.693}{k}$$

as before. The values of $t_\frac{1}{2}$ vary very widely for different radioactive isotopes. For example, the half-life of ^{238}U is 4.51×10^9 year; for ^{212}Po, $t_\frac{1}{2}$ is 3×10^{-7} s.

It is often convenient to follow the course of a reaction by studying the change in some physical property of the reacting system. In the decomposition of dinitrogen pentoxide there is an increase in pressure:

$$2N_2O_5 \rightarrow 4NO_2 + O_2$$

and thus the reaction may be followed by the indications of a manometer. Suppose the initial pressure of dinitrogen pentoxide (or its initial *partial pressure* if there is some other gas present, eg air) is p_0, the pressure after time t is p_t, and the final pressure after complete reaction is p_∞. Then the initial concentration of dinitrogen pentoxide is proportional to the total change in pressure, $p_\infty - p_0$, and the concentration of dinitrogen pentoxide present after time t is proportional to the further change in pressure, $p_\infty - p_t$. Substituting these expressions for the concentration terms in (8.5) and then integrating

$$\ln\{(p_\infty - p_0)/(p_\infty - p_t)\} = kt$$

and the plot of $\ln(p_\infty - p_t)$ against time gives a straight line of slope $-k$. From the data in Table 8.1, plotted in Figure 8.2, k is determined.

The value of k gives the half-time of the reaction at this temperature as $0.693/0.0927$, or 7.17 minutes. The thermal decompositions (pyrolyses) of a number of organic compounds in the gas phase proceed by similar unimolecular processes.

A reaction such as ester hydrolysis:

$$RCO_2R'(l) + H_2O(l) \rightarrow RCO_2H(l) + R'OH(l)$$

is often bimolecular, and is represented by a second-order rate equation:

$$-\frac{d}{dt}[RCO_2R'] = k[RCO_2R'][H_2O] \tag{8.15}$$

If, however, one reactant, usually water, is present in large excess, its change in concentration during the reaction is negligible, and its constant concentration may be included in k, ie (8.15) may be written:

$$-\frac{d}{dt}[RCO_2R'] = k'[RCO_2R'] \tag{8.16}$$

where $k' = k[H_2O]$.

Equation (8.15) is now of first-order form; under these conditions of excess of one reagent, the reaction is said to be *pseudo* first-order.

Second-order rate equations are of two kinds; for the decomposition of hydrogen iodide,

$$-\frac{d}{dt}[HI] = k[HI]^2 \tag{8.6}$$

and for the hydrolysis of an ester by hydroxide ions,

$$RCO_2R' + OH^- \rightarrow RCO_2^- + R'OH$$

$$-\frac{d}{dt}[RCO_2R'] = k[RCO_2R'][OH^-] \tag{8.17}$$

Table 8.1. *Kinetics of decomposition of dinitrogen pentoxide at 329 K*

t/min	p_t/mm	$(p_\infty - p_t)$/mm	$\ln\{(p_\infty - p_t)$/mm$\}$
0	100	331	5.802
3	178	253	5.533
5	220	211	5.352
10	297	134	4.898
15	345	86	4.454
20	377	54	3.989
30	409	22	3.091
40	421	10	2.303
∞	431	0	—

Figure 8.2. Evaluation of the rate constant for the thermal decomposition of dinitrogen pentoxide at 329 K.

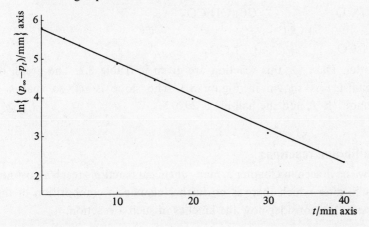

Integration of (8.6) gives:

$$kt = \frac{x}{a(a-x)} \tag{8.18}$$

where x is the number of moles of hydrogen iodide decomposed after time t.

If $x = a/2$,

$$kt_{\frac{1}{2}} = 1/a \tag{8.19}$$

ie $t_{\frac{1}{2}} = 1/ak$, so that the half-time is *inversely* proportional to a (cf (8.14)). The dimensions of k, for a *second-order* reaction, are concentration^{-1} time^{-1}, eg dm^3 mol^{-1} s^{-1}.

In terms of some physical property, eg pressure p,

$$kt = \frac{(p_t - p_0)}{(p_\infty - p_0)(p_\infty - p_t)} \tag{8.20}$$

since, if the amount of hydrogen iodide left after time t is proportional to $p_\infty - p_t$, the amount dissociated is proportional to $(p_\infty - p_0) - (p_\infty - p_t)$, ie to $p_t - p_0$.

Integration of (8.17) in the form:

$$\frac{-d[A]}{dt} = k[A][B]$$

where $[A]$ and $[B]$ are the concentrations of the reacting species, leads (Appendix A19) to:

$$kt = \frac{1}{(a-b)} \ln \left\{ \frac{a(b-x)}{b(a-x)} \right\} \tag{8.21}$$

where a and b are the initial concentrations of reactants A and B, respectively, and x is the amount of A, and of B, which has reacted after time t.

There is no unique half-life of this reaction, unless $a = b$, in which case (8.6) and (8.19) are applicable. The dimensions of k, from Equation (8.21), are again time^{-1} concentration^{-1}.

For example, the hydrolysis of the tetrachlorsuccinate ion

$$\begin{array}{ccc} \text{CCl}_2\text{CO}_2^- & & \text{CCl(OH)CO}_2^- \\ | & +\text{OH}^- \rightarrow & | & +\text{Cl}^- \\ \text{CCl}_2\text{CO}_2^- & & \text{CCl}_2\text{CO}_2^- \end{array}$$

is second-order. Data for this reaction are given in Table 8.2. The graph of $x/(a-x)$ against t is shown in Figure 8.3. The slope is ak, so that $k = 0.0160$ dm^3 mol^{-1}h^{-1}, and the half-life is 2920 h.

8.2 Equilibrium reactions

As we have seen in Chapter 5, many chemical reactions reach a position of equilibrium, after which there is no further change in composition of the reacting system. We consider now the kinetics of such a reaction.

Suppose that both the forward and reverse reactions are first-order processes, as represented by

$$A \underset{k_{-1}}{\overset{k_1}{\rightleftharpoons}} B \qquad (8.22)$$

where k_1 is the rate-constant for the forward reaction ($A \rightarrow B$), and k_{-1} the rate constant for the reverse reaction ($B \rightarrow A$). From the law of mass action, (5.8)

$$\frac{k_1}{k_{-1}} = \left(\frac{[B]}{[A]}\right)_{t=\infty} = K \qquad (8.23)$$

where K is the equilibrium constant for (8.22).

If the initial numbers of moles at time $t=0$ are $[A]_{t=0}=a$, $[B]_{t=0}=b$, then after time t,

$$[A]_t = (a-x) \quad [B]_t = (b+x)$$

where x is the number of moles of A reacted, and hence the number of moles

Table 8.2. *Kinetics of alkaline hydrolysis of the tetrachlorosuccinate ion at 298 K* $(a=b=0.0214 \ mol \ dm^{-3})$

t/h	$10^2 x$	$x/(a-x)$
500	0.311	0.170
1000	0.543	0.340
1500	0.723	0.510
2000	0.866	0.682
2500	0.983	0.851

Figure 8.3. Kinetics of hydrolysis of the tetrachlorosuccinate ion at 298 K.

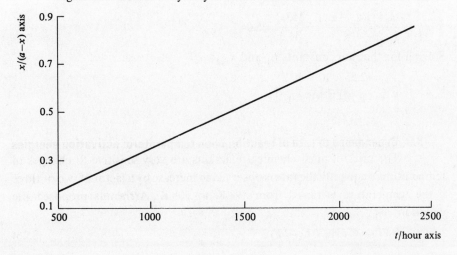

of B formed. The rate equation is thus:

$$\frac{dx}{dt} = k_1(a-x) + k_{-1}(b+x) \tag{8.24}$$

Integration of this equation (Appendix A19) leads to the form

$$(k_1 + k_{-1}) = \frac{1}{t} \ln\left(\frac{a - a_\infty}{a_t - a_\infty}\right) \tag{8.25}$$

This is a first-order integrated rate equation, from which the *sum* of the rate constants can be found. In order to determine k_1 and k_{-1} individually, their ratio (ie K, (8.23)) may be determined: alternatively, by studying the rate of reaction of pure component A in the initial stages, when the concentration of B is insufficient for the back reaction (B → A) to be of importance, k_1 is determined.

For example, an optically active ketone A slowly changes from the pure form A to an equilibrium mixture of A and its isomer B. The reaction can be followed by observing the change in optical rotation with time of a solution of the ketone. The results in Table 8.3 were obtained at 298 K; α_t is the optical rotation after time t h.

We can substitute the optical rotations in (8.25) from the relationships:

$$(a - a_\infty) \propto (\alpha_0 - \alpha_\infty)$$
$$(a_t - a_\infty) \propto (\alpha_t - \alpha_\infty)$$

The plot of $\ln(\alpha_t - \alpha_\infty)$ against time is shown in Figure 8.4. From the slope, $-k$, we obtain

$$(k_1 + k_{-1}) = 3.05 \times 10^{-2} \, \mathrm{h}^{-1}$$

From (8.23)

$$\frac{k_1}{k_{-1}} = \left(\frac{b+x}{a-x}\right)_{t=\infty}$$

Now $b = 0$, and x_∞ (the amount of A converted to B, at equilibrium) $= (a - a_\infty)$. Hence

$$\frac{k_1}{k_{-1}} = \frac{(\alpha_0 - \alpha_\infty)}{\alpha_\infty} = \frac{157.7}{31.3} = 5.04$$

Solving for the rate constants k_1 and k_{-1}:

$$k_1 = 2.55 \times 10^{-2} \, \mathrm{h}^{-1}$$
$$k_{-1} = 0.504 \times 10^{-2} \, \mathrm{h}^{-1}$$

8.3 Dependence of rate of reaction upon temperature: activation energies

The rates of most chemical reactions are very sensitive to changes in temperature; frequently the rate is observed to increase by a factor of two or three if the temperature is raised from 298 K to 308 K. Arrhenius proposed the equation

$$k_2(T) = A \exp(-E_a/RT) \tag{8.26}$$

to fit the experimental k_2 versus T data. In this equation, k_2 is the second-order rate constant, A the 'pre-exponential' factor, and E_a the activation energy of the reaction. The quantities A and E_a may be found by plotting $\ln k_2$ against $1/T$; A is the intercept and E_a/R the slope of this 'Arrhenius plot'.

The Arrhenius equation may be justified by the following argument. In order for molecules A and B to react in the gas phase, they must collide; the frequency of

Table 8.3. *Mutarotation of the ketone A at 298 K*

t/h	$\alpha_t/\deg$	$(\alpha_t - \alpha_\infty)/\deg$	$\ln\{(\alpha_t - \alpha_\infty)/\deg\}$
0	189.0	157.7	5.061
3	169.3	138.0	4.927
5	156.2	124.9	4.828
7	145.9	114.6	4.741
11	124.6	93.3	4.536
15	110.4	79.1	4.371
24	84.5	53.2	3.974
∞	31.3	0	—

Figure 8.4. Kinetics of mutarotation of ketone A at 298 K.

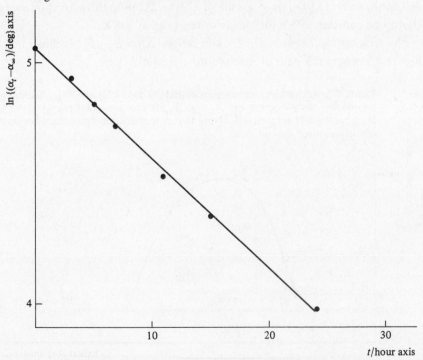

t/hour axis

such collisions is called the collision frequency, Z. For a gaseous mixture at 1 atm pressure and temperature 298 K, Z is about 10^{28} s^{-1} cm^{-3}, so that if every collision resulted in reaction, such gaseous reactions would be extremely fast, ie explosive. Again, from kinetic theory, Z is found to be proportional to $T^{-\frac{1}{2}}$, so that the result of a 10 K rise in temperature at 298 K would be to increase the rate of reaction by a factor of $(308/298)^{\frac{1}{2}}$, or about 1.02, not to produce the doubling in rate usually observed.

Thus only a very small fraction of the collisions lead to reaction; a limiting amount of energy, E_a, must be involved, so that the number of effective collisions is $Z \times P$, where P is the proportion of collisions in which the energy involved is E_a. From the Boltzmann distribution

$$P = \exp(-E_a/RT) \tag{8.27}$$

The Boltzmann distribution of molecular energies is shown, for two temperatures, in Figure 3.13; increase in temperature gives rise to a broader distribution, and to many more molecules with energies above the reaction threshold E_a.

P is obviously very much more sensitive to changes in temperature than is Z. For an increase in temperature from 298 K to 308 K:

$$\ln\{P(308)/P(298)\} = (-E_a/R \times 308) - (-E_a/R \times 298)$$

$$= (1/298 - 1/308)E_a/8.316 \tag{8.28}$$

and, for E_a about 53 kJ mole^{-1}, $k_2(308)/k_2(298) = 2.0$. With this activation energy, about one collision in 5×10^{10} leads to reaction, at 298 K.

The relationship between E_1, E_2 and ΔH ($-\Delta H = E_2 - E_1$) is illustrated in Figure 8.5, where the path of reaction for the equilibrium

Figure 8.5. Activation and reaction enthalpies for (8.29) at 670 K, assuming the mechanism in Section 8.3; Z_1 and Z_2 are the zero-point energy levels for $H_2 + I_2$ and 2HI respectively. (Note that E is measured upwards, whereas ΔH is measured downwards.)

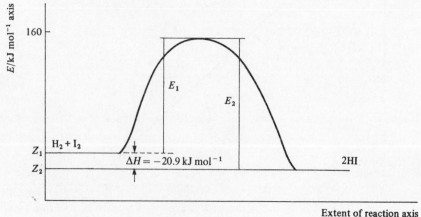

Extent of reaction axis

$$H_2 + I_2 \overset{k_1}{\underset{k_{-1}}{\rightleftharpoons}} 2HI \tag{8.29}$$

is traced. For the forward reaction $\Delta H = -20.9\,\text{kJ mol}^{-1}$ at 670 K. From the variation with temperature of the rate constant, Equation (8.27), the enthalpies E_1 and E_2 have the values $163\,\text{kJ mol}^{-1}$ and $184\,\text{kJ mol}^{-1}$ respectively.

E_1 and E_2 are the *activation enthalpies* of the forward and reverse reactions, (8.29). The activation enthalpy is often called the *activation energy*. E_1 may be considered to indicate the minimum kinetic energy which the two molecules, H_2 and I_2, must possess before, upon collision, the atoms react to produce two HI molecules; E_2 applies similarly to the collision process

$$HI + HI \overset{k_{-1}}{\rightarrow} H_2 + I_2$$

The probability that a molecule possesses enthalpy per mole in excess of E_1 is given by $\exp(-E_1/RT)$, where R is the gas constant per mole and T is the absolute temperature. If A_1 represents the total collision frequency of reactant molecules H_2 and I_2 whatever their energy, then the rate of reaction depends upon the product of A_1 and the factor $-E_1/RT$, or specifically

$$k_1 = A_1 \exp(-E_1/RT)$$

Thus

$$\ln k_1 = (-E_1/RT) + \ln A_1 \tag{8.30}$$

a similar equation to (8.26).

Since, as previously stated, A_1 varies as $T^{\frac{1}{2}}$ only, the exponential factor $\exp(-E_1/RT)$ clearly accounts for the large variation of k_1 with T, which implies that E_1 is quite large.

For example, if $E_1 = 58.6\,\text{kJ mol}^{-1}$ at $T = 300$ K,

$$k_{300} = A_1 \exp\left(\frac{-58\,600}{8.316 \times 300}\right)$$

and at $T = 310$ K,

$$k_{310} = A_1 \exp\left(\frac{-58\,600}{8.316 \times 310}\right)$$

ie

$$\frac{k_{310}}{k_{300}} = \exp\left\{\frac{58\,600}{8.316}\left(\frac{1}{300} - \frac{1}{310}\right)\right\} = 0.76$$

It should be noted that the exponential factor is very small, and that therefore only a very small fraction of the collisions, A_1 per second, lead to reaction. In the present example, $\exp(-E_1/RT)$ at 300 K has the value 6.2×10^{-11}.

The data in Table 8.4 refer to the decomposition of dibromosuccinic acid in aqueous solution. In Figure 8.6 the graph of $\ln k$ against $(1/T)$ is shown; the slope $(-E/R)$ is $-10\,131$ K, and the activation energy is therefore $84.3\,\text{kJ mol}^{-1}$.

The rates of nearly all chemical reactions *increase* with increase in temperature, but not by the same amount. Returning to the reaction (8.29), the rates of both the forward and backward reactions will increase with temperature, but not to the same extent. Thus the equilibrium constant K, (8.23), is dependent upon the temperature; eg in an exothermic reaction K decreases with an increase in temperature.

The rates of reactions are controlled largely by their activation energies. Thus, the reaction between hydrogen and chlorine, *in the dark*, is very slow – slower than that between hydrogen and iodine at the same temperature; both reactions proceed by the bimolecular collision process:

$$H_2 + Cl_2 \text{ (or } I_2) \rightarrow 2HCl \text{ (or 2HI)}$$

The activation energies are 209 kJ mol^{-1} and 192 kJ mol^{-1} respectively, so that the ratio of the rates is exp(-6.8) $= 1.1 \times 10^{-3}$ at 300 K (see Section 1.2). If a chain

Table 8.4. *The decomposition of dibromosuccinic acid in aqueous solution*

T/K	k/min^{-1}	$\ln k$	$1/T$
288	9.67×10^{-6}	-11.546	0.00347
308	7.08×10^{-5}	-9.556	0.00325
333	6.54×10^{-4}	-7.332	0.00300

Figure 8.6. Hydrolysis of dibromosuccinic acid in aqueous solution: variation of rate constant with temperature.

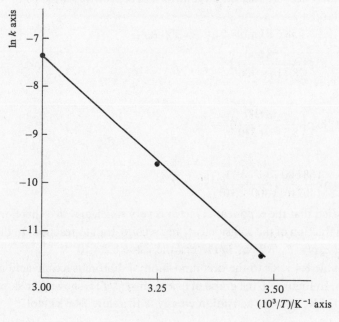

reaction is initiated (p. 295) the first reaction becomes very fast, because its effective activation energy is decreased considerably.

Where a number of parallel reactions can occur, the activation energy may determine the main product of reaction. Suppose a compound A may decompose in two ways, with the enthalpies shown:

$$B \quad \Delta H = b$$
$$\nearrow$$
$$A$$
$$\searrow$$
$$C \quad \Delta H = c$$

If $b < c$, the decomposition to form B is *thermodynamically* favoured. But suppose that the reaction paths are as shown in Figure 8.7. The activation energy for the formation of C is less than that for the alternative reaction, so that the formation of C is kinetically favoured. This is often the case in organic processes, where a number of different products could be formed; that produced in greatest yield, under particular conditions, is often determined kinetically rather than thermodynamically.

8.4 Unimolecular reactions

If molecules must collide with each other for reaction to occur, the existence of unimolecular reactions, for which the rate-determining step involves only one concentration term

$$-d[A]/dt = k[A]$$

requires explanation.

It was considered at one time that such reactions were initiated by radiation, as

Figure 8.7. Energy levels for the parallel reactions (*a*) A → B, (*b*) A → C.

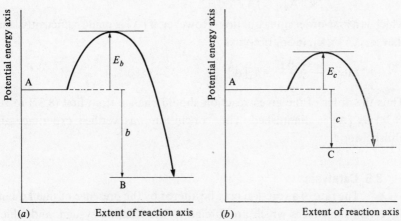

in photochemical processes: however, the correct explanation was put forward by A. E. Lindemann, and verified subsequently by C. N. Hinshelwood.

Lindemann assumed that a certain period of time elapses between the acquisition of activation energy by a molecule and its chemical reaction. If this period is long compared with the period between collisions, most molecules will lose their activation energy before they react. Thus we have

$$A + A \underset{k_{-1}}{\overset{k_1}{\rightleftharpoons}} A^* + A \quad \downarrow k_2 \tag{8.31}$$

$$\text{products}$$

where A represents a normal molecule of A, and A^* an activated molecule of the same species; k_1 and k_2 are the rate constants for the formation of A^*, and for its reaction process, respectively. Thus

$$d[A^*]/dt = k_1[A]^2 - k_{-1}[A^*][A] \tag{8.32}$$

$$-d[A^*]/dt = k_2[A^*] \tag{8.33}$$

If we postulate an equilibrium in which A^* is formed and deactivated or decomposed at equal rates, then

$$k_1[A]^2 = k_{-1}[A][A^*] + k_2[A^*] \tag{8.34}$$

and an equilibrium concentration of A^* is set up:

$$[A^*]_{eq} = \frac{k_1[A]^2}{(k_{-1}[A] + k_2)} \tag{8.35}$$

Hence the rate of formation of products is

$$k_2[A^*]_{eq} = \frac{k_2 k_1[A]^2}{(k_{-1}[A] + k_2)} \tag{8.36}$$

Thus if k_{-1} is sufficiently large compared with k_2, then at relatively large values of [A], such that $k_{-1}[A] \gg k_2$, (8.36) reduces to

$$\text{rate} = \frac{k_2 k_1[A]^2}{k_1[A]} = k_2 k_1[A] \tag{8.37}$$

which is a first-order rate equation. However, if [A] is made sufficiently small, so that $k_{-1}[A] \ll k_2$, (8.36) becomes

$$\text{rate} = \frac{k_2 k_1[A]^2}{k_2} = k_1[A]^2 \tag{8.38}$$

Thus the order of the given reaction should change from first (8.37) to second (8.38) as [A] is diminished. This conclusion was verified experimentally by Hinshelwood.

8.5 Catalysis

The rate of a reaction may be altered by the presence of small quantities of certain substances which are foreign to the reacting system, and which are called *catalysts*. The reaction rate may be increased, an example of *positive* catalysis (or just *catalysis*), or decreased, which is *negative* catalysis.

Catalysed systems may be either homogeneous or heterogeneous, but certain characteristics are common to both:

(a) The catalyst is unchanged chemically (ie in *chemical constitution* and *amount*) at the end of the reaction. For example, platinum gauze is used as a heterogeneous catalyst in the oxidation of ammonia; the gauze is eventually roughened after use, so that the platinum is involved in the mechanism of the reaction. The granular manganese dioxide added to catalyse the thermal decomposition of potassium chlorate may be recovered at the end of the reaction, but in a much more finely-divided state.

(b) A small amount of the catalyst affects a relatively enormous extent of reaction. For example, colloidal platinum, at a concentration of $10^{-6}\,\text{g dm}^{-3}$, catalyses the decomposition of hydrogen peroxide in alkaline solution to a very appreciable extent.

(c) The catalyst does not alter the position of equilibrium in a reversible reaction. Since it is unchanged chemically at the end of the reaction, the catalyst contributes nothing to the energy of the system, so that the same position of equilibrium should be obtained with or without a catalyst, at *constant temperature*. Thus, if reaction (8.22) is catalysed, both k_1 and k_{-1} must be increased (or decreased) to the same extent.

8.5.1 *Homogeneous catalysis*

This term implies that the catalyst is in the same phase as the reactants; e.g. the gaseous catalyst nitric oxide is used to promote the reaction between the gases sulphur dioxide and oxygen in the sulphuric acid chamber process, to give sulphur trioxide:

$$NO + \tfrac{1}{2}O_2 \rightarrow NO_2$$
$$NO_2 + SO_2 \rightarrow NO + SO_3$$

The *mutarotation* (change in optical activity) of an aqueous solution of glucose is catalysed by acids and bases; this is an example of *acid–base* catalysis. The varying efficiencies of different catalysts at the same concentration are revealed by the values of the rate constants in Table 8.5. The hydroxyl ion is clearly a very effective catalyst for this reaction.

The homogeneous catalyst provides an alternative path or mechanism for the reaction; in the case of a positive catalyst, a path of lower activation enthalpy. Thus, if C is a catalyst for the unimolecular decomposition of A, we can represent the two reaction paths

$$A \xrightarrow{k_1} \text{products} \tag{8.39}$$

$$A + C \xrightarrow{k_c} products + C \tag{8.40}$$

where (8.39) represents the uncatalysed reaction, and (8.40) the catalysed reaction. The rates are given by

$$-\frac{d}{dt}[A] = k_1[A] \tag{8.41}$$

and

$$-\frac{d}{dt}[A] = k_c[A][C] \tag{8.42}$$

respectively. Now, since C is a catalyst for this reaction, [C] does not change with time. Hence (8.42) may be written

$$-\frac{d}{dt}[A] = k'_c[A]$$

where

$$k'_c = k_c[C]$$

and the total rate is therefore

$$-\frac{d}{dt}[A] = (k_1 + k'_c[A]) \tag{8.43}$$

Now if $k'_c \gg k_1$, equation (8.43) becomes approximately

$$-\frac{d}{dt}[A] = k'_c[A] \tag{8.44}$$

Notice that, if (8.44) is obeyed, then the rate increases proportionately to the concentration of homogeneous catalyst C, since $k'_c = k_c[C]$. If [C] is varied, in a series of experiments, then k_c, and not k'_c, will remain constant (Table 8.6).

8.5.2 *Heterogeneous catalysis*

In this case the catalyst is present in a different phase from that of the reactants. For example, finely-divided platinum catalyses the reaction between hydrogen and oxygen. The gases may be mixed at room temperature and will

Table 8.5. *Efficiencies of catalysts in the mutarotation of glucose at 291 K*

Acid	k/s	Base	k/s
H_2O	9.5×10^{-5}	H_2O	9.5×10^{-5}
HEth	2.0×10^{-3}	Mandelate	6.1×10^{-2}
Mandelic acid	6.0×10^{-3}	Eth$^-$	2.7×10^{-2}
H_3O^+	1.4×10^{-1}	OH$^-$	6.0×10^3

remain unreacted indefinitely; if the platinum catalyst is introduced, combination takes place at the surface of the metal and water is formed. Platinum is also a positive catalyst for the decomposition of hydrogen peroxide:

$$2H_2O_2 \xrightarrow{Pt} 2H_2O + O_2$$

This decomposition can be retarded by the addition of small quantities of ethylene glycol (1,2-dihydroxyethane) or acetanilide; these substances are negative catalysts, or inhibitors, for this reaction. Platinum catalysts are very susceptible to poisoning by traces of impurities in the reactants, and are very expensive, so that catalysts which are less easily poisoned are frequently used, although they may show lower efficiency. For example, vanadium pentoxide is nowadays preferred to platinum as a catalyst for the formation of sulphur trioxide in the sulphuric acid contact process.

The heterogeneous catalyst also provides a reaction path of lower activation energy than that of the uncatalysed reaction. The reactants are first adsorbed on the catalyst surface; in this state they are more readily able to react.

$$A(g) + B(g) \rightarrow \text{product, activation enthalpy } E_u \qquad (8.45)$$

$$A(ads) + B(ads) \rightarrow \text{product, activation enthalpy } E_c \qquad (8.46)$$

If (8.45) refers to the uncatalysed reaction, and (8.46) to that predominating when the heterogeneous catalyst is present, then E_u is greater than E_c.

8.6 Chain reactions

Many chemical reactions occur by a series of relatively simple steps, some of which are repeated many times. In these cases the order of the reaction is very different from its stoichiometry.

8.6.1 *Hydrogen–chlorine reaction*

This chain reaction is photochemically initiated. Exposure to light of $\lambda \leqslant$ 546 nm (e.g. the bright white light given out on igniting magnesium ribbon)

Table 8.6. *Acid-catalysed hydrolysis of diazoethanoic ester at 298 K:*
$$N_2CHCO_2C_2H_5 + H_2O \xrightarrow{H_3O^+} HOCH_2CO_2C_2H_5 + N_2$$

pH	$10^3[H^+]/\text{mol dm}^{-3}$, $=c$	$k'_c/\text{min}^{-1}\,\text{mol dm}^{-3}$	$(k'_c/c)/\text{min}^{-1}$
2.50	3.25	20.8	6.40
2.74	1.82	11.7	6.45
3.05	0.90	5.8	6.33
3.44	0.36	2.7	6.38

causes rapid and essentially complete reaction. In the dark, the reaction may be initiated by traces of sodium vapour. The rate of reaction is markedly decreased by traces of oxygen or nitric oxide.

It is observed that the surface-to-volume ratio of the reaction vessel, and the nature of its surface (which changes as the reaction is repeated) also affect the rate of reaction once this has been initiated. It is further observed that there is a small initial increase in pressure, before any appreciable amount of hydrogen chloride has been formed.

These facts are explained by the postulated chain mechanism:

$$Cl_2 + h\nu \rightarrow Cl\cdot + Cl\cdot \qquad (8.47)$$

$$\left.\begin{array}{l} H_2 + Cl\cdot \rightarrow HCl + H\cdot \\ Cl_2 + H\cdot \rightarrow HCl + Cl\cdot \end{array}\right\} \qquad (8.48)$$

$$\left.\begin{array}{l} Cl\cdot + Cl\cdot \rightarrow Cl_2 \\ H\cdot + H\cdot \rightarrow H_2 \\ H\cdot + Cl\cdot \rightarrow HCl \end{array}\right\} \qquad (8.49)$$

Reaction (8.47) is the initiation step; the quanta must carry sufficient energy ($h\nu = 3.63 \times 10^{-19}$ J, or 218 kJ mol^{-1}) to dissociate the chlorine molecule into atoms – hence the upper limit of wavelength. The chlorine atoms propagate the reaction by the steps of (8.48); it is estimated that about 10^6 repetitions of (8.48) occur per chlorine atom, before the chain process is halted by one or more of the termination steps in (8.49). The chlorine radical is the chain carrier in this reaction series.

The termination steps in (8.49) are relatively slow, since each is really a three-body collision: for example

$$Cl\cdot + Cl\cdot + M \rightarrow Cl_2 + M^* \qquad (8.50)$$

where the excited M species, M*, carries away the enthalpy of reaction, which would otherwise lead to the immediate dissociation of the molecule formed, since this energy would be stored in the interatomic bond. M may be a mole-cule of the reaction vessel wall (which explains the importance of the surface and of the surface-to-volume ratio), or an added molecule such as oxygen or nitric oxide. These molecules decrease the overall rate of reaction by increasing the rate of the termination steps and so shortening the reaction chain.

The 'dark' initiation by traces of sodium vapour may be represented

$$Na\cdot + Cl_2 \rightarrow Na^+ + Cl^- + Cl\cdot \qquad (8.51)$$

and the small initial increase in pressure is due to the dissociation

$$Cl_2 \rightarrow Cl\cdot + Cl\cdot \qquad (8.52)$$

As we have seen, hydrogen and iodine react quite readily, at rather high temperatures, by a bimolecular collision mechanism; they can be induced to react by a similar photochemically-initiated mechanism ($\lambda < 875$ nm) at temperatures around 300 K.

8.6.2 *Hydrogen–bromine reaction*

The thermal ('dark') reaction between hydrogen and bromine in the gas phase at 470–570 K was studied by Bodenstein (1906); from his experimental results he deduced the rate equation

$$\frac{d[HBr]}{dt} = \frac{k_A[H_2][Br_2]^{\frac{1}{2}}}{k_B + ([HBr]/[Br_2])} \tag{8.53}$$

Equation (8.53) may be interpreted in terms of the following sequence of reactions:

(*a*) $Br_2 \rightarrow 2Br\cdot$

(*b*) $Br\cdot + H_2 \rightarrow HBr + H\cdot$

(*c*) $H\cdot + Br_2 \rightarrow HBr + Br\cdot$

(*d*) $H\cdot + HBr \rightarrow H_2 + Br$

(*e*) $Br\cdot + Br\cdot \rightarrow Br_2$

Stage (*d*) removes the chain carrier, the hydrogen radical, and explains the term $[HBr]/[Br_2]$ in the denominator of (8.53). Under given conditions this reaction is slower than that between hydrogen and chlorine, and this may be attributed to the relatively slow step (*b*). If rate constants $k_1 \ldots k_5$ are assigned to these steps of the reaction, then in (8.53)

$$k_A = (2k_2k_4/k_3)(k_1/k_5)^{\frac{1}{2}} \qquad k_B = k_3/k_4$$

8.6.3 *Pyrolytic reactions*

Many gas-phase decompositions of organic compounds – pyrolytic reactions, or pyrolyses – occur by chain mechanisms. The chain carrier is often a free radical, as in the decomposition of ethanol vapour

$$CH_3CHO \rightarrow CH_4 + CO$$

This reaction proceeds by the following stages:

$$CH_3CHO \rightarrow CH_3\cdot + CHO\cdot \tag{8.54}$$

$$CH_3\cdot + CH_3CHO \rightarrow CH_4 + CH_3CO\cdot \tag{8.55}$$

$$CH_3CO\cdot \rightarrow CH_3\cdot + CO \tag{8.56}$$

$$2CH_3\cdot \rightarrow C_2H_6 \tag{8.57}$$

Reaction (8.54) is the initiation step of the chain process. This is a relatively difficult step, with a high activation energy, because a strong carbon–carbon bond must be broken. As a result a few pairs of free radicals, fragments of molecules with incomplete electron shells, such as the methyl radical, $CH_3\cdot$, and the aldehyde radical, $CHO\cdot$, are formed. The aldehyde fragment plays no further part in the reaction, but the methyl radical reacts readily (ie at almost every collision) with aldehyde molecules, as in (8.55). One of the products of this reaction is methane; the other is the ethanoyl radical $CH_3CO\cdot$, which decomposes to form carbon monoxide and a new methyl radical, which reacts further. Thus the stages (8.55) and (8.56) are repeated many times for each methyl radical formed in the initiation step. These two stages constitute the propagation steps of the reaction, and virtually all the aldehyde is decomposed by (8.55). The methyl radical is the chain carrier, and the reaction rate is limited by the loss of methyl radicals from the system by such chain-terminating steps as (8.57).

An overall rate equation may be derived, based upon these stages (see Appendix A8):

$$(d/dt)[CH_4] = k[CH_3CHO]^{\frac{3}{2}} \tag{8.58}$$

so that the order of reaction is 1.5, and not unity as might have been expected from the stoichiometric equation. Fractional orders often arise for such chain processes.

8.6.4 *Reactions in solution: salt effects*

The rate of a reaction taking place in solution may be increased or decreased by the addition of a salt which does not take part directly in the reaction. The salt increases the ionic strength (p. 170) of the solution, and thus the activities of ions present; its effect upon reaction rate is referred to as the primary salt effect.

Consider a bimolecular reaction between species A and B in which the activation process leads to the formation of the species $AB^{\ddagger}$, the activated complex. The rate of the reaction is the rate of decomposition of this complex

$$dx/dt = k^{\ddagger}[AB^{\ddagger}] \tag{8.59}$$

In the *transition state theory* of reaction kinetics an equilibrium is postulated between the reactants and the activated complex:

$$A + B \rightleftharpoons AB^{\ddagger} \tag{8.60}$$

Treatment of this equilibrium by the methods of statistical thermodynamics shows that the rate constant in (8.59) is a function of temperature

$$k^{\ddagger} = kT/h \tag{8.61}$$

where k is Boltzmann's constant and h is Planck's constant; $k^{\ddagger}$ has units time^{-1}.

Writing $K^{\ddagger}$ for the equilibrium constant of (8.60)

$$K^{\ddagger} = \frac{[AB^{\ddagger}]f^{\ddagger}}{[A][B]f(A)f(B)} \tag{8.62}$$

where the activity coefficients will be important if some or all of the species A, B and $AB^{\ddagger}$ are charged, ie ionic. Then from (8.59)

$$\text{rate} = \frac{kT[AB^{\ddagger}]}{h} = \frac{kT([A][B])}{h} \frac{f(A)f(B)K^{\ddagger}}{f^{\ddagger}} \qquad (8.63)$$

so that the rate constant ($[A] = [B] = 1$) is

$$k^{\ddagger} = k' \frac{f(A)f(B)}{f^{\ddagger}} \qquad (8.64)$$

where

$$k' = kTK^{\ddagger}/h$$

Now the activity coefficients in (8.64) will be influenced by the ionic strength of the solution, which will in turn affect k; through (8.63) and (8.64), and the reaction rate. Writing (8.64) in logarithmic form

$$\ln k^{\ddagger} = \ln k' + \ln f(A) + \ln f(B) - \ln f^{\ddagger} \qquad (8.65)$$

If the solution is sufficiently dilute we may use the limiting form of the Debye–Hückel equation (p. 173)

$$-\ln f = Az(+)z(-)I^{\frac{1}{2}}$$

to express the activity coefficients:

$$\ln k^{\ddagger} = \ln k' - A\{z^2(A) + z^2(B)\}(I)^{\frac{1}{2}} + A(z(A) + z(B))^2(I)^{\frac{1}{2}}$$

since, if the charges on A and B are $z(A)$ and $z(B)$, that on $AB^{\ddagger}$ will be $z(A) + z(B)$ and

$$\ln k^{\ddagger} = \ln k' + 2Az(A)z(B)(I)^{\frac{1}{2}} \qquad (8.66)$$

This is the Brønsted–Bjerrum equation; k' is the rate constant $k^{\ddagger}$ at zero ionic strength.

Equation (8.66) shows that $k^{\ddagger}$ will vary with I according to the magnitude and sign of the term $z(A)z(B)$. This is illustrated in Figure 8.8. For the reaction

$$S_2O_8^{2-} + 2I^- \rightarrow I_2 + 2SO_4^{2-}$$

$z(A)z(B) = +2$, and the rate increases with increase in I; for the ester hydrolysis

$$RCO_2R' + OH^- \rightarrow RCO_2^- + R'OH$$

$z(A)z(B)$ is zero, and the rate is unchanged by change in ionic strength. In the reaction

$$[Co(NH_3)_5Br]^{2+} + OH^- \rightarrow [Co(NH_3)_5OH]^{2+} + Br^-$$

$z(A)z(B)$ is -2, and the rate decreases with increase in I. This dependence of the rate of reaction in solution upon the ionic strength is the *primary* salt effect.

The rate of a reaction in solution may also depend upon the ionic strength if the reaction is catalysed by an ion provided by dissociation of a weak electrolyte, for example hydroxonium ion catalysis by the weak acid HA. If the dissociation constant of the acid is K_a:

$$K_a = \frac{[H^+][A^-]}{[HA]} \frac{f(H^+)f(A^-)}{f(HA)} \qquad (8.67)$$

so that the amount of the catalyst formed is

$$[H^+] = \frac{K_a[HA]f(HA)}{[A^-]f(H^+)f(A^-)} \tag{8.68}$$

and will be dependent upon I because of the effect of I on the activity coefficients (see (5.52)). This is known as the *secondary* salt effect.

8.7 Enzyme catalysis

Many biological reactions are made possible by organic catalysts or enzymes. As in the case of inorganic catalysts, the quantity of enzyme present remains constant throughout the reaction. Thus, denoting the enzyme by E and the reactant or substrate by S, the reaction (enzymolysis) may be represented:

$$E + S \rightarrow P + E \tag{8.69}$$

where P is the product of reaction. The rate of reaction is generally found to be

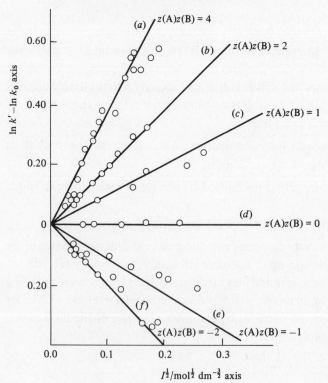

Figure 8.8. Variation of rates of typical ionic reactions with ionic strength: (a) $2Co(NH_3)_5Br^{2+} + Hg^{2+} \rightarrow$; (b) $S_2O_8^{2-} + 2I^- \rightarrow$; (c) $(NO_2NCO_2C_2H_5)^- + OH^- \rightarrow$; (d) $C_{12}H_{22}O_{11} + OH^- \rightarrow$; (e) $H_2O_2 + H^+ + Br^- \rightarrow$; (f) $[Co(NH_3)_5Br]^{2+} + OH^- \rightarrow$.

proportional to the enzyme concentration; Michaelis (1913) proposed the following general mechanism.

$$E+S \underset{k_{-1}}{\overset{k_1}{\rightleftarrows}} ES \overset{k_2}{\rightarrow} P+E \tag{8.70}$$

where ES is an active complex of enzyme and substrate, which may react further to give the product and free enzyme or may revert to the initial E and S. The rate of formation of product is

$$d[P]/dt = k_2[ES]$$

and

$$d[ES]/dt = k_1[E][S] - k_{-1}[ES] - k_2[ES]$$

Figure 8.9. Variation of the rate of enzymolysis with substrate concentration.

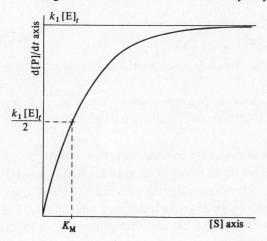

Figure 8.10. Relative affinity of enzyme and substrates A and B.

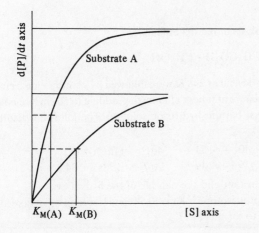

The steady-state hypothesis assumes that the major part of a reaction occurs with the reactive intermediates at constant concentration. Hence, using this hypothesis, we have

$$k_1[E][S] = (k_{-1} + k_2)[ES] \qquad (8.71)$$

If $[E]_t$ is the total enzyme concentration,

$$[E]_t = [E] + [ES], \quad \text{or} \quad [E] = [E]_t - [ES]$$

Since the concentration of the enzyme is much less than that of the substrate,

$$[S] - [ES] \approx [S]$$

$$[ES] \approx k_1 \frac{([E]_t - [ES])[S]}{k_{-1} + k_2}$$

$$\approx \frac{k_1[E]_t[S]}{k_{-1} + k_2 + k_1[S]} \qquad (8.72)$$

Thus

$$d[P]/dt = \frac{k_1 k_2 [E]_t[S]}{k_{-1} + k_2 + k_1[S]} = \frac{k_1[E]_t[S]}{K_M + [S]} \qquad (8.73)$$

where $K_M, = (k_{-1} + k_2)/k_1$, is the *Michaelis constant*; the enzymolysis (8.69) is thus first order in both $[E]$ and $[S]$.

From (8.73) the rate increases with $[S]$ to a maximum $k_1[E]_t$, Figure 8.9; K_M may therefore be defined as that substrate concentration at which the rate attains half its maximum value.

If the rates of reaction of a non-specific enzyme with two substrates are compared, Figure 8.10, it is seen that the enzyme has the greater affinity (gives the higher rate of reaction at a given concentration) for the substrate of *lower* K_M, ie a low K_M for a particular substrate–enzyme system indicates a high affinity. The constant K_M may be found by inverting (8.73) and then plotting $1/(d[P]/dt)$ against $1/[S]$ for a series of experiments.

Problems 8

8.1 The kinetics of the hydrolysis of methyl ethanoate

$$CH_3CO_2CH_3 + H_2O \xrightarrow{H^+} CH_3CO_2H + CH_3OH$$

in excess dilute hydrochloric acid at 298 K were followed by withdrawing 2 cm^3 portions of the reaction mixture at times t after mixing, adding to 50 cm^3 ice-cold water, and titrating against barium hydroxide solution. The following results were obtained:

t/min	0	10	21	40	115	∞
titre/cm^3	18.5	19.1	19.7	20.7	23.6	34.8

Determine the velocity constant and the half-life of the hydrolysis.

8.2 (a) The half-life of the krypton isotope ^{85}Kr is 10.6 year. How long will it take for 99% of a sample of ^{85}Kr to disintegrate?

(*b*) Calculate the mass of radon, ^{222}Rn, in equilibrium with 1 g of radium, ^{226}Ra, if the respective half-lives are:

^{222}Rn 3.83 day; ^{226}Ra 1622 year

8.3 The conversion of sucrose into glucose and fructose, in dilute hydrochloric acid solution, is a first-order reaction, and leads to a reversal of the sign of optical rotation (inversion).

The following polarimeter readings α were measured at times t:

t/min	5.0	20	44	90	140	175	∞
α/deg	12.2	9.95	6.95	2.70	0.10	-1.30	-4.00

Determine the first-order rate constant and the half-life for this reaction.

8.4 The following results refer to the decomposition of ammonia on a heated tungsten surface: $2NH_3 \rightarrow N_2 + 3H_2$

initial pressure/mm	65	105	150	185
half-life/s	290	460	670	820

Deduce the order of the reaction and determine the velocity constant.

8.5 In the homogeneous decomposition of nitrogen(I) oxide, it was shown that, at constant temperature, the time required for half the reaction to be completed, $t_{\frac{1}{2}}$, is inversely proportional to the initial pressure p_0. By varying the temperature the following results were obtained:

T/K	967	1030	1085
p_0/mmHg	294	360	345
$t_{\frac{1}{2}}$/s	1520	212	53

Deduce the order of reaction; calculate the velocity constant at 967 K, and the activation energy.

8.6 The rearrangement of N-chloroacetanilide to *p*-chloroacetanilide is catalysed by hydrogen ions:

$$C_6H_5N(Cl)COCH_3 \xrightarrow{H^+} ClC_6H_4N(H)COCH_3$$

N-chloroacetanilide liberates iodine from potassium iodide solutions, and the reaction can be followed by titrating with sodium thiosulphate. From the following data deduce the order of the reaction and determine the rate constant at the temperature of the experiment:

t/min	0	15	30	45	60	75
titre/cm^3	24.5	18.1	13.3	9.7	7.1	5.2

8.7 The mutarotation of α-*d*-glucose was followed by optical rotation, as measured by a polarimeter. The following rotations α_t were observed at times t from the start of the experiment:

t/min	0	10	20	30	40	50	60	∞
α_t/deg	130.7	110.6	97.6	85.5	77.4	72.0	55.3	47.5

Assuming the forward and reverse reactions to be kinetically of the first order, calculate the respective first-order rate constants.

8.8 It has been shown (R. A. Ogg, *J. Chem. Phys.* (1950), **18**, 573) that the observed rate of the gas-phase decomposition of dinitrogen pentoxide:

$$N_2O_5 \rightarrow 2NO_2 + \tfrac{1}{2}O_2$$

may be interpreted in terms of the sequence of reactions:

(i) $N_2O_5 \underset{k_2}{\overset{k_1}{\rightleftharpoons}} NO_2 \cdot + NO_3 \cdot$

(ii) $NO_2 \cdot + NO_3 \cdot \overset{k_3}{\rightarrow} NO_2 \cdot + O_2 + NO \cdot$

(iii) $NO \cdot + NO_3 \cdot \overset{k_4}{\rightarrow} 2NO_2$

in which the bimolecular reaction (ii) is rate determining. Verify that this mechanism leads to the first-order rate law

$$\frac{-d[N_2O_5]}{dt} = \frac{2k_1k_3[N_2O_5]}{k_2+k_3}$$

Appendices

A1 Stereoviews and stereoviewing

The representation of crystal and molecular structures by stereoscopic pairs of drawings has become commonplace in recent years. Indeed, some very sophisticated computer programs have been written which draw stereoviews from crystallographic data. Two diagrams of a given object are necessary, and they must correspond to the views seen by the eyes in normal vision. Correct viewing requires that each eye sees only the appropriate drawing, and there are several ways in which it can be accomplished.

(*a*) A stereoviewer can be purchased for a modest sum. Two suppliers are:

  (i) C. F. Casella and Company Limited, Regent House, Britannia Walk, London N1 7ND. This maker supplies two grades of stereoscope.

  (ii) Taylor–Merchant Corporation, 25 West 45th Street, New York, NY 10036, USA.

Stereoscopic pairs of drawings may then be viewed directly.

(*b*) The unaided eyes can be trained to defocus, so that each eye sees only the appropriate diagram. The eyes must be relaxed, and look straight ahead. This process may be aided by placing a white card edgeways between the drawings so as to act as an optical barrier. When viewed correctly, a third (stereoscopic) image is seen in the centre of the given two views.

(*c*) An inexpensive stereoviewer can be constructed with comparative ease. A pair of planoconvex or biconvex lenses each of focal length about 10 cm and diameter 2–3 cm are mounted in a framework of opaque material so that the centres of the lenses are about 60–65 mm apart. The frame must be so shaped that the lenses can be held close to the eyes. Two pieces of cardboard shaped as shown in Figure A1.1 and glued together with the lenses in position represents the simplest construction. This basic stereoviewer can be refined in various ways.

Figure A1.1. Simple stereoviewer. Cut out two pieces of card as shown and discard the shaded portions. Make cuts along the double lines. Glue the two cards together with the lenses E_L and E_R in position, fold the portions A and B backward, and fix P into the cut at Q. View from the side marked B. (A similar stereoviewer is marketed by the Taylor–Merchant Corporation, New York.)

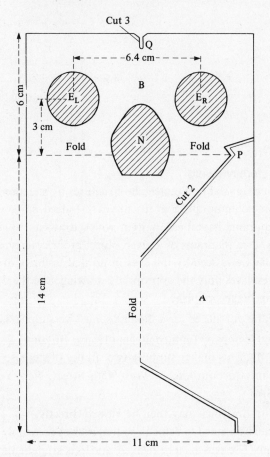

A2 Least-squares line

If it is desired to fit a straight-line relationship to a number N of observations in excess of two, it is often appropriate to use the method of least squares. Let the equation be of the form

$$y = ax + b \tag{A2.1}$$

where a and b are constants which have to be determined. For any observation i,

$$ax_i + b - y_i = e_i \tag{A2.2}$$

where e_i is an error which will be assumed both to be random and to reside in the value of the dependent variable y_i, the error in the independent variable x_i being relatively negligible. According to the principle of least squares, the best values of a and b are chosen such that the sum of the squares of the errors e_i is a minimum.

Thus

$$\text{Min}\left(\sum_i e_i^2\right) = \text{Min}\left\{\sum_i (ax_i + b - y_i)^2\right\} \tag{A2.3}$$

The required minimum value may be found by differentiating the right-hand side of (A2.3) partially with respect to both a and b, and setting each of the derivatives equal to zero.

Hence

$$\partial\left(\sum_i e_i^2\right)\Big/\partial a = 2\sum_i (ax_i^2 + bx_i - x_i y_i) = 0 \tag{A2.4}$$

and

$$\partial\left(\sum_i e_i^2\right)\Big/\partial b = 2\sum_i (ax_i + b - y_i) = 0 \tag{A2.5}$$

Thus, we may derive

$$a[x^2] + b[x] - [xy] = 0 \tag{A2.6}$$

and

$$a[x] + bN - [y] = 0 \tag{A2.7}$$

where $[x] = \sum_i x_i$ and (A2.6) and (A2.7) are known as the normal equations and $[x]$, for example, means $\sum_i x_i$ over the number N of observations. If each observation has a weight w then the normal equations become

$$a[wx^2] + b[wx] - [wxy] = 0 \tag{A2.8}$$

and

$$a[wx] + b[w] - [wy] = 0 \tag{A2.9}$$

Solving for a and b,

$$a = ([w][wxy] - [wx][wy])/\Delta \tag{A2.10}$$

and

$$b = ([wx^2][wy] - [wx][wxy])/\Delta \tag{A2.11}$$

where Δ is given by

$$\Delta = [w][wx^2] - [wx][wx] \tag{A2.12}$$

If all of the weights are unity, $[w] = N$.

The standard deviations in a and b may be estimated by the following procedure. From (A2.2),

$$[e^2] = \sum_i w_i(ax_i + b - y_i)^2 \tag{A2.13}$$

Then, we write, without proof here,

$$\sigma^2(a) = \{[e^2]/(N-2)\}[w]/\Delta \tag{A2.14}$$

and

$$\sigma^2(b) = \{[e^2]/(N-2)\}[wx^2]/\Delta \tag{A2.15}$$

σ being an estimated standard deviation and σ^2 the corresponding variance.

It is recommended that the least-squares line be compared, where feasible, with a plot of the experimental x, y values. In the light of this inspection, certain observations may be reasoned to be unreliable. It must be remembered that a least-squares procedure will always give the best fit to the observations, including the bad ones.

The least-squares technique may be extended to functions of higher degree. Thus, the quadratic function

$$y = ax^2 + bx + c \tag{A2.16}$$

may be fitted to experimental data in excess of three by this method, starting from an equation similar to (A2.3).

A3 Gamma function

The gamma function is useful in handling integrals of the type

$$\int_0^\infty x^n \exp(-ax^2)\,dx \tag{A3.1}$$

which occur in several areas of chemistry and chemical physics. The gamma function $\Gamma(n)$ may be represented by the integral equation

$$\Gamma(n) = \int_0^\infty t^{n-1} \exp(-t)\,dt \tag{A3.2}$$

The following particular results are important:

(a) For $n>0$ and integral,

$$\Gamma(n) = (n-1)! \tag{A3.3}$$

Note that $\Gamma(1) = 1$, because $0! = 1$.

(b) For $n>0$,

$$\Gamma(n+1) = n\Gamma(n) \tag{A3.4}$$

and if n is also integral,

$$\Gamma(n+1) = n! \tag{A3.5}$$

(c) $\Gamma(\tfrac{1}{2}) = \pi^{\frac{1}{2}}$ \hfill (A3.6)

As an example, we shall consider the solution of the integral

$$I = \int_0^\infty x^4 \exp(-x^2/2)\,dx \tag{A3.7}$$

Let $x^2/2 = t$, so that $x = (2t)^{\frac{1}{2}}$ and $dx = (2t)^{-\frac{1}{2}}\,dt$. Then,

$$I = 2\sqrt{2} \int_0^\infty t^{\frac{3}{2}} \exp(-t)\,dt \tag{A3.8}$$

Hence from (A3.2), (A3.4) and (A3.6)

$$I = 2\sqrt{2}\,\Gamma(\tfrac{5}{2}) \quad \text{or} \quad 3(\pi/2)^{\frac{1}{2}} \tag{A3.9}$$

A4 Propagation of errors

The number of significant figures in a result is not necessarily similar to the number of significant figures in the data. Consider $y = p^n$, where $p = 2.0 \pm 0.1$. For $n = 0.1$, y lies between 1.066 and 1.077, whereas for $n = 4$, y lies between 13.0 and 19.4.

Consider any function $y = f(p)$ (Figure A4.1). In the small interval δp, the change δy in y is given with good accuracy by

$$\delta y = \left(\frac{\mathrm{d}y}{\mathrm{d}p} \right) \delta p \tag{A4.1}$$

Consider next any function $y = f(p_1, p_2)$ where p_1 and p_2 are independent variables. For two small independent changes δp_1 and δp_2, the changes in y are by analogy with (A4.1)

$$(\delta y)_{p_1} = \left(\frac{\partial y}{\partial p_1} \right) \delta p_1 \tag{A4.2}$$

and

$$(\delta y)_{p_2} = \left(\frac{\partial y}{\partial p_2} \right) \delta p_2 \tag{A4.3}$$

Since we have assumed that these two variations in y are uncorrelated, they can be represented along two rectangular axes (Figure A4.2). Hence

$$(\delta y)^2 = (\delta y)_{p_1}^2 + (\delta y)_{p_2}^2 = \left(\frac{\partial y}{\partial p_1} \right)^2 (\delta_{p_1})^2 + \left(\frac{\partial y}{\partial p_2} \right)^2 (\delta_{p_2})^2 \tag{A4.4}$$

Generalizing for a function $y = f(p_j)(j = 1, 2, 3, \ldots n)$

$$(\delta y)^2 = \sum_{j=1}^{n} \left(\frac{\partial y}{\partial p_j} \right)^2 (\delta p_j)^2 \tag{A4.5}$$

The quantity δy can be equated to the standard deviation in y, $\sigma(y)$.

Figure A4.1. A function $y = f(p)$.

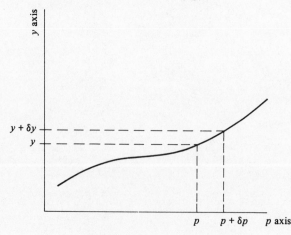

Figure A4.2. Representation of the uncorrelated errors δy_{p_1} and δy_{p_2}.

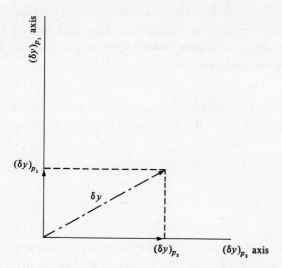

A5 Reduced mass

The calculation of a reduced mass is a central force problem in mechanics. We shall consider two atoms of masses m_1 and m_2, separated by an equilibrium interatomic distance r_e, oscillating about their stationary centre of mass. The centre of mass is determined by

$$m_1 r_1 = m_2 r_2 \tag{A5.1}$$

where r_1 and r_2 are the distances of masses m_1 and m_2 respectively from the centre of mass at any instant (C, Figure A5.1). If r is the bond length at any instant,

$$r = r_1 + r_2 \tag{A5.2}$$

Using (A5.1)

$$r_1 = m_2 r / (m_1 + m_2) \tag{A5.3}$$

and

$$r_2 = m_1 r / (m_1 + m_2) \tag{A5.4}$$

The atoms experience a restoring force proportional to the amount that the bond has been stretched $r - r_e$ at the given instant. The acceleration of the atoms will obey the classical laws, leading to

$$m_1 \ddot{r}_1 = -k(r - r_e) \tag{A5.5}$$

and

$$m_2 \ddot{r}_2 = -k(r - r_e) \tag{A5.6}$$

Using (A5.3) or (A5.4), we obtain

$$m_1 m_2 \ddot{r} / (m_1 + m_2) = -k(r - r_e) \tag{A5.7}$$

Equation (A5.7) is the classical equation for the motion of a particle of mass μ, where

$$\mu = m_1 m_2 / (m_1 + m_2) \tag{A5.8}$$

executing simple harmonic motion along the direction of r with a frequency given by

$$\omega = (k/\mu)^{\frac{1}{2}} \tag{A5.9}$$

that is

$$\mu \ddot{r} = -k(r - r_e) \tag{A5.10}$$

The expression for μ can be re-arranged as

$$\frac{1}{\mu} = \frac{1}{m_1} + \frac{1}{m_2} \tag{A5.11}$$

Figure A5.1. Two-atom system vibrating about a centre of mass, C.

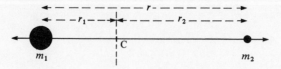

μ is called the reduced mass of the system. If one of the atoms is very much heavier than the other, say $m_1 \gg m_2$, then the frequency of oscillation is determined by the smaller mass m_2, just as though the mass m_1 represents a rigid body.

Example

The relative molar masses of hydrogen and iodine are 1.0079 and 126.9045 respectively (Appendix A20). Hence

$$\mu = \frac{1.00794u \times 126.9045u}{(1.0079u + 126.9045u)}$$

$$= 0.99958u$$

where u is the atomic mass unit.

A similar calculation can be carried out for the system (proton + electron) – see Problem 2.3.

A6 Volume of molecules in a gas

The constant b in the van der Waals' equation, (4.55), is related to the volume occupied by the gas molecules. More precisely, it is an *excluded* volume, or covolume, as the following analysis shows.

Let the molecules of a gas be spherical and of diameter σ. Two molecules cannot approach more closely than the sum of their van der Waals' radii, $2 \times \sigma/2$. Figure A6.1 shows the situation of two molecules in closest contact. The spherical volume of space in which the centres of the molecules cannot move is shaded, and the radius of *this* sphere is σ.

Thus the volume *excluded* per pair of molecules is $4\pi\sigma^3/3$, and that per single molecule $4\pi\sigma^3/6$. The actual volume of a single molecule is $4\pi(\sigma/2)^3/3$, or $\pi\sigma^3/6$. Hence, the excluded volume for a single molecule is four times its own volume.

Figure A6.1. Excluded volume in a gas.

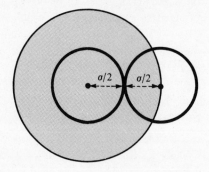

A7 Hypsometric formula — a special case of the Boltzmann distribution

Consider a rectangular column of an ideal gas of cross-sectional area A and height z, with reference to an origin O at ground level, at a uniform temperature T (Figure A7.1). The mass of a gas molecule is m and, at the height z, let the pressure of the gas be P and let there be N molecules per unit volume. We need to determine how N varies with z. Since the gas is assumed to be ideal, no intermolecular attractions need be considered. Hence, for one mole of gas

$$PV = RT \tag{A7.1}$$

The gas constant R is $N_A k$, where N_A is the Avogadro constant and k is the Boltzmann constant, and V is the volume containing one mole of gas. Since $N_A / V = N$,

$$P = NkT \tag{A7.2}$$

At a height $(z + \mathrm{d}z)$ the pressure is $(P + \mathrm{d}P)$. The gravitational force on the segment of width $\mathrm{d}z$ is $A\rho g\,\mathrm{d}z$, where ρ is the density of the gas and g is the gravitational acceleration. The pressure difference across the segment is $(-\mathrm{d}P/\mathrm{d}z)\,\mathrm{d}z$, the negative sign indicating that the gas pressure decreases in the positive direction of z. The hypsometric force on the segment is $-A(\mathrm{d}P/\mathrm{d}z)\,\mathrm{d}z$, and at equilibrium the two forces are balanced:

$$-A(\mathrm{d}P/\mathrm{d}z)\,\mathrm{d}z = A\rho g\,\mathrm{d}z \tag{A7.3}$$

or

$$\mathrm{d}P = -\rho g\,\mathrm{d}z \tag{A7.4}$$

Equation (A7.4) might be obtained also from a definition of pressure.

Figure A7.1. Construction for the hypsometric formula.

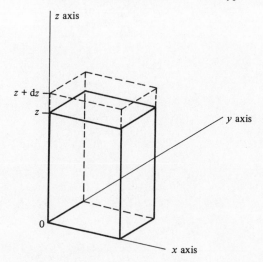

From (A7.2)

$$dP = -kT\,dN \tag{A7.5}$$

and using the fact that $\rho = mN$ we obtain,

$$dN/N = -mg\,dz/kT \tag{A7.5}$$

On integration, we obtain

$$\ln N = -mgz/kT + \text{constant} \tag{A7.7}$$

At $z = 0$ let $N = N_0$. Thus the constant becomes $\ln N_0$, and

$$N = N_0 \exp(-mgz/kT) \tag{A7.8}$$

Now, mgz is the gravitational potential energy per molecule of gas. Let U represent this potential energy per mole of gas. Then $U = N_A mgz$, and

$$N = N_0 \exp(-U/RT) \tag{A7.9}$$

As an example, consider air, of mean molar mass $0.030 \text{ kg mol}^{-1}$, at a height of 30 km above ground level at 298 K. From (A7.9), N/N_0 is approximately 1/35 times its value at ground level; at 300 km it is only about 3×10^{-16} times its value at ground level.

A more general derivation of (A7.9) may be obtained through statistical mechanics but this topic is beyond the scope of this book.

A8 Decomposition of ethanal: a chain reaction

The vapour phase decomposition of ethanal occurs by a series of steps, some of which may be repeated many times: the order of reaction is very different from that implied by the stoichiometry,

$$CH_3CHO \rightarrow CH_4 + CO \tag{A8.1}$$

The steps involved are

$$CH_3CHO \overset{k_1}{\rightarrow} CH_3\cdot + CHO\cdot \tag{A8.2}$$

$$CH_3\cdot + CH_3CHO \overset{k_2}{\rightarrow} CH_4 + CH_3CO\cdot \tag{A8.3}$$

$$CH_3CO\cdot \overset{k_3}{\rightarrow} CH_3\cdot + CO \tag{A8.4}$$

$$CH_3\cdot + CH_3\cdot \overset{k_4}{\rightarrow} C_2H_6 \tag{A8.5}$$

where $CH_3\cdot$, $CHO\cdot$ and $CH_3CO\cdot$ are free radicals, each having an unpaired electron, and k_1–k_4 are the rate constants for the steps. Consider the transient species $CH_3\cdot$. At equilibrium, its rate of formation

$$k_1[CH_3CHO] + k_3[CH_3CO\cdot] \tag{A8.6}$$

is equal to its rate of removal

$$k_2[CH_3\cdot][CH_3CHO] + k_4[CH_3\cdot]^2 \tag{A8.7}$$

Similarly, for the species $CH_3CO\cdot$ we have

$$k_2[CH_3\cdot][CH_3CHO] = k_3[CH_3CO\cdot] \tag{A8.8}$$

Hence, the steady-state concentration of $CH_3CO\cdot$ is given by

$$[CH_3CO\cdot] = k_2[CH_3\cdot][CH_3CHO]/k_3 \tag{A8.9}$$

Substituting for $[CH_3CO\cdot]$ in (A8.6) and equating to (A8.7) gives

$$k_1[CH_3CHO] + k_2[CH_3\cdot][CH_3CHO]$$
$$= k_2[CH_3\cdot][CH_2CHO] + k_4[CH_3\cdot]^2 \tag{A8.10}$$

or

$$[CH_3\cdot]^2 = k_1[CH_3CHO]/k_4 \tag{A8.11}$$

The rate of formation of methane is given by

$$\frac{d[CH_4]}{dt} = k_2[CH_3\cdot][CH_3CHO] = k_1^{\frac{1}{2}}k_2k_4^{-\frac{1}{2}}[CH_3CHO]^{\frac{3}{2}} \tag{A8.12}$$

so that the order of reaction is 1.5, and not unity as suggested by (A8.1).

A9 Selected ionization energies

The ionization energy (sometimes called ionization potential) I_z represents the energy change accompanying the reaction

$$M^{(z-1)+}(g) \to M^{z+}(g) + e^-$$

at 0 K. At any other temperature the kinetic energies of the gaseous species are involved. In this appendix, I_z is listed for z ranging from one to four. For $z = 1$, $M^{(z-1)+}$ is, of course, the unionized gaseous atom. The values are given in $kJ\,mol^{-1}$, although the unit is not a part of the definition.

	Atomic number	I_1	I_2	I_3	I_4
H	1	1312			
He	2	2371	5247		
Li	3	520.1	7297	11811	
Be	4	899.1	1757	14820	20999
B	5	800.4	1462	3642	25016
C	6	1087	2352	4561	6510
N	7	1403	2856	4577	7473
O	8	1316	3391	5301	7468
F	9	1681	3375	6046	8318
Ne	10	2080	3963	6176	9376
Na	11	495.8	4565	6912	9540
Mg	12	737.6	1450	7732	10543
Al	13	577.4	1816	2744	11577
Si	14	819.2	1577	3228	4356
P	15	1061	1896	2910	4950
S	16	999.6	2258	3333	4565
Cl	17	1255	2296	3850	5163
Ar	18	1520	2665	3946	5577
K	19	418.8	3068	4439	5874
Ca	20	589.5	1145	4941	6435
Sc	21	632.6	1243	2388	7130
Ti	22	659.0	1309	2715	4181
V	23	650.2	1370	2866	4669
Cr	24	652.7	1591	2991	4845
Mn	25	717.1	1509	3251	5113
Fe	26	762.3	1561	2956	5402
Co	27	758.6	1645	3231	5113
Ni	28	736.4	1751	3489	5402
Cu	29	745.2	1958	3666	5694
Zn	30	906.3	1733	3827	5983
Se	34	941.0	2075	2902	4139
Br	35	1142	2082	3463	4845
Kr	36	1350	2371	3564	5017
Rb	37	402.9	2653	3828	5113
Sr	38	549.4	1064	4149	5498
Ag	47	730.9	2072	3483	5017
Cd	48	867.3	1631	3377	5305
In	49	559.4	1820	2705	5594

	Atomic number	I_1	I_2	I_3	I_4
Sn	50	707.5	1412	2958	3936
Sb	51	833.5	1738	2287	4238
Te	52	869.4	2079	2953	3649
I	53	1007	1834	2991	4052
Xe	54	1170	2046	3100	4498
Cs	55	375.7	2264	3377	4920
Ba	56	560.7	962.7	3570	4728
La	57	541.4	1103	1850	5017
Ce	58	666.5	1187	1939	3540
Hg	80	1007	1809	3310	6945
Tl	81	589.1	1970	2875	4874
Pb	82	715.5	1450	3095	4076
Bi	83	702.9	1862	2470	4381
Rn	84	1037	1930	2894	4247

A10 Selected dissociation enthalpies and electron affinities

The dissociation enthalpies $D^{\ominus}(X_2)$ refer to the process

$$X_2(g) \rightarrow 2X(g)$$

at 298 K. The electron affinities $E(X^-)$ refer to the process

$$X(g) + e^- \rightarrow X^-(g)$$

at 0 K. Selected values for both these quantities are listed below. The electron affinities $E(X^{2-})$ are somewhat less precise as they are derived from lattice energy calculations rather than by experiment.

	$D^{\ominus}$/kJ mol^{-1}		$E(X^{z-})$/kJ mol^{-1}
F_2	157	F^-	-238.0
Cl_2	243.0	Cl^-	-348.6
Br_2	223.8	Br^-	-324.5
I_2	213.7	I^-	-295.5
H_2	431.8	H^-	-71
O_2	489.9	O^-	-142
		O^{2-}	749
		S^-	-206
		S^{2-}	414
		Se^-	-213
		Se^{2-}	490
		Te^-	-222
		Te^{2-}	406
		OH^-	-272
		CN^-	-339

A11 Solubility products

The data listed in this table are thermodynamic solubility products at 298 K. They are tabulated in the form $\alpha\ \beta$, where β is the exponent of 10 which multiplies α. Thus $1.7 - 10$ must be interpreted as 1.7×10^{-10}.

	α	β		α	β
$AgCl$	1.7	-10	$KClO_4$	8.9	-3
$AgBr$	5.0	-13	$La(OH)_3$	1.0	-19
AgI	8.5	-17	$Lu(OH)_3$	2.5	-24
$AgBrO_3$	5.8	-5			
$AgCN$	1.6	-14	$Mg(OH)_2$	8.9	-12
$AgCNS$	1.0	-12	MgF_2	8.0	-8
Ag_2CrO_4	1.9	-12	$MgCO_3$	8.0	-9
$Al(OH)_3$	5.0	-33	MgC_2O_4	8.6	-5
			$Mn(OH)_2$	2.0	-13
$Ba(OH)_2$	5.0	-3	MnS	7.0	-16
BaF_2	2.4	-5	$MnCO_3$	8.8	-11
$BaCO_3$	1.6	-9			
$BaSO_4$	1.1	-10	$Ni(OH)_2$	1.6	-16
$BaCrO_4$	8.5	-11	NiS, α	3.0	-21
BaC_2O_4	1.5	-8	NiS, β	1.0	-26
			$NiCO_3$	1.4	-7
$Ca(OH)_2$	1.3	-6			
CaF_2	1.7	-10	$Pb(OH)_2$	4.2	-15
$CaCO_3$	4.7	-9	PbF_2	4.0	-8
$CaSO_4$	2.4	-5	$PbCl_2$	1.6	-5
$Ca_3(PO_4)_2$	1.3	-32	$PbBr_2$	4.6	-6
CaC_2O_4	1.3	-9	PbI_2	8.3	-9
$Cd(OH)_2$	2.0	-14	$Pb(IO_3)$	1.2	-13
CdS	1.0	-28	PbS	7.0	-29
$CdCO_3$	5.2	-12	$PbCO_3$	1.5	-13
$Co(OH)_2$	2.5	-16	$PbSO_4$	1.3	-8
CoS, α	5.0	-22	$Pb_3(PO_4)_2$	1.0	-54
CoS, β	1.9	-27	$PbCrO_4$	2.0	-16
$CoCO_3$	8.0	-13			
$CsClO_4$	3.2	-3	$RbClO_4$	3.8	-3
$Cu(OH)_2$	1.6	-19	$Sc(OH)_3$	1.0	-27
$CuCl$	3.2	-7	$Sn(OH)_2$	3.0	-27
$CuBr$	5.9	-9	SnS	1.0	-26
CuI	1.1	-12	$Sr(OH)_2$	3.2	-4
Cu_2S	1.2	-45	SrF_2	7.9	-10
CuS	8.0	-37	$SrCO_3$	7.0	-10
$CuCO_3$	2.5	-10	$SrSO_4$	7.6	-7
$CuCNS$	4.0	-14	$SrCrO_4$	3.6	-5
			SrC_2O_4	5.6	-8
$Fe(OH)_2$	1.8	-15			
$Fe(OH)_3$	6.0	-38	$Th(OH)_4$	1.0	-39
FeS	4.0	-19	ThF_4	7.0	-12
Fe_2S_3	1.0	-88	$TlCl$	1.9	-4
$FeCO_3$	2.1	-11	$TlBr$	3.6	-6
$FePO_4$	1.5	-18	TlI	8.9	-8
			Tl_2S	1.2	-24
Hg_2Cl_2	1.1	-18			
Hg_2Br_2	1.3	-22	$Y(OH)_3$	8.1	-23
Hg_2I_2	4.5	-29	$Zn(OH)_2$	4.5	-17
Hg_2S	1.0	-45	ZnS	7.0	-26
HgS	1.6	-54	$ZnCO_3$	2.0	-10
$Hg_2(CNS)_2$	3.0	-20			

A12 Selected dissociation constants for acids and bases

The data in this appendix are thermodynamic dissociation constants at 298 K. They are tabulated in the form described in the table of solubility products; K_1, K_2 and K_3 represent the first, second and third dissociation constants respectively.

			K_1		K_2		K_3	
			α	β	α	β	α	β
Acids	Boric acid	H_3BO_3	7.3	-10	1.8	-13	1.6	-14
	Carbonic acid	H_2CO_3	4.3	-7	5.6	-11		
	Hydrofluoric acid	HF	3.5	-4				
	Phosphoric acid	H_3PO_4	7.5	-3	6.2	-8	2.2	-13
	Water	H_2O	1.0	-14				
	Ethanoic acid	CH_3CO_2H	1.8	-5				
	Benzoic acid	$C_6H_5CO_2H$	6.5	-5				
	Monochlorethanoic acid	$ClCH_2CO_2H$	1.4	-3				
	Dichlorethanoic acid	Cl_2CHCO_2H	3.3	-2				
	Trichlorethanoic acid	Cl_3CCO_2H	2.0	-1				
	Methanoic acid	HCO_2H	1.8	-4				
	Malonic acid	$CH_2(CO_2H)_2$	1.5	-3	2.0	-6		
	Ethandicarboxylic acid	$(CO_2H)_2$	5.9	-2	6.4	-5		
	Phenol	C_6H_5OH	1.3	-10				
Bases	Ammonia	$NH_3 \cdot H_2O$	1.8	-5				
	Water	H_2O	1.0	-14				
	Aniline	$C_6H_5NH_2$	3.8	-10				
	Ethylamine	$C_2H_5NH_2$	5.6	-4				
	Diethylamine	$(C_2H_5)_2NH$	9.6	-4				
	Triethylamine	$(C_2H_5)_3N$	5.7	-4				
	p-Phenylenediamine	$C_6H_4(NH_2)_2$	1.1	-8				
	Piperidine	$(CH_2)_5NH$	1.6	-3				
	Pyridine	C_5H_5N	1.7	-9				
	Hydrazine[a]	NH_2NH_2	1.7	-6				
	Hydroxylamine[a]	NH_2OH	1.1	-8				

[a] At 293 K.

A13 Conductances of selected ions

The conductivity κ, earlier called the specific conductance, of an electrolyte is defined as the conductance (reciprocal of resistance) between opposite faces of a unit cube of the solution or melt. In SI units the cube has a side of 1 m, and in cgs units it is a 1 cm cube. Hence,

$$\kappa_{SI}/\Omega^{-1}\,m^{-1} = 100\kappa_{cgs}/\Omega^{-1}\,cm^{-1}$$

The molar conductance Λ is defined by $\Lambda = \kappa/c$, where c is the concentration in the units appropriate to κ. In SI units, c is in $mol\,m^{-3}$ and in cgs units c is in $mol\,cm^{-3}$. Neither of these concentration units is in general use. To retain c in the common units of $mol\,dm^{-3}$, the following formulae are used:

$$\Lambda_{SI}/\Omega^{-1}\,m^2\,mol^{-1} = \frac{10^{-2}\kappa_{SI}/\Omega^{-1}\,m^{-1}}{c/mol\,dm^3}$$

$$\Lambda_{cgs}/\Omega^{-1}\,cm^2\,mol^{-1} = \frac{10^3\kappa_{cgs}/\Omega^{-1}\,cm^{-1}}{c/mol\,dm^{-3}}$$

Hence

$$\Lambda_{SI}/\Omega^{-1}\,m^2\,mol^{-1} = 10^{-4}\Lambda_{cgs}/\Omega^{-1}\,cm^2\,mol^{-1}$$

The data below serve to emphasize these relationships.

Selected molar conductances of some ions

	$\lambda_{SI}(291\,K)/$ $\Omega^{-1}\,m^2\,mol^{-1}$	$\lambda_{SI}(298\,K)/$ $\Omega^{-1}\,m^2\,mol^{-1}$	$\lambda_{cgs}(291\,K)/$ $\Omega^{-1}\,cm^2\,mol^{-1}$	$\lambda_{cgs}(298\,K)/$ $\Omega^{-1}\,cm^2\,mol^{-1}$
H^+	0.0314	0.0350	314	350
K^+	0.00646	0.00745	64.6	74.5
Na^+	0.00435	0.00509	43.5	50.9
NH_4^+	0.00645	0.00745	64.5	74.5
Ag^+	0.00543	0.00635	54.3	63.5
$\frac{1}{2}Ba^{2+}$	0.0055	0.0065	55	65
$\frac{1}{2}Ca^{2+}$	0.0051	0.0060	51	60
$\frac{1}{3}La^{3+}$	0.0061	0.0072	61	72
OH^-	0.0172	0.0192	172	192
Cl^-	0.00655	0.00755	65.5	75.5
NO_3^-	0.00617	0.00706	61.7	70.6
$CH_3CO_2^-$	0.00346	0.00408	34.6	40.8
$\frac{1}{2}C_2O_4^{2-}$	0.0063	0.0073	63	73
$\frac{1}{3}[Fe(CN)_6]^{3-}$	0.0099	0.0101	99	101
$\frac{1}{4}[Fe(CN)_6]^{4-}$	0.0095	0.0111	95	111

A14 Selected cryoscopic and ebullioscopic constants

These constants have the units $K \, mol \, kg^{-1}$. Their values must be determined from measurements on *dilute* solutions. Usually the units are given simply as K.

Molal freezing point depression constants

Ethanoic acid	3.90
Benzene	5.11
Tribromomethane	14.3
Camphene	35
Camphor[a]	40
Cyclohexane	20.2
Naphthalene	6.9
Nitrobenzene	6.9
Water	1.86

[a] Commercial camphor is not usually a pure chemical compound. The cryoscopic constant should be determined for the given sample by means of a solute of known molecular weight.

Molal boiling point elevation constants

Ethanoic acid	3.07
Propanone	1.71
Benzene	2.65
Tetrachloromethane	5.0
Ethanol	1.2
Water	0.52

A15 Selected electrode potentials

This appendix lists the standard electrode (reduction) potentials $\pi^{\ominus}$ at 298 K and unit activity. All electrode reactions are written as reductions with electrons on the left-hand side of the equation:

$$\tfrac{1}{2}Zn^{2+} + e^- \rightleftharpoons \tfrac{1}{2}Zn \quad \pi^{\ominus} = -0.763 \text{ V}$$

A negative value for the potential means that the reduced form (Zn) is a better reducing agent than is hydrogen. A positive value indicates that the oxidized form is a better oxidizing agent than is hydrogen, for example:

$$\tfrac{1}{2}Cu^{2+} + e^- \rightleftharpoons \tfrac{1}{2}Cu \quad \pi^{\ominus} = +0.337 \text{ V}$$

By convention, the standard electrode potential for the hydrogen electrode is defined to be zero.

Values for some standard electrode potentials at 298 K, $\pi^{\ominus}/V$

$Li^+ + e^- \rightleftharpoons Li$	−3.045	$AgBr + e^- \rightleftharpoons Ag + Br^-$	0.095
$K^+ + e^- \rightleftharpoons K$	−2.925	$\tfrac{1}{2}Sn^{4+} + e^- \rightleftharpoons \tfrac{1}{2}Sn^{2+}$	0.15
$Rb^+ + e^- \rightleftharpoons Rb$	−2.925	$Cu^{2+} + e^- \rightleftharpoons Cu^+$	0.153
$\tfrac{1}{2}Ba^{2+} + e^- \rightleftharpoons \tfrac{1}{2}Ba$	−2.90	$AgCl + e^- \rightleftharpoons Ag + Cl^-$	0.2223
$\tfrac{1}{2}Sr^{2+} + e^- \rightleftharpoons \tfrac{1}{2}Sr$	−2.89	$\tfrac{1}{2}Cu^{2+} + e^- \rightleftharpoons \tfrac{1}{2}Cu$	0.337
$\tfrac{1}{2}Ca^{2+} + e^- \rightleftharpoons \tfrac{1}{2}Ca$	−2.87	$Cu^+ + e^- \rightleftharpoons Cu$	0.521
$Na^+ + e^- \rightleftharpoons Na$	−2.714	$\tfrac{1}{2}I_2 + e^- \rightleftharpoons I^-$	0.5355
$\tfrac{1}{2}Mg^{2+} + e^- \rightleftharpoons \tfrac{1}{2}Mg$	−2.37	$\tfrac{1}{2}I_3^- + e^- \rightleftharpoons \tfrac{3}{2}I^-$	0.536
$\tfrac{1}{2}Be^{2+} + e^- \rightleftharpoons \tfrac{1}{2}Be$	−1.85	$Fe^{3+} + e^- \rightleftharpoons Fe^{2+}$	0.771
$\tfrac{1}{3}Al^{3+} + e^- \rightleftharpoons \tfrac{1}{3}Al$	−1.66	$\tfrac{1}{2}Hg_2^{2+} + e^- \rightleftharpoons Hg(l)$	0.789
$\tfrac{1}{2}Zn^{2+} + e^- \rightleftharpoons \tfrac{1}{2}Zn$	−0.763	$Ag^+ + e^- \rightleftharpoons Ag$	0.7991
$\tfrac{1}{2}Fe^{2+} + e^- \rightleftharpoons \tfrac{1}{2}Fe$	−0.440	$Hg^{2+} + e^- \rightleftharpoons \tfrac{1}{2}Hg_2^{2+}$	0.920
$\tfrac{1}{2}Cd^{2+} + e^- \rightleftharpoons \tfrac{1}{2}Cd$	−0.403	$\tfrac{1}{2}Br_2(l) + e^- \rightleftharpoons Br^-$	1.0652
$Tl^+ + e^- \rightleftharpoons Tl$	−0.3363	$H_3O^+ + \tfrac{1}{4}O_2(g) + e^-$	
$\tfrac{1}{2}Ni^{2+} + e^- \rightleftharpoons \tfrac{1}{2}Ni$	−0.250	$\rightleftharpoons \tfrac{3}{2}H_2O(l)$	1.229
$AgI + e^- \rightleftharpoons Ag + I^-$	−0.151	$\tfrac{7}{3}H_3O^+ + \tfrac{1}{6}Cr_2O_7^{2-} + e^-$	
$\tfrac{1}{2}Sn^{2+} + e^- \rightleftharpoons \tfrac{1}{2}Sn$	−0.136	$\rightleftharpoons \tfrac{7}{12}H_2O(l) + \tfrac{1}{3}Cr^{3+}$	1.33
$\tfrac{1}{2}Pb^{2+} + e^- \rightleftharpoons \tfrac{1}{2}Pb$	−0.126	$\tfrac{1}{2}Cl_2(g) + e^- \rightleftharpoons Cl^-$	1.3595
$H_3O^+ + e^- \rightleftharpoons \tfrac{1}{2}H_2(g) +$		$\tfrac{1}{3}Au^{3+} + e^- \rightleftharpoons Au$	1.50
$\quad H_2O(l)$	0	$\tfrac{8}{5}H_3O^+ + \tfrac{1}{5}MnO_4^- + e^-$	
		$\rightleftharpoons \tfrac{12}{5}H_2O + \tfrac{1}{5}Mn^{2+}$	1.51
		$Ce^{4+} + e^- \rightleftharpoons Ce^{3+}$	1.61
		$\tfrac{1}{2}S_2O_8^{2-} + e^- \rightleftharpoons SO_4^{2-}$	2.01

The notation $Li^+/Li = -3.045$, for example, is often used to indicate electrode reactions of the type tabulated above.

A16 Selected constant boiling-point (azeotropic) binary mixtures at 760 Torr

1. *Minimum boiling-point systems*

A	B	Mole% A	Wt% A	bp/K
H_2O	C_2H_5OH	10.6	4.43	351.4
H_2O	$(C_2H_5)_2O$	5.0	1.26	357.4
H_2O	C_6H_6	44.4	15.6	342.6
CH_3OH	$(CH_3)_2CO$	20.0	12.2	328.9
CH_3OH	C_6H_6	61.4	39.5	331.5
CH_3CO_2H	C_6H_6	97.5	96.8	353.3
C_2H_5OH	C_6H_6	44.8	32.4	341.4
C_2H_5OH	C_6H_{12}	33.2	21.4	331.9

2. *Maximum boiling-point systems*

A	B	Mole% A	Wt% A	bp/K
H_2O	HF	65.4	62.9	384.6
H_2O	HCl	88.9	79.8	381.8
H_2O	HBr	83.1	96.7	399.2
H_2O	HI	84.3	43.1	400.2
H_2O	$HClO_4$	32.0	7.79	476.2
H_2O	HNO_3 (735 Torr)	62.2	32.0	393.7
H_2O	HCO_2H	43.3	15.6	380.3
HCl	$(CH_3)_2O$	65.0	59.5	271.7
$CHCl_3$	$(CH_3)_2CO$	65.5	79.6	378.6
HCO_2H	$(C_2H_5)_2CO$	48.0	33.0	337.7
C_6H_5OH	$C_6H_5CH_2OH$	8.0	7.04	479.2
C_6H_5OH	C_6H_5CHO	54.0	51.0	458.8

A17 Selected enthalpies of atomization of solids

The enthalpy of atomization $\Delta H_a^\ominus$ of a solid refers to the process

$$X(c) \rightarrow X(g)$$

at 298 K. The following data, also known as sublimation enthalpies, are given in kJ mol^{-1}

Li	161	Cu	339
Be	326	Zn	131
B	407	Se	233
C	718	Rb	82.2
Na	108	Sr	164
Mg	149	Ag	286
Al	325	Cd	112
Si	368	In	244
S	101	Sn	301
K	89.5	Sb	254
Ca	177	Te	199
Sc	343	Cs	78.2
Ti	471	Ba	174
V	514	La	417
Cr	397	Hg(l)	60.7
Mn	279	Tl	180
Fe	418	Pb	196
Co	425	Bi	199
Ni	424	U	490

A18 Enthalpies of formation of alkali-metal halides

The enthalpy of formation $\Delta H_f^{\ominus}$ of an alkali-metal halide corresponds to the process

$$M(c) + \tfrac{1}{2}X_2(g) \rightarrow MX(c)$$

at 298 K and 760 Torr. The following values are given in kJ mol^{-1}.

	Li	Na	K	Rb	Cs
F	612	571	563	549	531
Cl	405	413	436	431	433
Br	349	362	392	389	395
I	271	290	328	328	337

A19 Integrated rate equations

1 *Second-order rate equation*

$$\frac{dx}{dt} = k(a-x)(b-x)$$

$$\int \frac{dx}{(a-x)(b-x)} = k \int dt$$

From partial fractions,

$$\frac{1}{(a-x)(b-x)} = \frac{1/(b-a)}{(a-x)} - \frac{1/(b-a)}{(b-x)}$$

Therefore

$$\int \frac{dx}{(a-x)(b-x)} = \frac{1}{(b-a)} \int \frac{dx}{(a-x)} - \frac{1}{(b-a)} \int \frac{dx}{(b-x)}$$

ie

$$kt = \frac{1}{(b-a)} \left[-\ln(a-x) + \ln(b-x) \right] + \text{constant}$$

$$= \frac{1}{(b-a)} \left[\ln\left(\frac{b-x}{a-x}\right) \right] + \text{constant}$$

At $t=0$, $x=0$, ie

$$0 = \frac{1}{(b-a)} \ln\left(\frac{b}{a}\right) + \text{constant}$$

hence

$$kt = \frac{1}{(b-a)} \ln \frac{a(b-x)}{b(a-x)}$$

2 *Equilibrium reactions (both reactions of first order)*

$$\frac{dx}{dt} = k_1(a-x) - k_{-1}(b+x)$$

$$= k_1 a - k_{-1} b - (k_1 + k_{-1})x$$

$$= (k_1 + k_{-1}) \left(\frac{k_1 a - k_{-1} b}{k_1 + k_{-1}} - x \right)$$

$$= (k_1 + k_{-1})(A - x)$$

where

$$A = \frac{k_1 a - k_{-1} b}{k_1 + k_{-1}}$$

therefore

$$\int \frac{dx}{A-x} = (k_1 + k_{-1}) \int dt$$

$$-\ln(A-x) = (k_1 + k_{-1})t + \text{constant}$$

At $t=0$, $x=0$,

$$(k_1+k_{-1})t=\ln\left(\frac{A}{A-x}\right)=\ln\left(\frac{a-a_\infty}{a_t-a_\infty}\right)$$

At equilibrium, k_1/k_{-1} = equilibrium constant K

$$K=\left(\frac{b+x}{a-x}\right)_{t=\infty}$$

If b (ie [B] at $t=0$) is zero,

$$K=\frac{x_\infty}{a-x_\infty}$$

A20 Periodic table of the elements

Each box contains the chemical symbol of the element, its atomic number, atomic weight (see below) and outermost electronic configuration. The elements are arranged by classical group number and period (principal quantum number of outermost electron/s). Some of the electronic configurations, particularly for elements of high atomic number, are the subject of current investigation and debate.

The atomic weights are those recommended by IUPAC 1975†: they are relative values, being scaled to $^{12}C = 12$. The atomic weights of certain elements depend on the history of the material. The values given apply to terrestrial elements and to certain artificial elements. The precision is ± 1 in the last digit quoted, or ± 3 where the element is marked with an *. Values in parentheses refer to the mass number of the isotope of longest half-life for certain radioactive elements whose atomic weights cannot be quoted precisely without reference to their origin.

† *Pure & Applied Chemistry*, Vol. 47, pp. 75–95 (1975).

Periodic Table of the Elements

Period	IA	IIA	IIIA	IVA	VA	VIA	VIIA	VIII	VIII	VIII	IB	IIB	IIIB	IVB	VB	VIB	VIIB	O
1	1 H 1.0079 $(1s)^1$																	2 He 4.00260 $(1s)^2$
2	3 Li* 6.941 $(2s)^1$	4 Be 9.01218 $(2s)^2$											5 B 10.81 $(2s)^2(2p)^1$	6 C 12.011 $(2s)^2(2p)^2$	7 N 14.0067 $(2s)^2(2p)^3$	8 O* 15.9994 $(2s)^2(2p)^4$	9 F 18.998403 $(2s)^2(2p)^5$	10 Ne* 20.179 $(2s)^2(2p)^6$
3	11 Na 22.98977 $(3s)^1$	12 Mg 24.305 $(3s)^2$											13 Al 26.98154 $(3s)^2(3p)^1$	14 Si* 28.0855 $(3s)^2(3p)^2$	15 P 30.97376 $(3s)^2(3p)^3$	16 S* 32.06 $(3s)^2(3p)^4$	17 Cl 35.453 $(3s)^2(3p)^5$	18 Ar* 39.948 $(3s)^2(3p)^6$
4	19 K* 39.0983 $(4s)^1$	20 Ca 40.08 $(4s)^2$	21 Sc 44.9559 $(3d)^1(4s)^2$	22 Ti* 47.90 $(3d)^2(4s)^2$	23 V* 50.9414 $(3d)^3(4s)^2$	24 Cr 51.996 $(3d)^5(4s)^1$	25 Mn 54.9380 $(3d)^5(4s)^2$	26 Fe* 55.847 $(3d)^6(4s)^2$	27 Co 58.9332 $(3d)^7(4s)^2$	28 Ni 58.70 $(3d)^8(4s)^2$	29 Cu* 63.546 $(3d)^{10}(4s)^1$	30 Zn 65.38 $(3d)^{10}(4s)^2$	31 Ga 69.72 $(4s)^2(4p)^1$	32 Ge* 72.59 $(4s)^2(4p)^2$	33 As 74.9216 $(4s)^2(4p)^3$	34 Se* 78.96 $(4s)^2(4p)^4$	35 Br 79.904 $(4s)^2(4p)^5$	36 Kr 83.80 $(4s)^2(4p)^6$
5	37 Rb* 85.4678 $(5s)^1$	38 Sr* 87.62 $(5s)^2$	39 Y 88.9059 $(4d)^1(5s)^2$	40 Zr 91.22 $(4d)^2(5s)^2$	41 Nb 92.9064 $(4d)^4(5s)^1$	42 Mo 95.94 $(4d)^5(5s)^1$	43 Tc (97) $(4d)^5(5s)^2$	44 Ru* 101.07 $(4d)^7(5s)^1$	45 Rh 102.9055 $(4d)^8(5s)^1$	46 Pd* 106.4 $(4d)^{10}(5s)^0$	47 Ag 107.868 $(4d)^{10}(5s)^1$	48 Cd 112.41 $(4d)^{10}(5s)^2$	49 In 114.82 $(5s)^2(5p)^1$	50 Sn* 118.69 $(5s)^2(5p)^2$	51 Sb 121.75 $(5s)^2(5p)^3$	52 Te* 127.60 $(5s)^2(5p)^4$	53 I 126.9045 $(5s)^2(5p)^5$	54 Xe 131.30 $(5s)^2(5p)^6$
6	55 Cs 132.9054 $(6s)^1$	56 Ba 137.33 $(6s)^2$	57 La† 138.9055 $(4f)^0(5d)^1(6s)^2$	72 Hf 178.49 $(4f)^{14}(5d)^2(6s)^2$	73 Ta* 180.9479 $(5d)^3(6s)^2$	74 W* 183.85 $(5d)^4(6s)^2$	75 Re 186.207 $(5d)^5(6s)^2$	76 Os* 190.2 $(5d)^6(6s)^2$	77 Ir* 192.22 $(5d)^7(6s)^2$	78 Pt* 195.09 $(5d)^9(6s)^1$	79 Au 196.9665 $(5d)^{10}(6s)^1$	80 Hg* 200.59 $(5d)^{10}(6s)^2$	81 Tl* 204.37 $(6s)^2(6p)^1$	82 Pb 207.2 $(6s)^2(6p)^2$	83 Bi 208.9804 $(6s)^2(6p)^3$	84 Po (209) $(6s)^2(6p)^4$	85 At (210) $(6s)^2(6p)^5$	86 Rn (222) $(6s)^2(6p)^6$
7	87 Fr (223) $(7s)^1$	88 Ra 226.0254 $(7s)^2$	89 Ac‡ 227.0278 $(5f)^0(6d)^1(7s)^2$															

† Lanthanides

57 La* 138.9055 $(4f)^0(5d)^1(6s)^2$	58 Ce 140.12 $(4f)^1(5d)^1(6s)^2$	59 Pr 140.9077 $(4f)^3(5d)^0(6s)^2$	60 Nd* 144.24 $(4f)^4(5d)^0(6s)^2$	61 Pm 145 $(4f)^5(5d)^0(6s)^2$	62 Sm* 150.4 $(4f)^6(5d)^0(6s)^2$	63 Eu 151.96 $(4f)^7(5d)^0(6s)^2$	64 Gd* 157.25 $(4f)^7(5d)^1(6s)^2$	65 Tb 158.9254 $(4f)^9(5d)^0(6s)^2$	66 Dy* 162.50 $(4f)^{10}(5d)^0(6s)^2$	67 Ho 164.9304 $(4f)^{11}(5d)^0(6s)^2$	68 Er* 167.26 $(4f)^{12}(5d)^0(6s)^2$	69 Tm 168.9342 $(4f)^{13}(5d)^0(6s)^2$	70 Yb* 173.04 $(4f)^{14}(5d)^0(6s)^2$	71 Lu 174.97 $(4f)^{14}(5d)^1(6s)^2$

‡ Actinides

89 Ac 227.0278 $(5f)^0(6d)^1(7s)^2$	90 Th 232.0381 $(5f)^0(6d)^2(7s)^2$	91 Pa 231.0359 $(5f)^2(6d)^1(7s)^2$	92 U 238.029 $(5f)^3(6d)^1(7s)^2$	93 Np 237.0482 $(5f)^4(6d)^1(7s)^2$	94 Pu (244) $(5f)^6(6d)^0(7s)^2$	95 Am (243) $(5f)^7(6d)^0(7s)^2$	96 Cm (247) $(5f)^7(6d)^1(7s)^2$	97 Bk (247) $(5f)^9(6d)^0(7s)^2$	98 Cf (251) $(5f)^{10}(6d)^0(7s)^2$	99 Es (254) $(5f)^{11}(6d)^0(7s)^2$	100 Fm (257) $(5f)^{12}(6d)^0(7s)^2$	101 Md (258) $(5f)^{13}(6d)^0(7s)^2$	102 No (259) $(5f)^{14}(6d)^0(7s)^2$	103 Lr (260) $(5f)^{14}(6d)^1(7s)^2$

Notes on atomic weights:

[a] Variations in isotopic composition of terrestrial material limits the precision to that quoted.

[b] Geological specimens reveal such variations in isotopic composition that the difference in implied atomic weight exceeds the precision given in the table.

[c] Variations in atomic weight may occur in commercial material because of unknown or unsuspected changes in isotopic composition.

[d] Atomic weight is for radioisotope of longest half-life.

A21 Slater's rules

In wave-mechanical calculations it is often sufficiently accurate to use the approximate analytical wave functions which have been derived by Slater. These functions apply to a single electron in a central field, that is, in a field in which the potential energy is a function of only a radial parameter r and provided by an effective charge ζe. The difference between ζ and the atomic number Z is the screening constant S for the particular electron. It measures the degree to which the other electrons in the atom screen the electron under consideration from the nuclear charge. Hence,

$$\zeta = Z' \text{ (effective)} = Z - S, \tag{A21.1}$$

and the values of S are found by the following rules.

First, the atomic orbitals which are occupied are divided into the groups $1s/2s,2p/3s,3p/3d/4s,4p/4d/4f/$ and so on. Then S is formed by summing the following contributions:

(a) from any orbital of energy *higher* than that of the group considered, zero;
(b) from each other electron in the group considered, 0.35 per electron, or 0.30 per electron if the group considered is 1s;
(c) from the electron group of next lowest energy to that of the group considered, 0.85 per electron if the electron being considered is s or p, and 1.00 per electron for all lower energy groups. If the electron being considered is d (or f), 1.00 per electron for all *lower* energy electron groups.

Examples

Atom	Z	Electron con-sidered	S		ζ
He	2	1s	(1×0.30)	$= 0.30$	1.70
Be	4	2s	$(1 \times 0.35) + (2 \times 0.85)$	$= 2.05$	1.90
C	6	1s	(1×0.30)	$= 0.30$	5.70
C	6	2s,2p	$(3 \times 0.35) + (2 \times 0.85)$	$= 2.75$	3.25
Na	11	3s	$(8 \times 0.85) + (2 \times 1.00)$	$= 8.30$	2.20
Na$^+$	11	2s,2p	$(7 \times 0.35) + (2 \times 0.85)$	$= 4.15$	6.85
Ni	28	3d	$(7 \times 0.35) + (8 \times 1.00) + (8 \times 1.00) + (2 \times 1.00)$	$= 20.45$	7.55

Solutions

Solutions 1

1.1 $O–Cl–O = 116.9°$.

1.2   Assuming a simple relationship with molecular mass, melting point would be 175–180 K and boiling point 205–210 K.

1.3 $+1.8e$.

1.4 $0.282\sqrt{3} = 0.488$ nm.

1.5   400 kJ mol^{-1} is equivalent to 6.6422×10^{-19} J per molecule which, divided by h, gives a frequency of 1.00×10^{15} Hz, or a wavelength of 300 nm; this radiation is in the near uv region.

1.6   0.043 cm^3 helium contains 116×10^{16} molecules. The molar volume of helium, under the given conditions, is $0.08207 \times 273/1$ dm^3 mol^{-1}. Hence, $N_A = 116 \times 10^{16} \times 0.08207 \times 273/(0.043 \times 10^{-3}) = 6.04 \times 10^{23}$ mol^{-1}.

Solutions 2

2.1   Rearrangement, remembering that $\lambda = 1/\tilde{v}$, leads to (2.1). $K = 4/R_H$; $E_{red} = 3.026 \times 10^{-19}$ J.

2.2   In order, from left to right, n_2 is 3, 4, 5 and 6.

2.3 $\mu = 9.1046 \times 10^{-31}$ kg; $a_0 = 52.947$ pm.

2.4 $\Delta p_x = h/10^{-15} \approx 6.6 \times 10^{-19}$ N s, and the mean kinetic energy is $3(\Delta p_x^2/2m_e) \approx 7 \times 10^{-7}$ J. The potential energy is $z_1 z_2 e^2/(4\pi\varepsilon_0 r) \approx -2 \times 10^{-13}$ J. The potential energy does not balance the kinetic energy associated with the uncertainty principle, and the system is unstable. A similar result will be obtained if the uncertainty principle is given in the energy–time formulation: $\Delta E\, \Delta t \approx h$.

2.5 $4\pi N^2 \int_0^\infty r^2 \exp(-2r/a_0)\, dr = 1$, whence $N = (\pi a_0^3)^{-\frac{1}{2}}$, see Appendix A3.

2.6   K $(Ar)(4s)^1$
       Ca $(Ar)(4s)^2$
       Sc $(Ar)(4s)^2(3d)^1$
       Ti $(Ar)(4s)^2(3d)^2$

V   (Ar)(4s)^2(3d)^3
Cr (Ar)(4s)^1(3d)^5
Mn (Ar)(4s)^2(3d)^5
Fe (Ar)(4s)^2(3d)^6
Co (Ar)(4s)^2(3d)^7
Ni (Ar)(4s)^2(3d)^8
Cu (Ar)(4s)^1(3d)^{10}
Zn (Ar)(4s)^2(3d)^{10}
Ga (Ar)(4s)^2(3d)^{10}(4p)^1
Ge (Ar)(4s)^2(3d)^{10}(4p)^2
As (Ar)(4s)^2(3d)^{10}(4p)^3
Se (Ar)(4s)^2(3d)^{10}(4p)^4
Br (Ar)(4s)^2(3d)^{10}(4p)^5
Kr (Ar)(4s)^2(3d)^{10}(4p)^6

For chromium and copper the half-full and full d electron configurations have extra stability, and are preferred to (Ar)(4s)^2(3d)^4 and (Ar)(4s)^2(3d)^9 respectively.

2.7 Be$_2$ Valence bond: no electrons are available for pairing, and so no canonical structures can be constructed.

Molecular orbital: $(1s\sigma)^2(1s\sigma^*)^2(2s\sigma)^2(2s\sigma^*)^2$, an overall slightly antibonding configuration.

LiH   Valence bond: Li—H, Li$^+$H$^-$ and Li$^-$H$^+$ are possible canonical forms, and a resonance hybrid of them is postulated.

Molecular orbital: $(1s\sigma)^2(2s\sigma)^2$; the lithium (1s)^2 electrons form the $(1s\sigma)^2$ 'core', and a σ-bond is formed from lithium 2s and hydrogen 1s.

2.8 With $x=3$, $\psi(sp^3)=\frac{1}{2}[\psi(s)+\sqrt{3}\psi(p)]$. Given $\psi(sp^3)_{1,1,1}=\frac{1}{2}[s+p_x+p_y+p_z]$, where s is used for $\psi(s)$ and so on, it follows that the other three functions are

$$\psi(sp^3)_{-1,-1,1}=\tfrac{1}{2}[s-p_x-p_y+p_z]$$
$$\psi(sp^3)_{1,-1,-1}=\tfrac{1}{2}[s+p_x-p_y-p_z]$$
$$\psi(sp^3)_{-1,1,-1}=\tfrac{1}{2}[s-p_x+p_y-p_z]$$

2.9 Cr^{2+} 4
     V^{3+} 2
     [Mn(CN)$_6$]$^{4-}$ 1
     Co(NH$_3$)$_3$F$_3$ 0

2.10 See Figure S2.1.

2.11 From the dipole moment $q=0.12$, and from electronegativities $q=0.15$. Using $q=0.12$, $\lambda=0.37$, so that $\psi(HBr)=\psi_{covalent}+0.37\psi_{ionic}$.

2.12 Using $-e^2/(4\pi\varepsilon_0 r)$, the electrostatic energy is -8.98×10^{-19} J.

2.13 Using (2.33), $f=1177$ N m^{-1}.

2.14 Using Figure 2.25, the following assignments may be made:

755 cm^{-1} CH$_3$—CH$_2$—CH$_2$—
945 cm^{-1} CH$_3$—CH$_2$—CH$_2$—

1080 cm^{-1} CH$_3$—CH$_2$—CH$_2$—

1180 cm^{-1} CH$_3$—C=O
                          |
                          CH$_2$—

1370 cm^{-1} CH$_3$–CH$_2$–CH$_2$—

1425 cm^{-1} —CH$_2$–C=O

1725 cm^{-1} CH$_3$—C=O
                          |
                          CH$_2$—

2950 cm^{-1} CH$_3$—C—
                          |

Combining this information, we arrive at the structure

CH$_3$—CH$_2$—C=O   butan-2-one
              |
              CH$_3$

Solutions 3

3.1 -4.763×10^8 kJ mol^{-1}.

3.2 $\Delta U = -34.5$ kJ mol^{-1}; $q = 22.4$ kJ mol^{-1}; endothermic.

3.3 -1414 J.

3.4   (a) 28.20 kJ mol^{-1}; (b) 28.12 kJ mol^{-1}; express P as $nRT/(V-nb) - an^2/V^2$.

3.5   At constant volume, $(\partial U/\partial S)_V = T$; at constant entropy, $(\partial U/\partial V)_S = -P$. Thus, $(\partial T/\partial V)_S = [(\partial/\partial V)(\partial U/\partial S)_V]_S$ and $-(\partial P/\partial S)_V = [(\partial/\partial S)(\partial U/\partial V)_S]_V$; the required result now follows.

3.6   3.10 kJ.

3.7   From Section 3.8.3, $w = C_V(T_2 - T_1) = C_V T_1[(P_2/P_1)^{1-1/\gamma} - 1]$; then, (a) $w = +8.36$ kJ mol^{-1}, and (b) $w = +7.11$ kJ mol^{-1}.

Figure S2.1. Trifluorotriamminocobalt (III) Co(NH$_3$)^{3}F^3 structure: (a) *cis*, (b) *trans*.

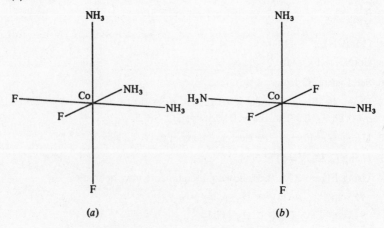

(a)   (b)

3.8 $\sum \Delta H_{cycle} = 0$ (see also p. 64). Experimental verification by calorimetric measurements. For example:

$$\Delta H / kJ\, mol^{-1}$$

$$\tfrac{1}{2}P_4(s) + \tfrac{5}{2}O_2(g) \rightarrow P_2O_5 \quad \Delta H_1$$

$$P_2O_3(s) + O_2(g) \rightarrow P_2O_5 \quad \Delta H_2$$

$$\tfrac{1}{2}P_4(s) + \tfrac{3}{2}O_2(g) \rightarrow P_2O_3 \quad \Delta H_3$$

Then, from Hess's law, $\Delta H_2 + \Delta H_3 - \Delta H_1 = 0$. In practice, we would be able to determine ΔH_1 and ΔH_2, and use them to find ΔH_3. For CuCl, see Figure S3.1, hence, $\Delta H_f^{\ominus}$ (CuCl, s) $= -134\, kJ\, mol^{-1}$.

3.9 $C_{10}H_8(s) + 12O_2(g) \rightarrow 10CO_2(g) + 4H_2O(l)$. Hence, $\Delta H_c = -5171\, kJ\, mol^{-1}$. By applying Hess's law, we find $\Delta H_f^{\ominus}(C_{10}H_8, s) = 87\, kJ\, mol^{-1}$.

3.10 $\Delta C_P = -11.62 + 3.965 \times 10^{-3}T + 12.33 \times 10^{-7}T^2$, whence $\Delta H_c(1000\,K) = -242.8 - 4.9 = -247.7\, kJ\, mol^{-1}$.

3.11 $-1426\, kJ\, mol^{-1}$.

3.12 $87\, J\, K^{-1}\, mol^{-1}$. A similar value is found for the entropy of vaporization of many liquids at their boiling points, which is the basis of an approximation known as Trouton's law. It occurs because the molar volumes of gases are approximately constant and the entropy of the gaseous state is much greater than that of the corresponding condensed state.

3.13 $+5.63\, kJ\, mol^{-1}$.

3.14 $\Delta H_f(C_3H_8, g)$ $-113\, kJ\, mol^{-1}$ (true value $-103.8\, kJ\, mol^{-1}$)

    $\Delta H_f(C_3H_6, g)$ $17\, kJ\, mol^{-1}$ (true value $20.4\, kJ\, mol^{-1}$)

    ΔH_{hydrog} $130\, kJ\, mol^{-1}$ (true value $125.1\, kJ\, mol^{-1}$)

3.15 The enthalpy of sublimation of a substance A is the enthalpy change (increase) accompanying the process

A(s) $\rightarrow$ A(g)

at 298.15 K and 1 atm.

3.16 $14.70\, J\, K^{-1}\, mol^{-1}$.

Figure S3.1. Thermochemical cycle for the formation of CuCl(s) and CuCl$_2$(s).

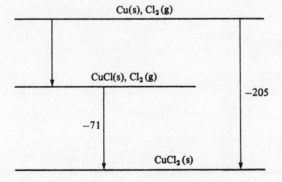

3.17 (a) At constant volume $dV=0$, and $dU=\d q=dq$.

  (b) (i) Under adiabatic conditions $\d q=0$, and $dU=\d w=dw$.

  (ii) At constant pressure and when only mechanical work of expansion is occurring, since $dH=dU+P\,dV+V\,dP$, $dH=\d q=dq$.

3.18 From (3.16), $\d w(=-P\,dV)=-R\,dT+RT/P\,dP$. Using (3.19), $\{(\partial/\partial P)(-R)\}_T=0$ and $\{(\partial/\partial T)(RT/P)\}_P=R/P$; hence $\d w$ is inexact.

3.19 $\Delta H_b(C\!-\!C,\ \text{arom})=511\,\text{kJ mol}^{-1}$; $\Delta H_f(C_{10}H_8,s)=-118\,\text{kJ mol}^{-1}$. The discrepancy is an indication of the unreliability of the method; the value from the combustion experiment is to be preferred.

3.20 0.11 K.

Solutions 4

4.1 From (4.10) $T=29.8$ K.

From (4.11) $P=2$ atm.

4.2 From (4.13) and $D=M_m/V$, where M_m is the molar mass,

$P=DRT/M_m$

From (4.65) $P=nRT/V(1+BP/RT)$, so $P/D=RT/M_m+BP/M_m$. Hence:

$10^{-5}P/\text{N m}^{-2}$	0.1223	0.2519	0.3697	0.6037	0.8522	1.013
$10^{-5}(P/D)/\text{m}^2\text{ s}^{-2}$	0.5373	0.5365	0.5360	0.5347	0.5333	0.5323

Figure S4.1 shows the plot of P/D against P. Assuming the linearity implied by the above equation, the data may be fitted by the method of least squares (see Appendix A2). The equation obtained is $y=-0.0055422x+0.53798$, where y represents $10^{-5}(P/D)/\text{m}^2\text{ s}^{-2}$ and x represents $10^{-5}P/\text{N m}^{-2}$. At $P=0$, the intercept, 0.53798, may be equated to RT/M_m, whence $M_m=0.04606\,\text{kg mol}^{-1}$ and $M_r=46.06$.

4.3 (a) $n(N_2)+n(H_2)=n_{total}=6$ mol. $P=6RT/50=3.00$ atm. (Did you remember to use R in $\text{dm}^3\text{ atm mol}^{-1}\text{ K}^{-1}$?) Hence, $p(H)=(4/6)3=2$ atm, $p(N_2)=(1/3)3=1$ atm.

Figure S4.1. Plot of P/D against P for dimethyl ether.

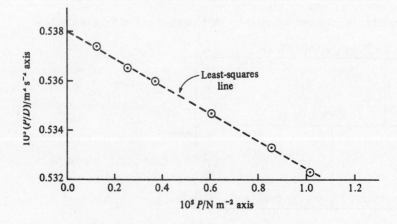

$10^5\ (P/D)/\text{m}^4\text{ s}^{-4}$ axis

Least-squares line

$10^5\ P/\text{N m}^{-2}$ axis

(b) Consider the reaction $H_2 + \frac{1}{3}N_2 \to \frac{2}{3}NH_3$

Initially:	4 mol	2 mol	0 mol
Finally:	0 mol	$2 - \frac{4}{3}$ mol	$\frac{8}{3}$ mol

Total pressure $P = (10/3)RT/50 = 1.667$ atm, whence

$p(N_2) = 0.333$ atm, $\quad p(NH_3) = 1.334$ atm, $\quad p(H_2) = 0$

4.4 $\int_0^\infty P(v)\,dv = 4\pi(a/\pi)^{\frac{3}{2}}\int_0^\infty v^2 \exp(-av^2)\,dv$, where $a = M/2RT$. Let $t = av^2$, so that $v = (t/a)^{\frac{1}{2}}$ and $dv = dt/(2a^{\frac{1}{2}}t^{\frac{1}{2}})$. Then the integral becomes

$$2\pi^{-\frac{1}{2}}\int_0^\infty t^{\frac{1}{2}}\exp(-t)\,dt = 2\pi^{-\frac{1}{2}}\Gamma(3/2) = 2\pi^{-\frac{1}{2}}\tfrac{1}{2}\pi^{\frac{1}{2}} = 1$$

where Γ represents the gamma function (see Appendix A3). Thus $|\bar{v}| = 4\pi(a/\pi)^{\frac{3}{2}}\int_0^\infty v^3 \exp(-av^2)\,dv$. Making the same substitution $|\bar{v}| = 2/(\pi a)^{\frac{1}{2}}\int_0^\infty t \exp(-t)\,dt = 2/(\pi a)^{\frac{1}{2}}\Gamma(2) = 2/(\pi a)^{\frac{1}{2}} = (8RT/\pi M)^{\frac{1}{2}}$. Using (4.38), $|\bar{v}|/(\overline{v^2})^{\frac{1}{2}} = (8/3\pi)^{\frac{1}{2}}$.

4.5 (a) $\mu_{JT} = (2a/RT - b)/C_P = 0.88$ K atm^{-1}, $T_i = 2739$ K.

(b) By differentiating PV with respect to P at constant temperature and setting the derivative to zero, we obtain $RT_{Boyle} = (a/b)[(V-b)/V]^2$. As the minimum is to occur at vanishingly low pressures, $V \to \infty$, so that $RT_{Boyle} = a/b$; hence $T_{Boyle}(NH_3) = 1369$ K.

4.6 Using (4.65) and Table 4.3, $PV/RT = Z = 0.992$.

4.7 Taking $p = 17$ J K^{-1} mol^{-1} in (4.66) and (4.67), $C_P/C_V = 1.28$.

4.8 See Figure S4.2.

4.9 See Figure S4.3.

4.10 (a) $(21\bar{3})$ (b) (104) (c) $(00\bar{1})$

      (d) $(1\bar{2}4)$ (e) $(\bar{3}\bar{1}2)$ (f) $(12, 3\bar{4})$

Conventions: in writing Miller indices, fractions should always be cleared and two-digit indices followed by a comma.

4.11

Plane	Area	Lattice points	Reticular density
(100)	a^2	1	$1/a^2$
(110)	$a^2\sqrt{2}$	2	$\sqrt{2}/a^2$
(111)	$a^2\sqrt{3}$	2	$2/a^2\sqrt{3}$
(230)	$a^2\sqrt{13}$	1	$1/a^2\sqrt{13}$

Ratio $1:\sqrt{2}:2/\sqrt{3}:1/\sqrt{13}$

4.12 Integrating (4.77) gives $\ln P = -\Delta H/RT + $ constant. By a least-squares fit of $\ln P$ to $1/T$, we obtain

$\ln P = 4563.0/T + 12.80$

Hence, $\Delta H = 37.9$ kJ mol^{-1}. At the boiling point $\ln P/$atm is zero, whence $T_{bp} = 37.9 \times 10^3/12.80R = 356$ K.

4.13 From (4.91) and (4.92), $d_{123} = 0.168$ nm, $\theta_{123} = 27.3°$.

Figure S4.2. Stacking of two cubes: tetragonal system, five *m* planes –
IJKL, HDPT, BFRN, AEQM, GCOS; one four-fold axis – along *z*; four
two-fold axes – along *x*, *y*, IK, JL; one centre of symmetry – *.

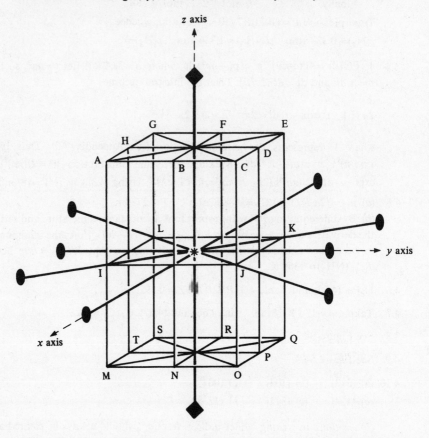

Figure S4.3. Structure of sodium chloride: thin lines show the conventional
face-centred cubic unit cell of side a_C, thick lines show the rhombohedral
unit cell of side $a_R = a_C / \sqrt{2}$; X–0–Z = 60° since Δ0XZ is equilateral.

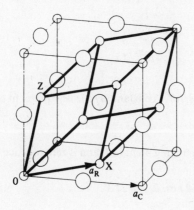

4.14 The following table may be set up:

$\sin^2 \theta$	$\sin^2 \theta/0.0188$	N	hkl	a/nm
0.0376	2.00	2	110	0.6528
0.0556	2.96	3	111	0.6575
0.0744	3.96	4	200	0.6563
0.0926	4.93	5	210	0.6577
0.112	5.96	6	211	0.6551
0.167	8.88	9	300, 221	0.6571
0.186	9.89	10	310	0.6563
0.205	10.90	11	311	0.6557
0.261	13.88	14	321	0.6556
0.316	16.81	17	410, 322	0.6565

(The common factor of 0.0188 is 0.0376/2)

Unit-cell type P; not all lines that are possible for a cubic P unit cell are present (see Example in Section 4.3.5)

Mean value ($\bar{a}$) for the unit-cell dimension 0.6561 nm

Standard deviation of the mean ($\sigma_{\bar{a}}$) 0.0004 nm

4.15 Tetrahedral holes 8:

$$\tfrac{1}{4}, \tfrac{1}{4}, \tfrac{1}{4}; \quad \tfrac{1}{4}, \tfrac{3}{4}, \tfrac{1}{4}; \quad \tfrac{3}{4}, \tfrac{1}{4}, \tfrac{1}{4}; \quad \tfrac{3}{4}, \tfrac{3}{4}, \tfrac{1}{4}; \quad \tfrac{1}{4}, \tfrac{1}{4}, \tfrac{3}{4}; \quad \tfrac{1}{4}, \tfrac{3}{4}, \tfrac{3}{4}; \quad \tfrac{3}{4}, \tfrac{1}{4}, \tfrac{3}{4}; \quad \tfrac{3}{4}, \tfrac{3}{4}, \tfrac{3}{4}$$

Octahedral holes 4:

$$0, 0, \tfrac{1}{2}; \quad 0, \tfrac{1}{2}, 0; \quad \tfrac{1}{2}, 0, 0; \quad \tfrac{1}{2}, \tfrac{1}{2}, \tfrac{1}{2}$$

4.16 Volume of hexagonal unit cell (Figure 4.14(m)) is $a^2 c \sin(120°)$. From Figure 4.27(b) there are two atoms per unit cell, and $a = 2r = 2$ nm, where r is the radius of the atom. Hence, the volume V occupied per atom is $c\sqrt{3}$ nm^3. An atom in the second layer and the three in contact with it from the first layer form a tetrahedron. Thus, the angle from the atom at the origin in the first layer to the atom in contact in the second layer and back to that distant c above the origin is the tetrahedral angle 109.47°. Hence $c/2 = 2r \sin(109.47/2°)$ so that $V = 5.66$ nm^3.

4.17

Compound	Electrons	Atoms	Electron/atom	Phase
AuZn	3	2	3/2	β
Au$_5$Al$_3$	14	8	7/4	ε
Cu$_{31}$Si$_8$	63	39	21/13	γ
Ag$_5$Al$_3$	14	8	7/4	ε

4.18 Unit-cell type P

Ca coordinated by 8 F at the corners of a cube

F coordinated by 4 Ca at the corners of a regular tetrahedron
4CaF$_2$ per unit cell

Ca: $0, 0, 0; \quad 0, \tfrac{1}{2}, \tfrac{1}{2}; \quad \tfrac{1}{2}, 0, \tfrac{1}{2}; \quad \tfrac{1}{2}, \tfrac{1}{2}, 0$

F: $\tfrac{1}{4}, \tfrac{1}{4}, \tfrac{1}{4}; \quad \tfrac{1}{4}, \tfrac{3}{4}, \tfrac{1}{4}; \quad \tfrac{3}{4}, \tfrac{1}{4}, \tfrac{1}{4}; \quad \tfrac{3}{4}, \tfrac{3}{4}, \tfrac{1}{4};$

$\tfrac{1}{4}, \tfrac{1}{4}, \tfrac{3}{4}; \quad \tfrac{1}{4}, \tfrac{3}{4}, \tfrac{3}{4}; \quad \tfrac{3}{4}, \tfrac{1}{4}, \tfrac{3}{4}; \quad \tfrac{3}{4}, \tfrac{3}{4}, \tfrac{3}{4}$

4.19 In the cubic unit cell, $a\sqrt{3} = 2r(+) + (2r(-))$, whence $a = 0.41508$ nm, making allowance for the eight-fold coordination. Hence,

$$D = \frac{1 \times (132.905 + 35.453) \times 1.66057 \times 10^{-27}}{(0.41508 \times 10^{-9})^3} = 3909 \text{ kg m}^{-3}$$

4.20

	Ether	Benzene	Propyl ethanoate	Water	Mercury
$\Delta H_v/RT$	10.6	10.7	11.1	13.1	11.3

Four of the liquids behave well according to Trouton's rule. Hydrogen bonding in water increases its structure, which decreases its entropy below that expected. Since all gases have very similar molar entropies, it follows that the entropy of vaporization $(\Delta H_v/RT)$ for water will be greater than that implied by the Trouton rule.

Solutions 5

5.1 Mean $K_p = 0.141$ atm (298 K) and 0.309 (308 K);
$\Delta H = 59.6 \text{ kJ mol}^{-1}$

5.2 $\Delta H/\text{kJ mol}^{-1}$ 177.9,   175.9,   175.2,   163.1,   162.8

5.3 (a) (i) Position of equilibrium moves L → R; K_p unchanged.
(ii) Position of equilibrium moves R → L; K_p decreases.
(iii) Position of equilibrium moves R → L; K_p unchanged.
(iv) Position of equilibrium moves R → L; K_p unchanged.
(b) (i) Rate of forward and reverse reactions decrease.
(ii) Rate of forward and reverse reactions increase.
(iii) Rate of forward and reverse reactions increase.
(iv) Rate of forward and reverse reactions increase.

5.4 Initial pH = 9.12
pH at equivalence point = 3.27
Final pH = 2.04

5.5 (a) Solubility in water = 2.45×10^{-3} g dm^{-3} (2.50×10^{-3} g dm^{-3} with $f_\pm = 0.98$)
(b) Solubility in 10^{-3} mol dm^{-3} sodium sulphate = 4.29×10^{-5} g dm^{-3}
(c) Solubility in 10^{-2} mol dm^{-3} sodium hydroxide = 2.75×10^{-3} g dm^{-3}

5.6 $x = 1.6051, -124.6051$, for both (a) and (b).

5.7 (a) 12.00,   (b) 2.30,   (c) 2.40,   (d) 7.01.

5.8 pH = 7.46 at 273 K; 7.00 at 298 K; 6.51 at 333 K. $\Delta H = 55.2$ kJ mol^{-1}.

5.9 (a) 3.45,   (b) 8.77,   (c) 5.13,   (d) 9.09.

5.10 pH = 4.05; ΔpH = +0.14.

5.11 1.8.

5.12 See Figure 5.11.

Solutions 6

6.1 (a) $x_i = n_i/\sum_i n_i$;
(b) property associated with number of particles present;
(c) $p_i = (n_i/\sum_i n_i)p$;

(*d*) a system of two components;

(*e*) temperature above (or below) which miscibility becomes partial;

(*f*) a solution for which the vapour obeys Raoult's law;

(*g*) composition of a mixture which has the minimum boiling point;

(*h*) distillation of a mixture in a current of steam, so that for a volatile component i, $p(i) + p(H_2O) = $ ambient pressure, at the boiling point.

6.2   33.52 mmHg; 153.04 mmHg.

6.3 $M_r = 44.1$.

6.4 $M_r = 45.9$.

6.5   Calcium chloride, unlike glucose, dissociates in aqueous solution to give Ca^{2+} and $2Cl^-$ ions. The apparent degree of dissociation α is found to be 0.80.

6.6 $M_r = 60.7$.

6.7 $k_f = 5.09$ K; 5.03 K; 4.90 K.

6.8 R varies from $0.0894 \, dm^3 \, atm \, mol^{-1} \, K^{-1}$ to $0.0816 \, dm^3 \, atm \, mol^{-1} \, K^{-1}$ as c decreases. Extrapolation to $c = 0$ gives $R = 0.0806 \, dm^3 \, atm \, mol^{-1} \, K^{-1}$.

6.9 $T = -0.153$ K.

6.10   (*a*) 0.069 K,   (*b*) 185.9 mmHg.

6.11 $\alpha_{assoc} = 0.088$; 0.942.

6.12   (*a*) 1.33 g,   (*b*) 1.44 g.

6.13   59.85% by weight.

6.14   (*a*) 4,   (*b*) 337 ± 2.

Solutions 7

7.1   0.7052 g silver.

7.2   Platinum anode: $\frac{1}{2}H_2O \rightarrow H^+ + \frac{1}{4}O_2 + e^-$

Copper anode: $\frac{1}{2}Cu \rightarrow \frac{1}{2}Cu^{2+} + e^-$

Dilute sodium chloride: $\frac{1}{2}H_2O \rightarrow H^+ + \frac{1}{4}O_2 + e^-$

Concentrated sodium chloride: $Cl^- \rightarrow \frac{1}{2}Cl_2 + e^-$

7.3   (*a*) Cell constant $= 3.965 \, m^{-1}$

(*b*) $\Lambda = 0.01093 \, \Omega^{-1} \, mol^{-1} \, m^2$

7.4 $\kappa = 4.467 \times 10^{-2} \Omega^{-1} m^{-1}$.

7.5   (*a*) $\Lambda_0(NaCl) = 0.01085 \, \Omega^{-1} \, mol^{-1} \, m^2$

(*b*) $\Lambda_0(NaEth) = 0.00765 \, \Omega^{-1} \, mol^{-1} \, m^2$

(*c*) $\Lambda_0(HCl) = 0.03782 \, \Omega^{-1} \, mol^{-1} \, m^2$

(*d*) $\Lambda_0(HEth) = 0.03462 \, \Omega^{-1} \, mol^{-1} \, m^2$

7.6 $S = 1.06 \times 10^{-10} \, ml.^2 \, dm^{-6}$; $f(\pm) = 0.985$; S (allowing for $f(\pm)$) $= 1.03 \times 10^{-10}$ $mol^2 \, dm^{-6}$.

7.7   (*a*) 0.177,   (*b*) 0.597,   (*c*) 0.503,   (*d*) 0.557

$t(Cl^-)$ is much smaller in hydrochloric acid because of the large mobility of the hydroxonium ion.

7.8   (*a*), (*b*) For the cell $Zn|Zn^{2+}, H^+|H_2$

Anode reaction $\frac{1}{2}Zn \to \frac{1}{2}Zn^{2+} + e^-$

Cathode $H^+ + e^- \to \frac{1}{2}H_2$

Overall $\frac{1}{2}Zn + H^+ \to \frac{1}{2}Zn^{2+} + \frac{1}{2}H_2$

For the cell $H_2|H^+, Ag^+|Ag$

Anode reaction $\frac{1}{2}H_2 \to H^+ + e^-$

Cathode $Ag^+ + e^- \to Ag$

Overall $\frac{1}{2}H_2 + Ag^+ \to H^+ + Ag$

For the cell $Zn|Zn^{2+}, Ag^+|Ag$

Anode reaction $\frac{1}{2}Zn \to \frac{1}{2}Zn^{2+} + e^-$

Cathode $Ag^+ + e^- \to Ag$

Overall $\frac{1}{2}Zn^{2+} + Ag^+ \to \frac{1}{2}Zn^{2+} + Ag$

(*c*), (*d*) $E^\ominus/V$: 0.763; 0.799; 1.562

$\Delta G^\ominus/kJ\,mol^{-1}$: -73.6; -77.1; -150.7

(*e*)   L → R in each case, because of the cell convention.

7.9   0.208 V; $dE/dT = -6.43 \times 10^{-4}\,V\,K^{-1}$.

7.10   (*a*) $\Delta G^\ominus = -233.7\,kJ\,mol^{-1}$, $K = 8.58 \times 10^{40}$

(*b*) Fe^{2+} is inherently unstable. $E^\ominus = 0.458$ V.

7.11   pH = 4.02.

7.12 $\Delta H = 5.38\,kJ\,mol^{-1}$.

Solutions 8

8.1 $k_1 = 3.35 \times 10^{-3}\,min^{-1}$; $t_{\frac{1}{2}} = 206$ min.

8.2   (*a*) 70.3 year

(*b*) 6.47×10^{-6} g

8.3 $k_1 = 0.01036\,min^{-1}$; $t_{\frac{1}{2}} = 66.9$ min.

8.4   Zeroth order; $k_0 = 0.232\,mm\,s^{-1}$.

8.5   Second order; $k_2 = 0.133\,dm^3\,mol^{-1}$, $E_A = 233.4\,kJ\,mol^{-1}$.

8.6   First order; $k_1 = 0.0206\,min^{-1}$.

8.7 $k_1 = 0.0161\,min^{-1}$, $k_{-1} = 0.00919\,min^{-1}$.

8.8 $-d[N_2O_5]/dt = 2d[NO\cdot]/dt = 2k_3[NO_2\cdot][NO_3\cdot]$

Substitute the expression for the steady-state concentration of $NO_3\cdot$.

$[NO_3\cdot] = k_1[N_2O_5]/(k_2 + k_3)[NO_2\cdot]$

Index